Knowledge Flows in a Global Age

Knowledge Flows in a Global Age

A Transnational Approach

Edited by John Krige

The University of Chicago Press
Chicago and London

The University of Chicago Press, Chicago 60637
The University of Chicago Press, Ltd., London

Published 2022
Printed in the United States of America

31 30 29 28 27 26 25 24 23 22 1 2 3 4 5

ISBN-13: 978-0-226-81994-5 (cloth)
ISBN-13: 978-0-226-82038-5 (paper)
ISBN-13: 978-0-226-82037-8 (e-book)

DOI: https://doi.org/10.7208/chicago/9780226820378.001.0001

Library of Congress Cataloging-in-Publication Data

Names: Krige, John, editor.
Title: Knowledge flows in a global age : a transnational approach / edited by John Krige.
Description: Chicago ; London : The University of Chicago Press, 2022. | Includes bibliographical references and index.
Identifiers: LCCN 2021061665 | ISBN 9780226819945 (cloth) | ISBN 9780226820385 (paperback) | ISBN 9780226820378 (ebook)
Subjects: LCSH: Technology transfer—History—20th century. | Technology transfer—Cross-cultural studies. | Diffusion of innovations—History—20th century. | Technology and international relations—History—20th century. | Globalization.
Classification: LCC T174.3 .K565 2022 | DDC 338.9/26—dc23/eng/20220110
LC record available at https://lccn.loc.gov/2021061665

♾ This paper meets the requirements of ANSI/NISO Z39.48-1992 (Permanence of Paper).

Contents

PART II

FACILITATING TRANSNATIONAL KNOWLEDGE FLOWS

Introduction

Writing the Transnational History of Knowledge Flows in a Global Age

JOHN KRIGE

The Global, the Transnational, and the National

Transnational history as method emerged along with a renewed interest in global history in the late twentieth century. It was one of a variety of intellectual responses to the convergence of two developments.[1] First, the process of economic globalization itself, manifest in changes in the organization of the world economy that exploited new technological infrastructures to increase the mobility of capital, goods, and people across national borders. Second, this tectonic shift was accompanied by frustration in academia with a genre of historical scholarship that was professionalized in the late nineteenth century along with the process of state formation and that took the territorial boundaries of the nation-state as framing its object of inquiry.[2] National borders were being dissolved in global imaginaries and rendered permeable in practice by the increased mobility of capital, goods, and people. The writing of history had to adapt to new realties and to cosmopolitan life experiences. Changes that occurred nationally began to be analyzed through the lens of transnational connectivities that intermingled people, goods, ideas, and cultures as they moved back and forth across receding national borders.[3] The history of science and technology was not indifferent to these developments. For historian of science Fa-ti Fan, "instead of looking at science and technology as products in a particular nation or civilization, the main focus of global history of science [was] on the transmission, exchange and circulation of knowledge, skills and material objects."[4] David Edgerton contrasted "technonational" historians of technology that took the state as the key unit of analysis with "technoglobalists," who regarded nation-states as "at best a temporary vehicle through which the forces of techno-globalism operate but are always about to disappear through the advance of globalizing technology."[5]

There is now an extensive body of literature that has enriched these framing principles. It is marked by a strong pushback against the implicit assumption that the economic globalization that inspired these new approaches was an irresistible historical force dissolving national boundaries and destroying national sovereignty. The subterranean impact of assumptions that "makes the ever-closer integration of the world appear a more or less natural development" has been thoroughly dissected.[6] Proponents of world history Charles Bright and Michael Geyer insist that, rather than thinking about transnational and global flows in terms that tend "to presume the (relative) openness of the world," we need to excavate "the structured networks and webs through which interconnections are made and maintained—as well as contested and renegotiated."[7] Africanist Frederick Cooper adds that detailed study of the "movement of people, as well as capital, reveals the lumpiness of cross-border connections, not a pattern of steadily increasing integration," obliging us to put "as much emphasis on nodes and blockages as on movement."[8] Fa-ti Fan warns us that when we use language that implies that the flow through networks is laminar and unperturbed, we need to be aware that a term like "circulation" can conceal what "may have been really a series of negotiations, pushes and pulls, struggles, and stops and starts."[9] He argues that to write a robust history of transnational knowledge flows it is "imperative to investigate the historical reasons and circumstances that fostered or hindered the movement of knowledge or material objects" and "to find out how and why certain mechanisms were introduced to control the coming and going of people and things."[10] Global imaginaries that dissolve boundaries evacuate a transnational history of content. There is no transnational knowledge flow without friction at national borders.

This book builds on the foundations laid in my previous volume on how knowledge moves, but consolidates them "theoretically" with these insights in mind.[11] The global is embraced not as an explicit object of analysis but as an extension of the spatial dimensions through which knowledge moves across national borders.[12] To sharpen, but also to simplify the distinctions at work here, we can say that national histories describe social changes that take place *within* the confines of national borders, that transnational history analyzes what happens *at* borders as knowledge moves *across* them, and that global histories explore social interconnections and interdependencies at a scale *beyond* national borders. The last are also often written within a contemporary imaginary that foresees the eventual *elimination* of national borders altogether by a relentless process of globalization, anticipated methodologically by decoupling the global from the national. The chapters in this volume resist that methodological move, seeking rather to analyze the factors that facilitate or impede the movement of knowledge-bearing people, ideas, and things across national borders. Patricia Clavin has written that the "challenge before historians interested in transnational phenomena . . . is whether, and how, to engage with the nation-state."[13] This book responds to that challenge by insisting on the centrality to the transnational project of transactions that occur at national borders.

The approach taken here occupies a quite different terrain to the majority of works on transnational history by the mainstream discipline.[14] These tend to limit the role of transnational flows of science and technology to providing the technological infrastructure that make global networking possible. For example, Ian Tyrrell, who has written an illuminating transnational history of the United States from 1789 to the present, defines transnational history as dealing with the "movement of people, ideas, technologies and institutions across national boundaries." However, technology enters his narrative as providing the communications systems that brought different nations and peoples together. Tyrrell settles on "satellite communications, ubiquitous jet transport, optic fiber cables and the internet" as the means of integration in the current global era.[15] Emily Rosenberg, who is engaged with science, technology, and expertise in the first age of globalization from 1870 to 1940, begins her book with a quote from the British missionary- explorer and agent of imperialism David Livingstone: "The extension and use of railroads, steamships, telegraphs, break down nationalities and bring people geographically remote into close connection. . . . They make the world one."[16] By contrast, in this volume, science and technology, or more generally "knowledge," do not simply provide the material infrastructure that makes transnational connectivity possible. The movement of knowledge embodied in people, ideas, institutions, and things across national borders is itself the object of transnational analysis.

In my previous edited collection of essays on this theme I regretted the rather heavy emphasis placed on both the United States and on the Cold War. The chapters presented here expand our field of vision beyond those in both space and time. They cover the broad sweep of the twentieth century up to the present day. They deal with major global powers like China, the United Kingdom, and the Soviet Union, as well as the United States, and also with colonial Portugal and the Palestinian authority on the West Bank. They cover high-performance computers, naval and space technologies, scientific fields like agriculture and statistics, and production and marketing systems for coffee and for penicillin. Moving knowledge across borders remains a social accomplishment. However, whereas in the previous volume I emphasized impediments to transnational knowledge flows, in this collection the inclusion of case studies that highlight the sharing of (agricultural) technologies—as distinct from the regulation of military and civil dual-use knowledge and know-how—introduces actors, agendas, and institutions that facilitate border crossings. They also enrich our understanding of borders as institutions by moving beyond nationally constructed legal regimes that inhibit knowledge flows to include international organizations, nongovernmental organizations (NGOs), and corporate and nonstate actors whose mission is to promote cross-border movement. And they provide material for me to briefly decenter the Global North in the conclusion to this book. Taking knowledge as the object of transnational history has not only thrown new light on the political economy of science and technology in an interconnected, interdependent world of nation-states. Through our collective work in this volume it has also forged personal and intellectual bonds between scholars with otherwise very different research agendas, enabling a truly multidisciplinary approach to a topic of considerable historical and contemporary interest.

Borders and/as Institutions

What is a border? For historian Charles Maier, a border is "partially a virtual construction. It is as much a site of the demonstrative extent of power as a real barrier. It regulates an exchange as much as it excludes."[17] Borders are not necessarily aligned with the geographical limits of the nation-state, though they presuppose them. (That said, much of the analysis developed here can apply to any centrally governed geographical space, or "territoriality," and not only that identified with the nation-state as such.) Their presence is made manifest, as Maier suggests, by an historically evolving assembly of institutions and policies mobilized by the state, and politically aligned social actors, that have the authority to impede or to facilitate traffic across them. "The concept and institution of the border," write sociologists Sandro Mezzadra and Brett Neilson, are constituted by "multiple (legal and cultural, social and economic) components."[18]

National borders are crucial variables in global studies of phenomena like migration but are otherwise often ignored in global history.[19] They are also sidelined in the global history of science and technology.[20] Yet this is the body of literature that I draw on extensively in this methodological introduction. The translation between the scales is made by treating the "local" sites of knowledge production that are major analytical units in global history as being nationally situated and by thinking about the networks through which knowledge "circulates" from one "locality" to another as making transnational connections across borders.[21]

As pointed out above, much transnational history, perhaps seduced by the aspirations of globalization to eliminate national borders altogether, is written as if people, ideas, and things circulated freely across them. By contrast, the studies in this volume draw our attention to the importance of national borders as sites of interventions that facilitate or resist transnational movement. Ironically, a mode of writing history that emerged along with the neoliberal, market-driven pursuit of building a borderless world in the late 1980s here reinserts national borders into the heart of the analysis when knowledge in motion is the object. In doing so it also problematizes the epistemic scale and scope of scientific internationalism.

Transnational Connnectivities: Scientific Internationalism

The factors that impede or facilitate the flow of knowledge across national borders depend on the nature of the knowledge that is shared between members of the research community, and on the political relationships between the collaborating states, that demarcate the scope of "internationalism" in response to changing national and geopolitical circumstances. Generalities are difficult to make. However, the role of scientific internationalism in facilitating transnational connectivities since at least the late nineteenth century needs particular attention.

Scientific internationalism is an ideology, and a practice, that promotes transnational cooperation between researchers who share new knowledge, regardless of national affiliation, in the name of scientific progress.[22] It is also readily invoked as an *explanation* for why knowledge moves across national borders. It has been given additional weight in the age of globalization. Indeed, the scale of its implementation today in "international conferences, international journals, and international visitors" led James Secord to suggest that modern scientific inquiry seems to many to be "the closest thing to a perfectly globalized system that we possess," a system in which it is easy to assume that "knowledge simply travels by itself" across borders.[23]

The transnational approach taken in this book resists a characterization of scientific (and technological) collaboration as a "perfectly globalized" practice that has abolished national borders. Beginning in the late nineteenth century, science and technology became increasingly associated with the state, "both because science needed the state and because the state could draw advantages from science."[24] This entanglement has led to ongoing negotiations between national authorities and corporations heavily engaged in research and development (R&D), as well as the academic research community, over the boundaries between knowledge and know-how that could circulate "freely" and that whose movement could be restricted by a state keen to protect a key resource in a competitive world order. In Cold War America in particular, the epistemic space for the uninhibited transnational flow of knowledge has been the subject of intense negotiations that define the "social contract" between the government and corporate and academic actors. My enthusiasm for the genre of transnational history performed in this and my previous edited collection is a consequence of the shift in my perspective that was needed when I made detailed studies of these negotiations over the terms of the transnational circulation of knowledge and know-how throughout the Cold War and beyond.[25]

The Cold War paradigm characterizing American research has bequeathed an administrative distinction between two categories of knowledge: classified (including the special category of restricted data in the nuclear domain) and unclassified.[26] It maps onto a dichotomy between knowledge that must be kept secret and remain protected behind a high wall and that cannot be shared transnationally (except under tightly controlled circumstances, as in the Manhattan Project) and knowledge that is "open" and that can circulate "freely" across national borders propelled by the ideology of scientific internationalism.[27] It also allows for the protection of commercializable knowledge by patents, though unless the patent is a "state" secret, in a liberal democratic society control over its circulation is vested in individuals, firms, and corporations that produced the knowledge, often in close consultation with the state and with its political and financial support.

This simple open/secret dichotomy needs to be revised fundamentally.[28] Perhaps because of an overemphasis on academic science, the bulk of the scholarship on knowledge in motion does not reckon with a gray zone of sensitive, unclassified knowledge that lies in the epistemic space between classified and open. It can be shared transnationally because it is unclassified, but only under strict conditions because it is sensitive, and is deemed a threat to national security.

Anthropologist Joseph Masco has recently drawn our attention to the vast scope of sensitive unclassified knowledge embedded in what he calls the US counterterrorist, regulatory state since the attacks on the World Trade Center in 2001.[29] Masco's focus is on the use of this epistemic category by the government as an instrument of domestic social control. The concept of sensitive unclassified knowledge also applies to restricting the scope of knowledge that can move freely across national borders in the name of national security. In fact, concerns that America's enemies were exploiting the nation's values of openness and freedom of information to acquire knowledge that could be exploited for nefarious ends have accompanied the national security state from its birth in the aftermath of World War II.[30] As early as 1947 the Truman administration was looking for ways to regulate information that was not deemed classifiable but that should not be allowed to circulate freely either. The Export Control Act of 1949 took a major step in this direction. It imposed regulations on the circulation of unclassified dual-use civil and military technology by corporations that included "advanced technical data, including industrial 'know-how' and strategically important scientific discoveries . . . [that] may have a war potential significance equal to, if not greater than, strategic commodities."[31] It soon also expanded the definition of the term "export" to cover sharing that data with foreign nationals *within the territory of the United States* on the grounds that they could take it back home. Universities protested. The regulation threatened to make it impossible to teach any foreign national in an American university without first acquiring an export license. To protect academic freedom, the vague epistemological distinction between basic science that was openly published and applied science that had practical implications was invoked to resolve a political dilemma.[32] The field of knowledge was carved up into three sectors: a domain of fundamental research that is still in force today, a space where knowledge can be shared internationally; a field where secret knowledge is classified and "nationalized"; and an intermediate gray zone of sensitive unclassified knowledge that is regulated by export controls.[33] Over subsequent decades a labyrinthine legal system embedded in social institutions and practices, and built on an immensely sophisticated and constantly negotiated concept of what "sensitive" knowledge is and how it travels, has been put in place. It inhibits or regulates transnational flows of sensitive unclassified knowledge (typically dual-use civil and military knowledge) in the name of national and economic security to ensure America's global scientific and technological preeminence or "leadership."

The introduction of this third, gray zone of knowledge has important implications for our understanding of the scope of scientific internationalism that facilitates transnational flows of unclassified knowledge. It implies that, when *sensitive* unclassified knowledge crosses borders to select countries, the range of scientific internationalism is bounded by regulatory instruments invoked by the national-security state. These impose epistemic limits on the scope of scientific internationalism, and its explanatory value, that have to be established empirically, along with the political scope of "international" collaboration itself. There are also political limits.

Scientific internationalism is premised on universalism, the epistemological conviction that scientific collaboration and consensus are possible despite national allegiances, and other ideological differences between people, because knowledge of the world is imposed by robust truths —"facts"—that all can agree on. Eliding the tensions between national competition and international collaboration, which is especially intense when *sensitive*, unclassified knowledge is in motion, scientific internationalism glosses over the changing alliances between nation-states that enable the ideology to gain purchase and to have material effects or, alternatively, to lose traction when international rivalry tears the collaborative network apart.

In this volume and elsewhere, Jessica Wang insists on the importance of recognizing the specific clusters of political relations at work "behind more general allusions to global or international phenomena." She has argued that "international science itself should be understood as arising from the heyday of high imperialism in the late nineteenth and early twentieth century," and treated as a contingent response to a particular configuration of global alliances that limited the scope of who was included under the umbrella of international cooperation.[34] I discuss her contribution to this volume in more detail below. For the present, the point to retain is that the politico-geographical scope of the international, and so of scientific internationalism, is a constantly (re)negotiated terrain whose scope fluctuates with the alliances that bind, and the centrifugal forces that dissolve, relationships between nation-states in an anarchic world system, and by the field and sensitivity of the knowledge in motion.

Knowledge, Power, and the Mangle of Movement

Knowledge empowers.[35] Indeed, ever since the Enlightenment multiple forms of knowledge (of astronomy, of cartography, of botany, of geography) have been yoked to the ambitions of monarchies or national governments to secure legitimacy at home and to project power abroad. Beginning in the late nineteenth century, state and corporate actors became systematically engaged in, and supportive of, scientific research and technological applications to both commercial and military ends. They were central to the first wave of globalization, they animated imperial rivalry, and they transformed manufacturing industry and the conduct of warfare. In 1945 they moved definitively to the heart of the political process in "advanced" industrialized economies. The ensuing political and ideological Cold War was also a competition for scientific and technological preeminence. Decolonization and economic modernization went hand in glove with faith in science and technology to lift millions out of poverty. Today the production and control of new knowledge is even more critical to the so-called fourth industrial revolution, in which a nation's capacity in 5G wireless communications, artificial intelligence, robotics, autonomous vehicles, and a range of other "critical" emerging technologies, and their underlying science, will define its relative position in the global distribution of power in the mid-twenty-first century.[36]

In the previous section I distinguished between different administrative categories of knowledge. Epistemologically speaking, the term "knowledge" in this book includes both science and technology. At the most general level it refers to both propositional knowledge (knowing that) and tacit knowledge (knowing how). It does not reduce technology to "applied science" but takes it to be a form of knowledge, including know-how, that is embedded in material objects and practices that are designed to transform the world around us.[37] In the case studies below, as in my previous volume, knowledge emerges as an historically and socially constructed assemblage of ideas, concepts, theories, techniques, practices, facts, information, and data that are appropriated and produced in response to local needs, aspirations, and agendas. It does not necessarily move as a bounded whole. It is best treated as an assemblage that can be disaggregated into different elements that travel on a variety of platforms. These elements can be embedded in commodities and "things," like genetically modified seeds, an inertial guidance system, or a semiconductor manufacturing plant. They can be carried in the heads and hands of "knowledgeable bodies"—as Robert Oppenheimer said, "The best way to send information is to wrap it up in a person."[38] And they can be inscripted in a material support like articles and preprints, technical reports and blueprints, or on a digital platform.

How does knowledge move through networks connecting distinct geographical spaces? The question long predates the transnational turn. It was posed in a classic paper by George Basalla in 1967 entitled "The Spread of Western Science" that adopted a three-stage model to explain how the values and institutions of science "diffused" from the West to the "Rest."[39] It was addressed by Bruno Latour, who suggested that data acquired at the "periphery" was stabilized at its source and conveyed as "immutable and combinable mobiles" to "metropolitan" centers of calculation, where it was integrated into the existing pool of knowledge.[40] It has been discussed extensively by historians of science and empire, and of scientific travel, and by scholars of science, technology, and society (STS).[41] The recognition that scientific knowledge is socially constructed and locally produced, and that making it universal and portable requires human intervention, gave the question of how knowledge moves an additional epistemological twist.[42] The chapters here show that the process whereby knowledge produced in one context moves across borders and is embedded in another varies with the kind of knowledge that it is, as well as its reproducibility and manipulability (that enable its selective appropriation at its new location), and the social and political interests at stake. The accounts of how knowledge and skills are produced locally and disseminated transnationally are so heterogeneous, and each case is so specific, that no single analytic framework can adequately explain *how* knowledge moves transnationally, and the changes in meaning that such knowledge and skills assume as their context changes. To take the transnational approach is essentially to stop thinking of knowledge circulating freely and immutably from one side of a border to another and to ask how its flow is facilitated, as well as "regulated, impeded, reshaped, and reconstituted in new settings."[43]

There is no "intrinsic" reason why knowledge should move transnationally.[44] Sujit Sivasundaram warns us that "mobility should not be stressed to the extent that immobility, disjuncture and the workings of the local are forgotten."[45] Some kinds of knowledge are labeled classified so as to restrict their circulation to a small, authorized group of people and to stop them moving across national borders at all. If knowledge moves, it is because people who build the network have the power and the motivation to set it in motion, to navigate the institutions that impede or facilitate its move across borders, and to creatively adapt it to local needs at the receiving end. The negotiations at border crossings are underpinned by the understanding that knowledge empowers, and that its acquisition and deployment by the "other" opens new possibilities (and imposes constraints) that can reshape the relationship between the "nodes" in the network and reconfigure the local settings in which it was produced and used. Knowledge that crosses borders is saturated with power by the very fact that it is set in motion to serve a transformative goal.

The concept of "contact zone" is widely used by historians of science and empire to characterize a local site of encounter in an asymmetrical field of power. It was coined by Mary Louise Pratt in her analysis of travel writing, "in an attempt to invoke the spatial and temporal copresence of subjects previously separated by geographical and historical disjunctures," "usually involving conditions of coercion, radical inequality and intractable conflict."[46] Kapil Raj took these contact zones to be sites of shared learning in his study of the encounter of British imperial actors with Indian literati. He described them as sites for "the making of scientific knowledge through co-constructive processes of negotiation between different skilled communities and individuals from both regions," that reconfigured existing knowledges, skilled practices, and subjectivities for both partners.[47] If contact zones are understood as local sites of spatial and temporal *copresence* and of knowledge *co-construction*, they can be used more generally to describe any face-to-face encounter in which knowledge, tacit skills, and intangible knowledge are transferred between the participants in an asymmetric field. As such, the concept serves as a useful bridge between transnational histories of knowledge in motion and global histories of science and technology.

A transnational history that takes border crossings as key sites of analysis does not readily engage with dyadic colonizer-colonized relationships in which the reach of imperial power is undisturbed by national boundaries. It gains traction rather in colonial and neocolonial projects dedicated to nation building, modernization, and "development," in which "peripheral" entities have acquired a degree of autonomy vis-à-vis metropolitan "centers" and affirm their national sovereignty in the face of neoimperial and global pressures.[48] Emily Rosenberg has suggested that "within the friction of contact zones" between transnationally connected sites "there was creativity as well as oppression, coproduction as well as imposition of 'imperial knowledge.'"[49] This formulation is a useful starting point but it needs to be unpacked. Creativity and coproduction have often marked the transnational flow of knowledge between elites at nodes of the networks (in an asymmetric field of force). By contrast, oppression and imposition frequently characterized the local mobilization of "useful" knowledge to refashion practices on the "ground" (at the expense of indigenous knowledge and ways of doing). Knowledge moved across two borders, "horizontally" between different national centers of power and vertically, top-down, between local social strata. We can separate them analytically; in practice they were often intertwined. The mission to reform, improve, and modernize was embodied in the knowledge that traveled, and facilitated its mobility across borders and into the new ecological niche that was constructed to appropriate it—not without local resistance—at the receiving end.[50]

The sources that historians use, and the absence of subaltern voices in them, complicate immensely the effort to recapture the transformation of traditional forms of knowledge and "best practices" that are imposed "from above," producing a lacuna that turns the transnational network "into a sort of iron cage through which no native can break."[51] That cage is sometimes constructed by transnational actors themselves, who devise policies that deliberately make no allowance for engaging with their indigenous "targets" at all. Matthew Connelly has described how population control experts strived to produce policy programs for needy countries "standardized like a Model T Ford." As one critic put it, "no need to speak the language, or even to meet a non-Ph.D.-holding native. Visits to the country, if required at all, could be confined to short stays in Western luxury hotels."[52] Historians concerned with knowledge flows can fruitfully collaborate with anthropologists, postcolonial studies, feminist studies, and STS scholars who have sought to break down the walls of that political and epistemological cage.[53]

Encountering Materiality: Networks and Travel

Transnational historians focus on networks that cross national borders. Communications technologies constitute the "hard" backbone of networks.[54] The physical interconnection of different geographic spaces on the globe has been made possible by increasingly sophisticated technological infrastructures, from steamships and undersea cables to commercial jet aircraft and the internet. Today's communication technologies, which connect handheld smartphones and personal computers via the internet and the World Wide Web using fiber-optic cables and telecommunications satellites, allow "instant" real-time communication between individuals on a planetary scale. They have rendered the global "an evocative shorthand for claims for the 'death' of distance and time."[55] All national borders have not vanished into thin air, however.[56] Indeed, historical sociologist Saskia Sassen points out that "the national is often one of the key enablers and enactors of the emergent global scale."[57] Historian Jeffrey Byrne agrees that "an ever more observant and pervasive sovereign state might even be the essential facilitator of globalization, rather than its victim."[58] The national and the global are not mutually exclusive.[59] On the contrary, the capacity of nation-states (and regions like the European Union) to close their borders to travel in an attempt to contain the spread of COVID-19 affirmed their power to impose limits on the reach of globalization. The national and the global are coproduced, to use Jessica Wang's felicitous formulation.[60]

Travel through global networks crosses national borders. It is not frictionless. To speak of friction is not only to speak of obstruction: friction also makes movement possible, as in the colloquial phrase "the rubber meets the road." Travel involves physical movement that successfully navigates "the sticky materiality of practical encounters" and that foregrounds the affirmation of national sovereignty and the policing of national borders in transnational narratives.[61] At the mundane, everyday level, multiple obstacles—"insufficient funds, late buses, security searches, and informal lines of segregation"—disrupt it, as Anna Tsing reminds us.[62] Passport and visa restrictions, complemented by export controls, are invoked at borders that are policed by immigration and customs officials who affirm a state's sovereignty within its territorial limits. They were particularly important as instruments of control and surveillance by the US national security state in the early Cold War, when they were used to target the movements of "knowledgeable bodies," as Mario Daniels calls them.[63] Nonhuman actors like the environment can also constrain movement and shape trajectories of travel, especially in regions of the globe that lack a sophisticated communications infrastructure. Gisela Mateos and Edna Suárez-Díaz have described the multiple obstacles (including environmental) faced by a truck driver tasked with transporting a mobile radioisotope exhibition through a number of countries in Latin America.[64] The truck was built at the Oak Ridge National Laboratory in Tennessee with the US federal highway system in mind. It was totally ill-adapted to the winding, hilly, narrow roads of some host countries, and was too wide to be loaded as cargo on local trains. The powerful earthquake in Valdivia, Chile, in 1960, measuring 9.5, made access to the country physically impossible. Travel during the globalization era of the early twentieth century was equally perilous. In 1915 David Fullaway, an American entomologist working in Hawai'i, took a ten-month ocean voyage in search of parasites that would kill melon flies on the islands. His journey was facilitated by the interimperial links established between the botanic stations that he visited in Java, India, and the Philippines and the local knowledge of government officials that he could tap into. It was disrupted by his detention for four days by the authorities in Dhanushkodi in India, by another delay in Colombo in British Ceylon, by poor-quality coal that slowed his steamship's voyage to Singapore, and by a typhoon from the Philippines that swept across his path on his way home.[65] His need to protect his specimens led him to prefer to travel in a rugged transport ship rather than an ocean liner. It avoided northerly ports of call where lower temperatures would threaten his specimens and it was equipped with cabins that were "more suited to handling parasites en route."[66]

Microstudies of the friction and resistance that accompany travel are an essential dimension of transnational histories that are concerned not only with why but also with how knowledge moves through networks. They do not simply remind us that knowledge does not "circulate" freely. Methodologically they oblige an engagement with the nation-state as transnational actor, with the negotiated boundary between national allegiance and "scientific internationalism," and with the coproduction of national border-making/-suppressing institutions and the global networks that they disrupt/facilitate. They also cast light on global inequalities that leave their mark on the construction and maintenance of transnational networks and the "epistemic communities of experts" that do not have access to the resources of the Global North. I address this issue at greater length in the Conclusion.

The Specificity of This Volume and a Summary of the Chapters

This volume shares with other forms of transnational history a conviction that national borders are porous. It decouples these "borders" from geographical territorial boundaries and understands them as "performed" by institutions that impede or facilitate the movement of knowledge across them. It resists any formulation that suggests that knowledge moves by itself and seeks instead to excavate the sites of friction knowledge encounters on its journey through a network.[67] Where it differs from other genres of transnational history is in taking knowledge itself, embodied in people, ideas, and things, and in knowledge-based practices, as well as tacit knowledge and know-how, as its object of analysis, asking not only how and why scientific and technological knowledge move across borders, but what transnational transactions take place *at* borders. In the remainder of this introduction I briefly describe the individual chapters in this book, which cover many scientific and technological fields, and a diversity of national constellations, over a century of transnational transactions. They amply demonstrate the new vistas opened up for historians of many stripes, and for historians of science and technology in particular, who make this methodological choice.

The chapters are divided into two main groups. Those in Part I deal with the transnational circulation of military or civil-military dual-use knowledge and technologies whose flow is *regulated* by nation-states in the name of national security or national economic competiveness in the Global North. Those in Part II deal with the transnational circulation of agricultural things, knowledge, and know-how and the factors that *foster* their movement across national borders between the Global North and the Global South. They are bookended by a survey by Jessica Wang, who historicizes the collegial transnational behavior expressed in "scientific internationalism," and by a Conclusion, in which I touch on some of the political and intellectual aspects of decentering the Global North in several of the chapters.

Wang (chapter 1) announces her aim as being to "to situate the rise of international science as both ideology and practice within the history of the international system as a whole." To that end, she traces the co-dependency of scientific nationalism and internationalism from the early modern period up to the eve of World War I. Wang emphasizes that, prior to 1815, rivalry between empires for control over information (especially cartographic) and to monopolize commercially valuable materials (notably botanical) put a premium on secrecy and the use of bureaucratic and legal instruments to regulate the circulation of knowledge that had often been acquired by theft from competitors or by pillage of indigenous resources. Denial coexisted with cooperation in specific situations that enhanced state power, while also fostering the cosmopolitan internationalism of the Republic of Letters, whose members corresponded more or less freely in a parallel denationalized space. These transnational collaborative activities gained momentum after the Congress of Vienna, which helped stabilize a balance of power between imperial rivals and gradually defused tensions between them. Communications revolutions—steam-powered transport on land and on sea, and the telegraph—paved the way to more intimate face-to-face interactions and increased interdependence and interimperial collaboration. State-sponsored networks were put in place to acquire accurate data on the oceans, which enhanced the projection of imperial power for seafaring nations and facilitated international commerce. Spurred on by what Helen Tilley describes as Europe's pursuit of "global colonialism," international scientific conferences proliferated and new permanent international committees, some of them staffed by civil servants, were established from the latter half of the nineteenth century.[68] Many of these bodies brought together scientists to set regulatory standards in nomenclature and measurement that were essential to sharing knowledge in collaborative ventures as well as in global trade. Wang concludes that the scientific internationalism of the late nineteenth and early twentieth centuries evolved thanks to progressive and egalitarian commitments to the universality of knowledge and its service to the common good, along with the creation of transnational epistemic communities and their professionalization. Scientific internationalism also expressed "a pragmatic realm of interstate practice, designed for an era of increasingly globalized and interdependent commercial relations."[69]

Part I begins with chapter 2, in which Katherine Epstein describes the efforts made by the British Admiralty to regulate the flow of knowledge produced by several inventors in the first two decades of the twentieth century across an internal border between the public and private sectors, from where it could more easily flow abroad, and an external border between Britain and foreign nations. She focuses on three important inventions that secured the technological advantage of the Royal Navy's onboard weaponry: a gunnery-targeting system, a torpedo propulsion system, and a gyrotechnology that served as master gun sight when situated high on the ship to maximize visibility. The new knowledge that could be controlled took several forms—it could be encoded in documents, embodied in individual inventors, or embedded in devices—leading to disputes over what could and should be regulated. The navy wanted to secure rights to both present and future patents, and to protect them with secrecy, arguing that upgrades and improvements drew on knowledge acquired at the navy's expense, for instance in sea trials. Inventors, for their part, were wary of granting the navy a monopoly on their discoveries and could threaten to sell their inventions to a foreign power. In an effort to ensure that the knowledge remained a national asset under its control, the navy developed two foundational "legal technologies" in peacetime, decades before the United States: secret patents and the Official Secrets Act. If these legal instruments failed to ensure control, the navy resorted to other means, like smearing an inventor's name or striking him off the list of government contractors. These measures were anything but the routine bureaucratic procedures that were put in place after World War II. The Admiralty discussed particular arrangements with each inventor as it thought fit, choosing whatever mix of carrots and sticks it deemed necessary to maintain its control over his knowledge. This involved an ongoing process of negotiation over the terms of the agreement between multiple stakeholders—including a sometimes recalcitrant Treasury that was responsible for compensating inventors financially.

In chapter 3, Michael Falcone describes the first halting steps taken by American officials during World War II to construct the bureaucratic apparatus needed to protect and exploit new knowledge of penicillin shared with them by British researchers. While penicillin has already been widely studied by historians, the chapter investigates in detail a somewhat ignored aspect of its biography, namely, "the ways that transnational technology flows led the US government to construct a knowledge regime that served the interests of both American capitalism and American hegemony . . . and the transition from Pax Britannica to Pax Americana." Falcone quotes Vannevar Bush, head of the wartime Office of Scientific Research and Development (OSRD), as saying in 1945 that "American and British scientists have worked so closely together that it will be utterly impossible, and a matter of no vital interest, to attempt to assign many explicit accomplishments to one or the other." In fact, the OSRD played a major role in ensuring that the transnational flow of knowledge from Britain to America was almost exclusively one-way, infuriating British researchers, pharmaceutical companies, and the government, who were expecting reciprocity. To incite skeptical American companies to invest in the production of an unknown new drug, Bush and his close collaborators, including MIT president Karl Compton, allowed the corporations to control most penicillin-related intellectual property (IP) and helped them gain near exclusive access to global markets. The American government sacrificed the harmony of the international US-UK political and military alliance to encourage and then to protect its domestic industry, and pursued commercial avenues for strategic penicillin sales, even if it meant driving a wedge between London and its crumbling empire. As war turned to peace, it also opened new markets to US corporations in the name of public health, using penicillin as a political instrument to further the implementation of the Marshall Plan and to consolidate American global hegemony in the postwar period.

Mario Daniels's study of the contested sales of high-performance computers (HPCs) to the Soviet Union in the early days of détente (chapter 4) throws entirely new light on the creative development of instruments by allied governments to regulate the circulation of sensitive technology and knowledge. Trade in HPCs to the Soviet bloc was strictly curtailed by export controls and had to be agreed to by the members of CoCom, the allied Coordinating Committee for Multilateral Export Controls. The Soviets wanted to use computers imported from the United States or the United Kingdom, which were far more powerful than anything of their own, at their renowned high-energy physics laboratory at Serpukhov. The US intelligence community feared that these imported computers would be diverted to military ends, such as weapons system design and testing. To prevent this, British and American officials devised, and the Soviets accepted, a package of safeguards that included on-site monitoring of the computer by Western personnel at Serpukhov. The idea of using safeguards to monitor Soviet compliance with nuclear nonproliferation agreements was very much in the air at the time—the nuclear nonproliferation treaty had just been signed—and was stoutly resisted by Moscow. By contrast, safeguards built into the contract for an imported computer were accepted no less than fifteen years before they were implemented for nuclear weapons tests. They gave "Western governments unprecedented rights of access on Soviet state territory" and combined "surveillance through the physical presence of specialists with credentials from Western governments" with "close technical monitoring of the potentially dangerous calculations the computers would execute on site." As the CEO of Control Data Corporation told Congress, the Soviets "bought and paid for the computer but we are telling them what they can and cannot do with it." This intrusive behavior was not restricted to limiting the use of computer hardware to ensure that there was no "diversion" of computing capacity from Serpukhov to the weapons program. It also covered software and the tacit knowledge of how to do things. The US authorities were not only worried that the Soviets might copy their HPCs. They were also concerned that, through face-to-face learning with Western experts, they would enhance their capacity to use them in new, unexpected, and uncontrollable ways.

In chapter 5 I describe the public outcry surrounding the risk that sensitive technical data and intangible know-how had been shared by Western aerospace engineers who assisted their Chinese homologues to get to the bottom of three accidents with Long March rockets that had failed to place into orbit American-built satellites in the 1990s. The companies were accused of improving the performance and reliability not only of China's civilian satellite launchers but also of its military ballistic missiles, without first seeking an export license to do so. Two government committees looked into what unclassified sensitive knowledge had been divulged and the mechanisms that facilitated its transfer to China in the joint accident inquiries. Their epistemologically sophisticated and prudent conclusions—which reflected the difficulty they had in tracking the transnational flow of intangible knowledge and of establishing its take-up—made little difference. In a domestic political climate dominated by conservative forces in Congress opposed to President Clinton's trade liberalization with China, new laws were passed that imposed tight restrictions on the use of the Long March launcher by American satellite manufacturers even before their findings were available. The Department of Defense insisted that the only way to control the flow of knowledge in such launch campaigns was to have trained American monitors present at every step of the process, "from the cradle to the grave"—a system of "safeguards" analogous to that put in place to stop HPCs sold to the Soviets being diverted to military ends. The Chinese authorities, for their part, rejected out of hand the charges made by the US government, and insisted that they did not learn anything new from Western experts—an exchange that was all the more acrimonious since they were accused of stealing sensitive nuclear and space technology.

While these four chapters describe the measures taken by national institutions to *impede* the transnational flow of sensitive knowledge to a rival power, Part II of the collection comprises five chapters that describe the ideological, political, commercial, personal, and material considerations that *facilitated* the transnational flow of knowledge and knowledge-based commodities. To sharpen the contrast all of them relate to agriculture. Furthermore whereas, in Part I, the flows occurred between technologically sophisticated countries with already strong knowledge bases in the Global North, here the focus is on North-South relations where asymmetries of knowledge and power are more palpable. Agricultural cooperation is a marker of that asymmetry. Indeed, four of these chapters engage with "North-South" technical assistance programs intended to improve the yields of a variety of products for vulnerable farmers.

Gabriela Soto Laveaga (chapter 6) casts a critical eye on the credit given to Norman Borlaug, the "father of the Green Revolution," who introduced Mexican high-yielding varieties of semidwarf wheat seeds (Sonora 64 and Lerma Rojo 64) into India in 1963. Multiple personal, ideological, national, and material ambitions and agendas were threaded together to produce the network through which technical assistance moved from agricultural research stations in the Yaqui valley in Mexico into Indian farms several decades later. In 1943 the Mexican authorities, with Rockefeller Foundation funding, established an Office for Special Studies that could capitalize on the indigenous agronomic infrastructure put in place in the 1930s to breed hybrid seeds for a national agrarian reform program. There, Norman Borlaug, who won the Nobel Prize for Peace in 1970, cooperated with experienced Mexican agronomists to advance his research. It was this pool of hybrid seeds and local skills that was tapped and "transferred" across national borders. The geopolitical ambitions of the Mexican and American governments, the established linkages between Indian and Mexican agricultural research institutes, the high visibility of the Rockefeller and Ford Foundations on the Indian subcontinent, Borlaug's notoriety and the prestige acquired by Indian agronomists working with him, and the fear of an impending famine together facilitated the transnational transfer of the seeds and the agricultural practices needed to enhance their yield (irrigation, fertilizer, etc.). The Mexican authorities sought to lever their researchers' experience in developing hybrids to garner international prestige and to support agrarian improvement programs in South Asia by sending diplomats and agronomists on demonstration missions for skeptical Indian farmers. Borlaug and the Rockefeller Foundation were vectors of a American modernizing effort simultaneously to challenge Soviet aid initiatives in the country and to promote India "as the free world's contender against China in a fateful race to set the course of the developing world," as Henry Kissinger put it.[70] The fanfare surrounding the American contribution, and the kudos attached to Indian agronomists working alongside the US Nobel Prize winner, had the nefarious effect of writing the Mexican contribution to knowledge transfer out of the official narrative. As Soto Laveaga points out, by calling seeds with Mexican names "Borlaug's seeds," such narratives "perform a seemingly impossible act" of explicitly referencing Mexico "yet at the same time shedding themselves of any affiliation with Mexican expertise and domestic science." Restoring the Mexican contribution to its rightful place not only disrupts the dominant North-South America-centric narrative of Indian modernization and the Green Revolution. It injects a "South-South" dimension that decenters the United States and opens an intellectual space for a global analysis of the circulation of seeds across multiple national borders.

In chapter 7, Maria Gago asks how the robusta variety of coffee produced locally in the Portuguese colony of Angola moved globally onto breakfast tables in main street America. She argues that its journey was possible only thanks to the successful negotiation of agreed quality standards between agricultural scientists employed by the colonial authorities with the US-based Green Coffee Association. Robusta was produced by European smallholders and indigenous growers in the so-called cloud forests, high in the mountains of the African colony. Its bitter taste and high caffeine content were disadvantages compared to the delicate arabica flavors—until after the war. The speed and convenience of using it to make instant and packed blend coffee made it attractive to American consumers and a major source of income for the Portuguese colonial regime. Access to the lucrative US market was expanded in successive rounds of negotiations between the Portuguese Coffee Export Board and American brokers, importers, industrialists, and other stakeholders in the coffee business represented in the Green Coffee Association. By gradually opening the US market to increasingly low-quality beans, as defined by standards agreed upon by both parties, "Portugal" became the world's fourth largest coffee producer. The transnational connectivity between the two ends of the supply chain that linked producers in Angola to consumers in America was consolidated by an international agreement that restricted market access to a group of coffee-producing countries that was signed in 1962 with US support, and that excluded Cuba.

Robusta coffee moved transnationally along channels that were built within the framework of a capitalist system that wedded the manufacture of taste by the market with the domination and exploitation of labor by a colonial regime. The standards that regulated the interface between these two worlds measured quality by form, size, and defects per 450 g, and by the percentage of beans damaged by pests like the berry borer. These quantifiable properties stripped away the social, political, and environmental specificities of the local site of production, the deep knowledge required to grow the plants in the rain forests, and the use of hulling, sorting, and polishing machines imported from another Portuguese (ex)colony, Brazil. By defining a grid that correlated morphological features like size and defects with dollars and cents, standardization rendered the output from the Angolan cloud forests legible in global commodity markets. The local knowledge that had transformed a plant into a crop disappeared from view; coffee was propelled into global markets by the evaluation of its profitability by agricultural scientists, economists, state officials, market analysts, and consumer organizations.

Tiago Saraiva (chapter 8) shows that a training program organized by the UN Food and Agriculture Organization (FAO) in 1953 was the transmission belt for the transnational flow of statistical methods developed in the United States in the 1930s to the Portuguese colony of Guinea-Bissau in the 1950s. Mordecai Ezekiel of FAO's Economics Division, and previously an economic adviser to Henry Wallace, President Roosevelt's secretary of agriculture during the New Deal, devised the programs. The statistical techniques he advocated had been honed in consultation with American farmers to plan optimum corn/hog production ratios during the Depression. Ezekiel was one of many American and colonial civil servants who, after the war, moved out of national administrations into the UN system, where they "played a leading role in managing the transition from an imperial to an international world system."[71] In 1947 the fascist Portuguese colonial government, eager to establish its international respectability in an emerging postcolonial world order, committed itself to participating in a world agricultural survey promoted by the FAO. The scheme was underwritten by Truman's Point Four program, announced at his inaugural speech in 1949, to launch "a bold new program for making the benefits of our scientific advances and industrial progress available for the improvement and growth of underdeveloped areas." Amílcar Cabral, an indigenous government official in Guinea-Bissau, was sent to a training course in Nigeria organized by the FAO and the British Colonial Office and including French colonial officials. Back home he deployed the skills he had learned to devise an agricultural survey of Guinea-Bissau that used data on soil erosion as an index of colonization: land left fallow to plant groundnuts for export, to the benefit of metropolitan commercial interests, eroded more rapidly than land farmed to satisfy the nutritional needs of the domestic population. A set of techniques developed to promote participatory democracy among the farmers of the American Corn Belt were used by Cabral in agricultural surveys that "helped make plausible a political project of national independence." The universalistic, putatively anticolonial rhetoric of the young UN and its agencies, the promotion of American science and technology by the Truman administration as a tool of development, and the establishment of educational programs to improve the scientific skills of educated people in Third World countries provided the ideological impulse and material resources to transfer useful knowledge from the rich industrialized "Global North" to the emerging "Global South," including Portugal's African colonies.

In chapter 9, Courtney Fullilove contrasts two modes of agricultural development in Palestine, one focused on achieving food security, the other on achieving food sovereignty for peasant farmers. The former is typically state-centered and backed by international R&D, and is fostered by ex situ seed banks that encourage farmers to produce high-yield staple crops for the market. The latter aspires to community control of resources for food and agriculture, and arose in the 1980s in opposition to corporate and multinational control of food systems that "took up the mantle of European colonial governments in shaping institutions and regulations to organize natural resources in occupied territory." Fullilove describes the initiatives taken by a Palestinian NGO, the Union of Agricultural Work Committees (UAWC), to tap into international technical assistance programs for preserving biodiversity and food security while encouraging Palestinian farmers to aspire to food sovereignty (that is inseparable from territorial sovereignty). The UAWC was the first Arab member of Via Campesina, an antiglobalization movement dedicated to peasant control over natural resources. It gave substance to its agenda by drawing on foreign aid that had been flowing since the Oslo Accords were signed in 1993 to establish the Community Seed Bank in Hebron in 2010. The facility is equipped with a Dutch-built state-of-the-art seed blower, a grading machine, and advanced breeding and reproduction facilities. It supports rainfed agriculture, seeing it as a solution to the problem of climate change and to the uncertainty of access to water for irrigation. The seed bank focuses on forty-five local vegetable crops that are unique to local markets. It also collects wild relatives of vegetables and medicinal plants. The transnational circulation of this equipment and know-how has gone along with training UAWC staff through technical assistance and capacity-building programs that have been selectively adapted to local needs but that also "establish a hierarchy of knowledge in the practice of seed saving." The challenge the UAWC faces is to navigate between the opportunities provided by internationally sponsored high-technology and advanced knowledge, and its own rootedness in local farming practices that are associated with environmental and political ambitions that can easily clash with the aims of its donors.

Knowledge moves transnationally in a plant phenotype database described by Sabina Leonelli (chapter 10) by uploading information acquired using the local expertise of multiple stakeholders—breeders, farmers, and researchers, many of them in the Global South—into a global data linkage tool (named Crop Ontology) whose structure and management are in the hands of scientists in the Global North. The system is constructed deliberately to facilitate border crossings by sidelining "political and social considerations when it comes to the circulation of plant data," and by making it available in a readily usable form to any stakeholder who has a smartphone. The denationalization and decommodification of Crop Ontology were facilitated by three factors: a shared commitment to a bottom-up approach to define a common semantic platform for plant traits; "the absence—so far—of a strong commercial interest from the West and of highly centralized governmental control over what is being farmed," in the case of a popular legume like the cassava root; and the advantage that, in their early stages of development, large digital plant data platforms that prize openness and diversity are rich with possibilities and creative uncertainties so that no one quite knows in which direction they will evolve and just what commercial opportunities they will produce. If this ideological and political context should change, national borders that have been technically effaced on the digital platform can spring back to constrain knowledge circulation in the name of "national rights, interests, and claims to ownership." A global regime structured on the principles of scientific internationalism would then be subverted and renationalized from within by the agents of corporate capitalism in collusion with national governments.

My concluding chapter reminds readers of the implications of taking knowledge and know-how as the objects of transnational history. It stresses that the knowledge/power/national-interest nexus imposes narratives that differ depending on the knowledge itself that is in movement. Because knowledge empowers, it engages the interest of the stakeholders and the institutions that manage transnational transactions at borders. Those institutions and their representatives are in turn embedded in multiple histories, including foreign policy agendas, that intersect, overlap, and clash at the transnational juncture. The transnational historian then must not only be sensitive to the peculiarities of the kind of knowledge that moves but must also be prepared to take a multidisciplinary approach to do justice to the complex interests that engage diverse stakeholders in its movement. The inclusion of a number of chapters on postwar agricultural development in the volume also allows for a brief consideration of what transnational history involves when we decenter the Global North. It highlights the different dynamics between strategies of inhibition in North-North relations and strategies that foster collaboration in North-South and South-South transnational relations. These are but preliminary reflections, yet they help foreground the political implications of transnational histories that insist on the entanglement of national modernizing agendas in emerging powers with the already-available resources in the global pool of knowledge generated by the advanced industrial countries.

Acknowledgments

Thanks to the three anonymous reviewers, and to Martin Collins, Mario Daniels, Kate Epstein, Sabina Leonelli, Jahnavi Phalkey, and Jessica Wang for their valuable feedback on early drafts of this Introduction.

Notes

1. Sven Beckert and Dominic Sachsenmeier, *Global History, Globally: Research and Practice around the World* (London: Bloomsbury Academic, 2018); Sebastian Conrad, *What Is Global History?* (Princeton, NJ: Princeton University Press, 2016); Akira Iriye and Pierre-Yves Saunier, eds., *The Palgrave Dictionary of Transnational History from the Mid-19th Century to the Present Day* (Basingstoke: Palgrave Macmillan, 2009).

2. For a more detailed analysis see Fa-ti Fan, "The Global Turn in the History of Science," *East Asian Science, Technology and Society* 6, no. 2 (2012): 249–258; Lissa Roberts, "Situating Science in Global History: Local Exchanges and Networks of Circulation," *Itinerario* 32, no. 1 (2009): 9–29; Simone Turchetti, Nestor Herran, and Soraya Boudia, "Introduction: Have We Ever Been 'Transnational'? Towards a History of Science across and beyond Borders," *British Journal for the History of Science* 45, no. 3 (2012): 319–336.

3. I owe the term "transnational connectivities" to Emily S. Rosenberg, *Transnational Currents in a Shrinking World, 1870–1945* (Cambridge, MA: Belknap Press of Harvard University Press, 2012).

4. Fan, "The Global Turn in the History of Science," 251.

5. David E. H. Edgerton, "The Contradictions of Techno-Nationalism and Techno-Globalism: A Historical Perspective," *New Global Studies* 1, no. 1 (2007): 1–32.

6. Conrad, *What Is Global History?*, 210. History repeats itself. Edgerton quotes George Orwell in 1944 as being tired of hearing that the radio and the aeroplane had abolished distance and frontiers, when in fact they had increased nationalism and made travel more difficult between one country and another, especially during the 1930s; Edgerton, "The Contradictions of Techno-Nationalism," 12.

7. Charles Bright and Michael Geyer, "Regimes of World Order: Global Integration and the Production of Difference in 20th Century World History," in *Interactions: Transregional Perspectives on World History*, ed. Jeremy H. Bentley, Renate Bridenthal, and Anand A. Young (Honolulu: University of Hawai'i Press, 2005), 202–237, quotations at 204. See also Michael Geyer and Charles Bright, "World History in a Global Age," *American Historical Review* 100, no. 4 (1995): 1034–1060.

8. Frederick Cooper, "What Is the Concept of Globalization Good For? An African Historian's Perspective," *African Affairs* 100 (2001): 189–213, quotations at 194 and 209.

9. Ann Curthoys and Marilyn Lake, eds., *Connected Worlds: History in Transnational Perspective* (Canberra: ANU Press, 2009), speak of transnational approaches that reach "for metaphors of fluidity, as in talk of circulation and flows (of people, discourses, and commodities) alongside metaphors of connection and relationship" (6), the kind of use that Fan, "The Global Turn in the History of Science," 252, is warning us about here. Kapil Raj, *Relocating Modern Science: Circulation and the Construction of Knowledge in South Asia and Europe, 1650–1900* (Basingstoke: Palgrave Macmillan, 2007), defines a form of encounter in which production and circulation occur simultaneously (see note 43 below).

10. Fan, "The Global Turn in the History of Science," 253.

11. John Krige, ed., *How Knowledge Moves: Writing the Transnational History of Science and Technology* (Chicago: University of Chicago Press, 2019).

12. Katharine N. Rankine, "Anthropologies and Geographies of Globalization," *Progress in Human Geography* 27, no. 6 (2003): 708–734, has been particularly useful here.

13. Patricia Clavin, "Defining Transnationalism," *Contemporary European History* 14, no. 4 (2005): 421–439, at 436. I have already answered that challenge implicitly in my earlier work: John Krige, *American Hegemony and the Postwar Reconstruction of Science in Europe* (Cambridge, MA: MIT Press, 2006); Krige, *Sharing Knowledge, Shaping Europe: US Technological Collaboration and Nonproliferation* (Cambridge, MA: MIT Press, 2016).

14. Historians of US foreign policy and of diplomatic history also tend to treat technology not as an actor but as the backdrop of the stage on which the drama of social life is played out. See John Krige, "Technodiplomacy: A Concept and Its Application to U.S.–France Nuclear Weapons Cooperation in the Nixon-Kissinger Era," *Federal History* 12 (2020): 99–116. See also Odd Arne Westad, "The New International History of the Cold War: Three (Possible) Paradigms," *Diplomatic History* 24, no. 4 (2000): 551–565.

15. Ian Tyrrell, *Transnational Nation: United States History in a Global World* (New York: Palgrave Macmillan, 2007), 3, 5.

16. Rosenberg, *Transnational Currents in a Shrinking World*, 1.

17. Charles S. Maier, *Among Empires: American Ascendancy and Its Predecessors* (Cambridge, MA: Harvard University Press, 2006), 106.

18. Sandro Mezzadra and Brett Neilson, *Border as Method, or, the Multiplication of Labor* (Durham, NC: Duke University Press, 2013), 3.

19. Andrew McKeown, "Global Migration, 1846–1940," *Journal of World History* 15, no. 2 (2004): 155–189.

20. As in many of the texts cited here. See also John Krige, review of *Technology and Globalisation: Networks of Experts in World History*, ed. David Pretel and Lino Camprubí, *Journal of History of Science and Technology* 14, no. 2 (2020): 115–119.

21. Lissa Roberts suggests that we "think about circulation, not as movement that has a designated center—that is, a clear and privileged point of origin and return—but as a continuous path whose formative trajectory is constituted out of multiple points of local contact and exchange." At the risk of oversimplification a transnational approach to knowledge flows has a distinct point of departure and of arrival, treats circulation as typically linking two units, and embeds point of local contact and exchange in national contexts, cf. Roberts, "Situating Science in Global History," 18.

22. Ronald E. Doel, "Scientists, Secrecy and Scientific Intelligence: The Challenges of International Science in Cold War America," in *Cold War Science and the Transatlantic Circulation of Knowledge*, ed. Jeroen van Dongen (Leiden: Brill, 2015), 11–35; Paul Forman, "Scientific Internationalism and the Weimar Physicists: The Ideology and its Manipulation in Germany after World War I," *Isis* 64, no. 2 (1973): 150–180; Jean-Jacques Salomon, "The Internationale of Science," *Minerva* 1, no. 1 (1971): 23–42; Brigitte Schroeder-Gudehus, "Nationalism and Internationalism," in *Companion to the History of Modern Science*, ed. R. C. Olby, G. N. Cantor, J. R. R. Christie, and M. J. S. Hodge (London: Routledge, 1990), 909–919; Geert J. Somsen, "A History of Universalism: Conceptions of the Internationality of Science from the Enlightenment to the Cold War," *Minerva* 46, no. 3 (2008): 361–379.

23. James Secord, "Knowledge in Transit," *Isis* 95, no. 4 (2004): 654–672, quotations at 670.

24. Salomon, "The Internationale of Science," 25.

25. One of its main results was Mario Daniels and John Krige, *Knowledge Regulation and National Security in Postwar America* (Chicago: University of Chicago Press, 2022). The whole question first arose when I attended a town hall meeting for my engineering colleagues called by the president of Georgia Tech and addressed by three senior officials of the FBI who described the threat of sensitive knowledge transfers to China in academic settings. I engaged with a new research problematic first described in John Krige, "Regulating the Academic 'Marketplace of Ideas': Commercialization, Export Controls, and Counterintelligence," *Engaging Science, Technology and Society* 1 (2015): 1–24.

26. On classification, see Peter Galison. "Removing Knowledge," *Critical Inquiry* 31, no. 1 (2004): 229–243.

27. Alex Wellerstein, *Restricted Data: The History of Nuclear Secrecy in the United States* (Chicago: University of Chicago Press, 2021).

28. I do not discuss here the role of agnotology, the deliberate attempt to suppress knowledge to further specific commercial or political interests. The public denial by executives of US tobacco companies that smoking could cause lung cancer, when their own in-house research had established a link, is a typical case in point. So too is the repression by European elites of knowledge that certain Caribbean herbal plants were abortifacients. See Naomi Oreskes and Eric M. Conway, "Challenging Knowledge: How Climate Science Became a Victim of the Cold War," and Londa Schiebinger, "West Indian Abortifacients and the Making of Ignorance," both in *Agnotology: The Making and Unmaking of Ignorance*, ed. Robert N. Proctor and Londa Schiebinger (Stanford, CA: Stanford University Press, 2008), 55–89 and 149–162, respectively. Nor do I deal with nonscientific forms of knowledge, extensively studied by anthropologists and ethnographers. See, for example, Helen Tilley, "Global Histories, Vernacular Science, and African Genealogies: or, Is the History of Science Ready for the World?," *Isis* 101, no. 1 (2010): 110–119. Neil Safier, "Global Knowledge on the Move: Itineraries, Amerindian Narratives, and the Deep Histories of Science," *Isis* 101, no. 1 (2010): 133–145, discusses the intersection of these forms of knowing with European forms of knowledge.

29. Joseph Masco, *The Theater of Operations: National Security Affect from the Cold War to the War on Terror* (Durham, NC: Duke University Press, 2014).

30. The history and changing motivations of these policies are described in Daniels and Krige, *Knowledge Regulation and National Security.*

31. The argument developed at length in Daniels and Krige, *Knowledge Regulation and National Security,* is summarized in Mario Daniels and John Krige, "Beyond the Reach of Regulation? 'Basic' and 'Applied' Research in Early Cold War America," *Technology and Culture* 59, no. 2 (2018): 226–250, at 237–238.

32. The classic text on this ambiguity is of course Paul Forman, "The Primacy of Science in Modernity, of Technology in Postmodernity, and of Ideology in the History of Technology," *History and Technology* 23, no. 1–2 (2007): 1–152.

33. Daniels and Krige, "Beyond the Reach of Regulation?"

34. Jessica Wang, "Plants, Insects, and the Biological Management of American Empire: Tropical Agriculture in Early Twentieth-Century Hawai'i," in "Empires of Knowledge," ed. Axel Jansen, John Krige, and Jessica Wang, special issue, *History and Technology* 35, no. 3 (2019): 203–236, at 205.

35. The intimate connection between knowledge and power is one answer to Chris Bayly's concern, raised during an AHA "conversation" in 2006, about how transnational histories incorporate the element of power into the concept of circulation; see C. A. Bayly, Sven Beckert, Matthew Connelly, Isabel Hofmeyr, Wendy Kozol, and Patricial Seed, "*AHR* Conversation: On Transnational History," *American Historical Review* 111, no. 5 (2006): 1441–1464, at 1452.

36. The President of the United States, *National Strategy for Critical and Emerging Technologies* (Washington, DC: Homeland Security Digital Library, October 2020), annex, https://www.hsdl.org/?abstract&did=845571.

37. Eric Schatzberg, *Technology: Critical History of a Concept* (Chicago: University of Chicago Press, 2018); Jon Agar, "What Is Technology?," *Annals of Science* 77, no. 3 (2020): 377–382.

38. Cited by David Kaiser, *Drawing Theories Apart: The Dispersion of Feynman Diagrams in Postwar Physics* (Chicago: University of Chicago Press, 2005), 60. See also Mario Daniels, "Controlling Knowledge, Controlling People: Travel Restrictions of U.S. Scientists and National Security," *Diplomatic History* 43, no. 1 (2019): 57–82.

39. George Basalla, "The Spread of Western Science," *Science* 156, no. 3775 (1967): 611–622. For a fine reassessment, see Warwick Anderson, "Remembering the Spread of Western Science," *Historical Records of Australian Science* 29, no. 2 (2018): 73–81, https://doi.org/10.1071/HR17027. See also Roy MacLeod, "Introduction," in *Osiris*, vol. 15, *Nature and Empire: Science and the Colonial Enterprise*, ed. McLeod (Chicago: University of Chicago Press, 2000), 1–13; Roberts, "Situating Science in Global History."

40. Bruno Latour, *Science in Action: How to Follow Scientists and Engineers through Society* (Milton Keynes: Open University Press, 1986), chap. 6. This was an element of Latour's and Michel Callon's Actor-Network Theory; see Bruno Latour, *Reassembling the Social: An Introduction to Actor-Network-Theory* (Oxford: Oxford University Press, 2005). See also Matthew Sargent, "Recentering Centers of Calculation: Reconfiguring Knowledge Networks within Global Empires of Trade," in *Empires of Knowledge: Scientific Networks in the Early Modern World*, ed. Paula Findlen (London: Routledge, 2018), 297–316.

41. Authors not cited elsewhere include Francesca Bray, "Only Connect: Comparative, National, and Global History as Frameworks for the History of Science and Technology in Asia," *East Asian Science, Technology and Society* 6, no. 2 (2012): 233–241; Francesca Bray, Barbara Hahn, John Bosco Lourdasamy, and Tiago Saraiva, "Cropscapes and History: Reflections on Rootedness and Mobility," *Transfers* 9, no. 1 (2019): 20–41; Jürgen Renn and Malcolm D. Hyman, "The Globalization of

Modern Science," in *The Globalization of Knowledge in History*, ed. Jürgen Renn (Germany: Edition Open Access, 2017), 561–604, https://www.mprl-series.mpg.de/media/studies/1/Studies1.pdf; Safier, "Global Knowledge on the Move"; Sanjay Subrahmanyam, "Between a Rock and a Hard Place: Some Afterthoughts," in *The Brokered World: Go-Betweens and Global Intelligence, 1770–1820*, ed. Simon Schaffer, Lissa Roberts, Kapil Raj, and James Delbourgo (Sagamore Beach, MA: Watson International, 2009), 429–440; Zuoyue Wang, "The Cold War and the Reshaping of Transnational Science in China," in *Science and Technology in the Global Cold War*, ed. Naomi Oreskes and John Krige (Cambridge, MA: MIT Press, 2014), 343–370. For historians of technology, in addition to David Edgerton (see note 43 below), see Martin Kohlrausch and Helmuth Trischler, *Building Europe on Expertise: Innovators, Organizers, Networkers* (London: Palgrave Macmillan, 2014); and David Pretel and Lino Camprubí, eds., *Technology and Globalisation: Networks of Experts in World History* (Cham: Springer, 2018).

42. Warwick Anderson, "Introduction: Postcolonial Technoscience," *Social Studies of Science* 32, no. 5–6 (2002): 643–658; Anderson, "From Subjugated Knowledge to Conjugated Subjects: Science and Globalisation, or Postcolonial Studies of Science?," *Postcolonial Studies* 12, no. 4 (2009): 389–400; Christophe Bonneuil, "Development as Experiment: Science and State Building in Late Colonial and Postcolonial Africa, 1930–1970," in MacLeod, *Osiris*, vol. 15, *Nature and Empire*, 258–281; David Wade Chambers and Richard Gillespie, "Locality in the History of Science: Colonial Science, Technoscience, and Indigenous Knowledge," in MacLeod, *Osiris*, vol. 15, *Nature and Empire*, 221–240; David Livingstone, *Putting Science in Its Place* (Chicago: University of Chicago Press, 20003); Raj, *Relocating Modern Science*.

43. Asif Siddiqi, "Dispersed Sites: San Marco and the Launch from Kenya," in Krige, *How Knowledge Moves*, 175– 200, at 175. David Edgerton gives this refashioning a very particular meaning in his emphasis on the transnational circulation of "old" technologies that "find a distinctive set of uses outside the time and place where [they were] first used on a significant scale." He calls these "creole" technologies. David Edgerton, "Creole Technologies and Global Histories: Rethinking How Things Travel in Space and Time," *Journal of the History of Science and Technology* 1, no. 1 (2007): 75–112, at 101. See also David Edgerton, *The Shock of the Old: Technology and Global History Since 1900* (Oxford: Oxford University Press, 2007).

44. This all the more true when we consider artifacts in motion, when we can easily lose sight of "the complex material-cultural embeddedness of things that must be loosened or pried out of their original matrices in order to set them in motion"; Bray et al., "Cropscapes and History," 21.

45. Sujit Sivasundaram, "Sciences and the Global: On Methods, Questions, and Theory," *Isis* 101, no. 1 (2010): 146–158, at 158.

46. Mary Louise Pratt, *Imperial Eyes: Travel Writing and Transculturation* (London: Routledge, 1993), 7, 6.

47. Raj, *Relocating Modern Science*, 223.

48. The literature is vast. For a valuable methodological overview, see Jeffrey James Byrne, "Reflecting on the Global Turn in International History or: How I Learned to Stop Worrying and Love Being a Historian of Nowhere," *Rivista Italiana di Storia Internazionale* 1, no. 1 (2018): 11–42. For a survey of recent books, see Artemy M. Kalinovsky, "Sorting out the Recent Historiography of Development Assistance: Consolidation and New Directions in the Field," *Journal of Contemporary History* 56, no. 1 (2021): 227–239. For now classical works, see Nick Cullather, "Development? Its History," *Diplomatic History* 24, no. 4 (2000): 641–653; David Engerman, Nils Gilman, Mark H. Haefele, and Michael Latham, *Staging Growth: Modernization, Development and the Global Cold War* (Amherst: University of Massachusetts Press, 2003); Daniel Immerwahr, *Thinking Small: The United States and the Lure of Community Development* (Cambridge, MA: Harvard University Press, 2015); Ricardo D. Salvatore, *Disciplinary Conquest: U.S. Scholars in South America, 1900–1945* (Durham, NC: Duke University Press, 2016); Helen Tilley, *Africa as a Living Laboratory: Empire, Development, and the Problem of Scientific Knowledge, 1870–1950* (Chicago: University of Chicago Press, 2011). Two collections that I have coedited are also useful here: John Krige and Jessica Wang, eds., "Nation, Knowledge, and Imagined Futures: Science, Technology, and Nation Building, Post 1945," special issue, *History and Technology* 31, no. 3 (2015); and Axel Jansen, John Krige, and Jessica Wang, eds., "Empires of Knowledge," special issue, *History and Technology* 35, no. 3 (2019).

49. Rosenberg, *Transnational Currents in a Shrinking World*, 7.

50. For the concept of ecological niche, see Bonneuil, "Development as Experiment."

51. Warwick Anderson criticizing Latour's book *Pandora's Hope*: Anderson, "From Subjugated Knowledge to Conjugated Subjects."

52. Matthew Connelly, *Fatal Misconception: The Struggle to Control World Population* (Cambridge MA: Belknap Press of Harvard University Press, 2008), 234, 235.

53. Warwick Anderson's work is a good starting point: Anderson, "From Subjugated Knowledge to Conjugated Subjects"; Gabrielle Hecht and Warwick Anderson, eds., "Postcolonial Technoscience," special issue, *Social Studies of Science* 32, no. 5–6 (2002). For feminist studies, Sandra Harding, *Sciences from Below: Feminisms, Postcolonialities, Modernities* (Durham, NC: Duke University Press, 2008).

54. Of course, we must not reduce networks to the technological infrastructure on which they are built; see Rachel Midura, "Conceptualizing Knowledge Networks: Agents and Patterns of Flow," in Findlen, *Empires of Knowledge*, 373–377.

55. Martin Collins, *A Telephone for the World: Iridium, Motorola, and the Making of a Global Age* (Baltimore, MD: Johns Hopkins University Press, 2018), 3.

56. Wendy Brown describes the tension between "increasingly liberalized borders, on the one hand, and the devotion of unprecedented funds, energies and technologies to border fortification, on the other": Brown, *Walled States, Waning Sovereignty* (New York: Zone Books, 2010), 20.

57. Saskia Sassen, *Territory, Authority, Rights: From Medieval to Global Assemblages* (Princeton, NJ: Princeton University Press, 2006), 1.

58. Byrne, "Reflecting on the Global Turn in International History," 37.

59. Sassen criticizes widespread "explanations and interpretations about the new global age [that] tend to posit the mutual exclusivity of the national and the global"; *Territory, Authority, Rights*, 405.

60. Jessica Wang, private communication.

61. Anna Lowenhaupt Tsing, *Friction: An Ethnography of Global Connection* (Princeton, NJ: Princeton University Press, 2005), 1.

62. Tsing, *Friction*, 5.

63. Daniels, "Controlling Knowledge, Controlling People"; Mario Daniels, "Restricting the Transnational Movement of 'Knowledgeable Bodies': The Interplay of US Visa Restrictions and Export Controls in the Cold War," in Krige, *How Knowledge Moves*, 35–61.

64. Gisela Mateos and Edna Suárez-Díaz, "Technical Assistance in Movement: Nuclear Knowledge Crosses Latin American Borders," in Krige, *How Knowledge Moves*, 345–367, at 346.

65. Wang, "Plants, Insects."

66. Wang, "Plants, Insects," 223.

67. For a comprehensive statement, see Curthoys and Lake, *Connected Worlds*. See also Bayly et al., "*AHR* Conversation on Transnational History."

68. Tilley, "Global Histories, Vernacular Science, and African Genealogies," 112.

69. See also Rosenberg, *Transnational Currents in a Shrinking World*, which devotes considerable attention to internationalism during the wave of globalization in the late nineteenth and early twentieth centuries.

70. Nick Cullather, *The Hungry World: America's Cold War Battle against Poverty in Asia* (Cambridge, MA: Harvard University Press, 2011), 138.

71. Perrin Selcer, *The Postwar Origins of the Global Environment: How the United Nations Built Spaceship Earth* (New York: Columbia University Press, 2018), 20. See also Jessica Wang, "Colonial Crossings: Social Science, Social Knowledge, and American Power from the Nineteenth Century to the Cold War," in van Dongen, *Cold War Science and the Transatlantic Circulation of Knowledge*, 184–213.

Chapter One

Knowledge, State Power, and the Invention of International Science

JESSICA WANG

How and when did science become international, and what kind of internationalism has defined scientific activities in different historical contexts? This question has not always made sense within the history of science, a discipline founded in the midst of the progressive and cosmopolitan internationalisms of the early twentieth century, which shaped the field's conceptual apparatus from the very beginning. When George Sarton launched the journal *Isis* in 1913, he insisted on the international character of science in language perfectly consistent with the progressive internationalist sensibilities of the early twentieth century, which imagined the quest for peace and human betterment as the self-evident objectives and attainable possibilities of the modern age. Science, Sarton declared, was "the most precious patrimony of humankind" and "the great peacemaker"; it united "the most elevated and knowledgeable minds of all nations, races, and beliefs." Although he conceded that individual scientists occasionally gave in to ignoble nationalist impulses, Sarton made clear that "chauvinistic and nationalistic tendencies" had no place in science.[1] Despite one world war and a second one in progress, internationalist hopes remained viable, and Robert K. Merton in 1942 could still plausibly postulate as normative qualities of science a communism (later famously amended as "communalism") and universalism antithetical to secrecy and unconstrained by borders.[2] In the post–World War II period American physicists involved in the political battles for civilian and international control of atomic energy also confidently touted the internationalism and universalism of science as reasons to minimize secrecy and maximize information exchange. Doing so, they argued, would allow humankind to forge a cooperative global order, one that could simultaneously tap into the beneficent potential of nuclear energy and guarantee a peaceful world, safe from the nuclear devastation that future conflicts would unleash if nations did not learn to live and work with each other.[3]

In the context of such assumptions, the persistence of nationalism within a scientific enterprise imagined as inherently international seemed puzzling. By the 1970s, however, historians Paul Forman and Brigitte Schroeder-Gudehus questioned the vaunted internationalism of science and reached for deeper, more incisive understandings of the problem and its complex realities in their respective studies of the period between the world wars. Forman suggested that scientific nationalism and internationalism ought to be understood not as antagonistic but as operating hand in hand, while Schroeder-Gudehus pointed out that state-led efforts to reintegrate Germany into the community of nations in the mid-1920s actually outpaced scientific organizations' willingness to restore supposed internationalist norms and lift their own sanctions against German science.[4] In more recent decades, scholars have increasingly accepted a codependent relationship between nationalism and internationalism as characteristic of modern science.[5] Meanwhile, international and global historians have produced an increasingly sophisticated literature that interrogates the basic conceptual apparatus and configurations of political power that undergird the nation-state form, the international system, and the fundamental nature of international relationships.

This chapter brings together approaches from the history of science and international history as a means of thinking through the historical evolution of international science and the role of the state in regulating or attempting to regulate flows of knowledge from the early modern period up to the eve of World War I. I seek to situate the rise of international science as both ideology and practice within the history of the international system as a whole, in order to explore how geopolitical circumstances shaped and reshaped the international politics surrounding knowledge about the natural world. Rather than persisting with an openness/secrecy binary that associates smooth flows of knowledge across borders with a normative science and barriers with an aberrant post–World War II national security state, my account here underscores a long history of states' roles in either promoting or discouraging international scientific cooperation and interchange. Ultimately, international science and its emphasis on collaboration and exchange needs to be understood as a product of a tangled science-state relationship that constantly shifted and pivoted in response to changes in the international system itself.

I begin by providing two brief snapshots from the history of early modern science—one of cosmography in sixteenth-century Spain and the other of botany and empire in the seventeenth and eighteenth centuries—in order to outline the workings of regimes of knowledge based on imperial powers' efforts to monopolize information and natural objects, as opposed to the Republic of Letters tradition and its cosmopolitan assumptions about the circulation of knowledge. Far from being incompatible with science, state-sponsored demands for secrecy constituted an established part of the social order that defined scientific inquiry in the early modern period. Controls over knowledge flows were never complete, however, and cross-border communications and collaborations coexisted uneasily with states' efforts to preserve their respective informational edges. Official piracy and other forms of state-sponsored theft also eroded the usefulness of secrecy, and as states lost their exclusive lock on information and species, demonstrations of openness became increasingly advantageous from a diplomatic perspective. From broken monopolies, revolutions in transportation and communications, and the "long peace" of 1815 to 1914 emerged a new, self-conscious age of internationalism in scientific relations, as well as international affairs more broadly, which I discuss in the final section. The internationalism of science and its idealized assumptions about the easy, borderless movement of knowledge thus emerged historically in tandem with global affairs, and through a constant process of negotiation and renegotiation between knowledge producers and state power.

Secrecy and Place: Cosmography, Regulation, and Imperial Power in Sixteenth-Century Spain

In early modern Europe, the production of knowledge about the natural world evolved amid not just the associational linkages that forged a "republic of letters" as both ideal and social network, but also cultures of secrecy, including states' interests in the monopolization of information and objects. Empire formed an intrinsic, indivisible part of inquiry into the natural world. As Antonio Lafuente and Nuria Valverde put it, "Science and empire are cause and effect of one another . . . each determines and is defined by the other."[6] In particular, imperial expansion went hand in hand with official secrecy, since competing European powers were precisely most lacking in detailed information about the natural bounty of distant places and how to travel to them. An age of empire rooted in exploration of the unknown seemed to offer endless opportunities to seek competitive advantages through exclusivity of information and access to valuable goods.

Clandestine understandings of knowledge permeated early modern European societies for both cultural and geopolitical reasons. Early modern occult philosophy facilitated a cultural outlook that associated value and status with secrecy and the power of mystery. Nature itself had its secrets, and authoritative knowledge rested upon claims of access to arcana.[7] In Elizabethan England, for example, Ralph Bauer observes that Walter Ralegh built his stature as an authority on the New World "not through a rhetoric of transparency and self-effacement . . . but rather through a rhetoric [of] secrecy, opacity, and self-investment." Ralegh's strategy of advancement rested upon strong cultural foundations. As Bauer points out, in Shakespeare's *The Tempest* and other literary representations, learned tracts such as Sebastian Münster's *Cosmographie*, or the lived experiences of indigenous-European encounters in New Spain and other theaters of exploration, powerful forms of natural knowledge emerged from the ability to master and extract secrets from exotic places. The acquisition of knowledge required wresting away secrets not just from other people but from nature itself, the ultimate source of mystery.[8] Craft traditions also emphasized the protection of information and knowledge, so that guilds could control the privilege of entry into skilled trades. When Benjamin Franklin published charts of the Gulf Stream in the 1780s, his endeavors were facilitated by a decline in craft secrecy that had been ongoing throughout the eighteenth century, which allowed him to tap into mariners' expertise.[9] States and their efforts to control information for instrumental purposes were hardly aberrational in cultural settings that deliberately shrouded nature with a sense of mystery or jealously guarded trade secrets.

Exploration and territorial expansion undergirded European powers' efforts to monopolize information and novel sources of wealth. Recent studies of science in the Spanish empire, for example, underscore the importance attached to producing and protecting knowledge about the New World. As Alison Sandman observes, contrary to present-day assumptions that knowledge tends to circulate, "In discussions of cartography . . . the importance of secrets is assumed almost without question," and Spanish imperial officialdom placed a high premium on navigational and cartographical furtiveness. Consequently, Spain's official cosmographers understood "knowledge of geography, cartography, and the details of navigation to the Indies as state secrets, damaging if revealed," and a welter of rules and laws oversaw the licensing of pilots and mandated prohibitions against their sharing information with foreigners.[10] More generally, as María M. Portuondo has shown in her close and incisive account of the Council of Indies, the House of Trade (Casa de la Contratación), and Spanish cosmography, a complex regime of laws, regulations, and administrative power surrounded knowledge about the region that Spain called "the Indies" and mandated the defense of navigational, cartographical, hydrographical, and natural-historical information that, from the Spanish monarchy's perspective, constituted state secrets critical to the interests of the Crown.[11]

Surveys of terrain, waterways, natural resources, and local peoples were all essential for the extension of Spanish power in the New World, as means to travel reliably, identify and exploit new sources of wealth, navigate unfamiliar social relationships, and claim power over territory.[12] With the allure of novel botanical and mineral riches came transformations in the culture of knowledge itself, as explorers who once aspired to follow traditions of classical learning refashioned themselves and their work toward the increasingly pragmatic objectives of an imperial age. In the 1580s, for example, Francisco Hernández's grand ambitions to publish an all-encompassing natural history of plants in Mexico—in Latin, Spanish, and Nahuatl, no less—meshed uneasily with the Council of Indies' desire for a much narrower and focused account of materia medica. The episode, Portuondo notes, offers "a fascinating example of how the dynamics of empire shaped the practice of a scientific discipline by directing its output toward utilitarian ends."[13] Within the Council of Indies, a move away from humanistic traditions and toward mathematical rigor and exact measurement, particularly in cartography, also indicated how imperial ambitions reshaped inquiry within Spain's cosmographical establishment.[14]

Concerns about information security weighed heavily throughout this era of state-sponsored surveying, map making, and knowledge production. Intense imperial rivalries with Britain and France to establish footholds in the New World, including raids on Spanish outposts in 1578–1579 that earned Francis Drake his knighthood from Elizabeth I, reinforced the Spanish Crown's determination to preserve whatever advantages it could through secrecy. Hence an elaborate system of information control governed and emanated from the Council of Indies and the House of Trade. By the mid-sixteenth century, the monarchy possessed a virtual state monopoly on the production of cartographic materials, regularly mandated that the Council of Indies maintain control over maps and rutters from recent expeditions, and employed book bans and other forms of censorship to restrict information flows. For example, a royal edict in September 1556 forbade the publication of any book on the Indies without prior examination and approval by the Council of Indies. By the early 1570s, further legal refinements directed the cosmographer-chronicler and his notary at the Council of Indies to keep a lid on sensitive documents, including navigational materials and ethnographic descriptions of conditions in the Indies. Both officials were expected to exercise utmost discretion in the public release of information, and the original records had to be deposited in the council's secret archive. Extensive rules and instructions on how to compile, record, and update hydrographical details about the Indies also advised keeping crucial navigational details out of official rutters and charts, where they could leak too easily to unwanted recipients.[15]

The House of Trade operated within a similar regulatory environment. When it came to mapping the Strait of Magellan for purposes of settlement and defense in the early 1580s, for example, the captain of the expedition as well as cosmographer Rodrigo Zamorano back in Seville both came under official secrecy restrictions, including storage of the newly rendered maps under lock and key at the House of Trade. The aftermath of Drake's attacks added extra urgency to the need for information control. When a summary of cosmographer-chronicler Juan López de Velasco's ambitious *Geografía y descripción universal de Indias* began to circulate in 1580, Velasco himself directed that his own copy be imprinted with what Portuondo describes as "the early modern version of a red 'top secret' label," and the king recalled all copies for safekeeping at the Council of Indies.[16] In a time period of unfamiliar coastlines and waterways, when inexperienced pilots sometimes guided their ships to the wrong ports even back in Europe, maps and charts constituted vital sources of strategic advantage. From the Crown's perspective, cartographical and navigational knowledge required firm control, lest its unwanted release tip the balance in favor of Spain's European rivals for exploration and settlement in the New World. Admittedly, early modern states never controlled information as tightly as they would have liked.[17] Nonetheless, the aspirational quality of state secrecy guided institutional practices and tied the flow of information about the natural world to the discretion of the state.

Botany, Empire, and Imperial Monopolies

Sixteenth-century Spanish cosmography offers a telling example of how states sometimes erected elaborate legal and bureaucratic architectures around the regulation of scientific knowledge long before the mid-twentieth-century era of security clearances and classified information. Seventeenth- and eighteenth-century botanical endeavors provide another perspective from which to contemplate the historical combination of science, state power, and empire. As natural resources whose potential cultivation promised a wealth of medicinals and other valuable commodities, plants signified vast economic opportunities, and botany therefore loomed large as a target of state-sponsored institution building. Londa Schiebinger and Claudia Swan have described succinctly the centrality of plants to empire: "Botany was 'big science' in the early modern world."[18]

Multiple circumstances testify to European powers' investment in botany during the late seventeenth and eighteenth centuries. In France, botany came second only to cartography in funding for scientific endeavors. The British deployed well over a hundred official collectors into the field between 1770 and 1820, and that figure does not include countless correspondents in the colonies who took it upon themselves to build reputations as learned men by sending ideas and samples to London. Spain mounted close to five dozen expeditions to the New World between 1760 and 1808, primarily to investigate plant resources. By 1800, the sixteen hundred or so botanical gardens in Europe or in European-claimed territories strewn throughout the world also demonstrated imperial powers' commitment to cultivating and transplanting new plant varieties, as part of a mission of colonial agricultural development.[19] With the exploration and development of botanical resources came transfers of knowledge between indigenous peoples and European arrivals, as well as an enormity of human suffering that beggars the imagination.[20] The massive atrocities that accompanied the rise of the Dutch spice trade in the seventeenth century, the nexus of slavery and plantation agriculture in rice, tobacco, and cotton in British North America in the seventeenth and eighteenth centuries and its postcolonial descendant, the United States, as well as sugar cultivation in the Spanish Caribbean and in Brazil, Cuba, and other independent Latin American nations quickly come to mind as exemplars of this global history of colonial violence.[21]

Beyond colonization, theft also provided a time-honored means of acquiring and exploiting new plant species. Well-known examples include the British Empire's penetration of China's tea monopoly and the botanical theft that made tea production in British India possible, as well as Henry Wickham's famed success in smuggling 70,000 rubber seeds out of Brazil after a long history of failed attempts to acquire and spread rubber in defiance of Brazilian law. In the 1850s and 1860s, during the early decades of Latin American independence, the Royal Botanic Gardens at Kew—the metropolitan seat of British imperial botany—also sponsored expeditions to acquire cinchona that openly violated the export laws of Peru, Bolivia, and Ecuador.[22] The era of high imperialism and its imbrication with global capitalism further expanded the reach, power, and costs—both human and environmental—of plantation agriculture, with tea in British India and rubber cultivation in the Netherlands Indies, British Ceylon, and French Indochina as direct outcomes of biopiracy. Abusive corporate labor practices on the plantations of colonial economies, along with parallel circumstances such as the semicolonialism of US-financed, corporate banana empires in Central America or the genocidal history of rubber in the Amazon, became part and parcel of nineteenth- and early twentieth-century commodity capitalism.[23]

As imperial powers identified new sources of botanical wealth—not infrequently by cajoling or threatening indigenous informants, or by stealing outright—they also prioritized secrecy and the monopolization of valuable seed and plant varieties, and they sought aggressively to break each other's monopolies.[24] Daniela Bleichmar adeptly describes "the fiercely competitive nature of Atlantic science" in the eighteenth century, and how governments on both sides of the Atlantic "scrambled to establish, maintain, or break trade monopolies in natural commodities."[25] Lucile Brockway's early and now indispensable study of Kew Gardens also emphasized the monopolistic impulses of colonial botany from the eighteenth century all the way into the early twentieth century. As Brockway observed, collecting expeditions aggressively sought botanical riches with little regard for local sovereignty, and "European governments competed with each other, each trying to establish botanical monopolies and to break the monopolies of their rivals."[26]

In the earlier period, mercantilist economic strategies, which presumed axiomatically the need to avoid reliance on imports through the development of intra-imperial sources of spices, materia medica, and other botanical goods, went hand in hand with the long eighteenth century of rivalries and recurrent warfare among England, France, and Spain, prior to the stabilization of continental relations that came out of the Congress of Vienna in 1815. In a geopolitical order that rested on regular bouts of warfare and autarkic economic tendencies, it made perfect sense to try to keep desirable plant varieties out of the hands of competitors, whether by implementing legal prohibitions on the exportation of critical plant species or by outright denying to ambitious naturalists entry into colonial possessions.[27] Alexander von Humboldt, for example, never made it into British India because the British East India Company, conscious of his French patronage, repeatedly refused him permission. Royal Spanish assent to his earlier five-year sojourn in South America from 1799 to 1804 came with strings attached, including a demand for reciprocity in the form of plant and animal samples from the expedition. Even so, such largesse from Carlos IV toward Humboldt caught contemporary Spanish observers by surprise, which suggests the normative quality of Spanish efforts to close off the New World colonies to foreign competitors, even if Humboldt also benefited from a moment of liberal reform aimed at trying to shore up Spanish power in the Americas.[28]

The policies of imperial powers and their chartered companies to limit access to colonized spaces reflected the intense, concerted efforts of competing powers to break their rivals' botanical monopolies.[29] In an early eighteenth-century case, watchful Spanish supervision accompanied La Condamine and his fellow French naturalists to the interior of Spanish Peru. La Condamine managed to steal some cinchona and rubber seedlings anyway, but their failure to propagate back in France kept Spain's botanical stronghold over these commodities intact.[30] Frequent military conflicts also offered ample opportunities for botanical piracy. During the American Revolution, which escalated to a large-scale confrontation between Britain and France with theaters of war across the globe, the British reaped botanical rewards in 1782, when they intercepted a French ship loaded with plant materials from the East Indies.[31] The renewal of military conflict during the Napoleonic Wars also disrupted botanical exchanges between transatlantic correspondents. Thomas Jefferson and other natural history enthusiasts and political elites eagerly hobnobbed with Alexander von Humboldt during the latter's visit to the United States in 1804, but wartime interdictions by British ships threatened the trade in natural history specimens between American and French savants. When Charles Willson Peale prepared a preserved alligator for Humboldt to take back to the naturalist Étienne Geoffroy Saint-Hilaire in Paris, he also relayed his hopes to eventually convey other natural riches. For the moment, however, Peale remained too nervous about "plundering war ships and Privateers" and thought it best to wait "untill better times."[32] The politically and economically disastrous Embargo Act of 1807, which Jefferson pushed in order to pressure the British navy to stop the impressment of sailors on American ships and British confiscation of US goods, disrupted the traffic in medicinals and other botanical products as well.[33] International tensions over trade and impressment eventually led the United States into the folly of the War of 1812, which, among its less noted effects, continued to threaten Franco-American scientific exchanges. Late in 1813, for example, Thomas Jefferson promised to send to Humboldt the first two volumes of the official account of the Lewis and Clark expedition, along with tobacco seeds, "if it be possible for them to escape the thousand ships of our enemies spread over the ocean."[34]

Such circumstances highlight how warfare offered states opportunities for botanical plunder. The frequency with which colonial territorial holdings changed hands from the seventeenth to the early nineteenth centuries, whether as a result of changing diplomatic alliances or peace agreements, also reshaped the imperial botanical landscape. In 1662, Charles II's royal marriage strengthened English ties to Portugal and brought Bombay's pepper trade under British rule. A century and a half later, the Napoleonic Wars transferred Mauritius, with its valuable spice and sugar trade, from French to British control. The Dutch similarly lost opportunities for botanically based wealth in Ceylon and South Africa to warfare with the

British in the late eighteenth and early nineteenth centuries, although the British returned Java to Dutch control after the War of 1812 in order to serve their desire for a regional counterweight to the potential expansion of French power.[35] The geopolitical machinations of European power thus defined botany as an object of intense competition for natural resources, economic opportunities, and sources of imperial advantage.

Collaboration amid Competition: From the Republic of Letters to State-Sponsored Openness

At the same time, the drive for monopolization was far from complete, and within this competitive global order, networks of correspondence and exchange also coexisted in an uneasy balance between secrecy and openness. Policies of secrecy in sixteenth-century Spain, for example, did not prevent the exchange of specimens from the respective gardens of Rodrigo Zamorano of the House of Trade and Charles de L'Écluse (Clusius) in the Netherlands.[36] Such courteous forms of interchange through correspondence gradually gained social and institutional form in the late seventeenth- and eighteenth-century Republic of Letters, which established an ethos of interchange among self-identified cosmopolitans, along with formal societies and their sober eagerness to welcome appropriately serious and reliable foreign correspondents. Expectations of reciprocity among literati who craved scientific contact, and who advanced their stature and credibility through their readiness to provide novel scientific insights and materials via regular epistolary contact, enabled an elite virtual social world that prized an ideal of open exchange without regard for national borders.[37]

Victoria Johnson's recent biography of the American physician, naturalist, and aspiring institution builder David Hosack offers an illustrative example of the Republic of Letters at work. Hosack crossed paths with an astonishing array of US political leaders as he built his medical career and forged his reputation in transatlantic natural history circles. The simple fact that he served as a physician to the families of both Alexander Hamilton and Aaron Burr says much about his elite connections; Hosack attended Hamilton during the latter's final hours after his fatal duel with Burr in 1804. As a young man, Hosack had studied medicine at the University of Pennsylvania with Benjamin Rush as a key mentor, before heading to Edinburgh and London in 1793 to burnish his medical credentials. At Edinburgh he developed a fascination for medical botany and the living plant resources that undergirded the materia medica, and when he moved on to London, Hosack sought out William Curtis and his Brompton Botanic Garden, where he enthusiastically immersed himself in learning to identify medicinal plants, work with samples, and analyze them according to Linnaean systematics. The American visitor's dedication so impressed Curtis that he nominated Hosack for foreign member status in the Linnean Society of London, which became just the first of Hosack's institutional ties to the Republic of Letters. As the years passed and his circle expanded, additional laurels testified to his standing within transatlantic networks. In 1816, for example, Hosack was honored with election to full membership in the Royal Society. A year later, he earned foreign membership in the Horticultural Society of London, an occasion that he celebrated by dispatching eighteen Seckel pear trees to London, which marked the variety's introduction to the British Isles.[38]

Hosack returned to the United States with the grand aspiration of building a great botanical garden of his own, with public funding, that he hoped could aid nation building by helping the United States to develop medical, agricultural, and industrial resources. His activities situated him within a transatlantic world of botanical exchange, defined by shipments of dried specimens and living plants between interested gentlemen who molded friendships through correspondence and travel, as well as acquisitions from commercial suppliers. As Hosack pursued his dream garden, he exploited his British network and French contacts to acquire seeds and plants from all over the world. Even as the aforementioned international tensions of the 1800s hampered botanical acquisitions, Hosack nonetheless amassed a collection of hundreds of different plants from diverse places, and at its height, the garden boasted more than two thousand distinct species. These included, as Johnson notes, samples not just from continental Europe, but from "botanical outposts in Australia, India, the Far East, Africa, and South America." Various fruit trees from China and Japan, avocados from South America, cotton from the West Indies, turmeric from the East Indies, and aloe species from the Cape of Good Hope were just a few of the plants that came to populate Hosack's Manhattan greenhouse.[39]

As Hosack's activities indicate, the Republic of Letters brought to life a rich transatlantic world of correspondence and botanical exchange. Hosack's ability to accumulate a vast array of living plant species contrasted markedly with the botanical monopolies sought by states. At the same time, however, one should not overestimate the extent to which national borders dissipated within the Republic of Letters. Although American naturalists in the early national period avidly sought specimens and reputations through European contacts across national boundaries, in other settings, the density of correspondence networks tended toward intra-imperial communications rather than fully internationalized relationships, as in the case of the creole naturalists of New Spain who sought to build reputations back in the metropole. Similarly, Hosack's forebears in the colonial period relied primarily on ties to savants in London, to whom they provided highly sought-after specimens in order to solicit hard-to-acquire instruments and publications back across the Atlantic.[40] In Dutch Indonesia, visiting savants—including Joseph Banks, James Cook, and Louis-Antoine de Bougainville—could expect a warm reception, but by the mid-nineteenth century, Dutch business interests were asking the metropolitan government to restrict the circulation of Javanese plants to the Rijksherbarium in Leiden, and plant exchanges from Buitenzorg in its early years occurred mainly within Dutch circles.[41] In addition, the work of Richard Drayton and Jim Endersby on British imperial botany suggests the extent to which botanical communications, at least those mediated by state institutions, tended to reinforce intra-imperial relationships over international ties well into the nineteenth century. Timothy Barnard has also highlighted the dependence of the Singapore Botanical Garden on the British Empire's Kew-centered network for advice, leading personnel, and political support during its first several decades of operation, in which "the vast majority of correspondence was directed to Kew Gardens." The facility was not confined completely to reliance on the metropole—colonial botanists in Singapore took the measure of their own institutional progress by following developments at Buitenzorg, and by the early twentieth century, they were exchanging samples with Dutch scientists and playing host to Robert Koch and other prominent visitors. Even so, only in the 1920s did the Singapore Botanical Garden develop a more independent identity as "an autonomous center, part of larger transnational networks, specializing in the collection and collation of botanical knowledge of the Malay Peninsula and surrounding regions."[42]

These examples indicate the limits of the Republic of Letters and the extent to which flows of scientific knowledge tended toward intra-imperial networks rather than fully cross-border, international, and interimperial relationships. At the same time, however, the Republic of Letters made it possible to imagine science as an enterprise defined by cosmopolitan elites, pursued in a literary social space that proclaimed its universalist character even as it still embodied national and imperial rivalries. In the early national period of the United States, for example, when Thomas Jefferson and other compatriots challenged European descriptions and Linnaean classifications of North American species, they did so as both a matter of national pride as well as a declaration of cosmopolitan belonging.[43]

Meanwhile, under the right circumstances, states also endorsed openness over secrecy. For example, governments were sometimes more willing to grant access to their colonies when requests came from nationals not perceived as dangerous rivals. In one instance, Linnaeus's student Pehr Kalm was able to make arrangements for his North American explorations in the midst of King George's War in part because of Swedish diplomacy and in part because of the strength of Linnaeus's own correspondence networks with naturalists in England and in British North America.[44] Hence diplomacy sometimes aligned with the Republic of Letters to facilitate international courtesies and make scientific travel possible. Similarly, as Moritz von Brescius has pointed out in a fascinating and important study, in the mid-nineteenth century, the Schlagintweit brothers received permission from the British East India Company long denied to Humboldt, which allowed them to undertake an expedition in India and in the Himalayas, precisely because their German origins meant the British did not see them as a competitive threat.[45]

Furthermore, active warfare did not necessarily preclude official endorsement of international scientific endeavors. Although wartime censorship tended to impede the circulation of journals, lettered communications among the scientific literati continued to flow, even when they occasionally contained cartographical and other information that warring states might have preferred to keep to themselves. More dramatically, states also sometimes mandated official protection for traveling savants in the midst of hostilities. Consider the transits of Venus of 1761 and 1769, for which the Republic of Letters enrolled hundreds of savants to take measurements. With the Seven Years' War (1756–1763) still in force for the first transit, the British nonetheless granted safe passage to at least one set of French astronomical travelers, although such permissions did not always hold in actual practice. When Anglo-French warfare broke out yet again after the French allied themselves with the American revolutionaries in 1778, the French government took care to carve out a zone of exception for Captain Cook's third voyage and even directed French ports to provide Cook "all necessary aid, despite the war," in recognition of him "as a man whose work interests all nations."[46] Scientific expeditions could still be a risky business when military conflicts broke out, as François Arago found out in 1808, when he was arrested and jailed for several months under suspicions of espionage while undertaking geodetic measurements in the mountains of Spain.[47] But states could also choose to try to build cultural capital by demonstrating their magnanimity toward science in times of war. Science served the pursuit of national glory, and as Lorraine Daston once commented, "Glory required an audience, preferably an international one."[48]

Lack of available expertise from within also encouraged governmental outreach at times. An early example comes from Antonio Barrera-Osorio's description of how the Spanish Crown decided in 1535 to grant a license and monopoly to two Germans in an attempt to develop pastel and saffron as agricultural commodities in New Spain after prior Spanish attempts had failed. The saffron initiative quickly came to naught when rodents consumed the roots, but pastel underwent a serious trial after the Germans brought in five skilled pastel makers from France. Charles V initially directed the House of Trade to deny the French masters entry to Spanish territory in the New World, but after appeals from the bankers who were funding the project, the Crown relented. In the end, however, the product from New Spain proved unable to match the quality of Toulouse pastel, and the entire effort went bust.[49]

In addition, the notional quality of colonial political authority kept knowledge and specimens beyond the tight control of imperial regulation, as much as governments might have hoped otherwise. In the case of cosmography and the early modern Spanish Atlantic, Alison Sandman has cautioned that the diverse backgrounds of ships' crews meant that written geographical knowledge did not remain secret. To the extent that monopolies on information held, it was in the form of tacit knowledge on the part of pilots.[50] Cameron Strang's recent study of the American Gulf South from the sixteenth to the early nineteenth centuries also shows how fuzzy and ever-shifting borders combined with loose political allegiances among ambitious naturalist-explorers to thwart ambitions for monopolization, especially as Spanish power in the region weakened and was ultimately replaced by the United States.[51] The work of entrepreneurs, such as the British trader and smuggler who sold seeds of *Cinchona ledgeriana* to Buitenzorg after being rebuffed by British institutions and thereby helped to establish Dutch Java's hold on the global quinine market, speaks as well to the cross-imperial linkages that formed despite imperial powers' hopes to monopolize resources.[52]

As these examples indicate, as much as empires sought to protect their cartographical and botanical resources, even in the absence of outright theft or espionage, knowledge about the natural world leaked through difficult-to-enforce barriers. As exclusive control over facts or access to botanical goods faded, the advantages of openness began to outweigh secrecy. María M. Portuondo's account of Spanish cosmography again provides crucial insights. By the end of the sixteenth century, as the success of English raids compromised Spain's regime of secrecy surrounding detailed geographical information, publicity became increasingly useful for two reasons. First, ongoing shifts in the structure of royal patronage lent a new importance to "*public* production" of knowledge as a means of demonstrating patrons' stature. Second, with the breaking of monopolies on cosmographical information, revelation showcased Spain's geopolitical prowess and shored up its territorial claims. According to Portuondo, "For Philip III, geographical knowledge was most valuable if it could be deployed, albeit properly contextualized, to create a public image of Spanish domination and prestige."[53] Hence the Crown's new willingness to publish Velasco's once-secret summary of the *Geografía*, the results of the Council of Indies' past projects to determine the longitude of various New World locales through simultaneous measurement during lunar eclipses, previously unreleased maps, and other geographical information. Portuondo concludes that such actions, taken together, "became one of the tools through which the Spanish monarchy announced to the world the extent of its empire, and thus cosmography ceased to be the secret science of the Spanish empire."[54] The assertion of territorial prerogatives required a delicate calibration between secrecy and openness—secrecy in order to defend privileged access to physical spaces, and openness in order to legitimate claims and warn off competitors by showcasing the extent and depth of the colonial gaze and presence on distant shores.

Brockway's study of Kew also sheds light on the forms of collaboration that existed alongside interimperial competition, and the circumstances in which it made sense for imperial powers to trade botanical information and plant varieties. Simultaneous forms of competition and collaboration, for example, defined British and Dutch colonial production of cinchona in the second half of the nineteenth century. The difficulty of cultivating cinchona beyond its native habitat maintained Spain's long monopoly over the precious bark, which newly independent nations in nineteenth-century Latin America attempted to reinforce through legal prohibitions on cinchona exportation. As noted above, the Dutch broke the Spanish monopoly thanks to the global trade in illicit seeds. Meanwhile, under Kew's sponsorship, the British also acquired other cinchona varieties and set about experimenting with them in Ceylon and India. During this period, William G. McIvor, the arborist and horticulturist in charge of cinchona development in British colonial India, corresponded regularly with J. E. DeVrij, the Dutch quinine expert at Buitenzorg Botanical Gardens in Java. Together, the two men determined the highest-yielding varieties of cinchona. This level of cooperation might seem surprising, but as Brockway pointed out, two very different political economies ultimately defined the two empires' respective interests in cinchona. From the Dutch perspective, it probably helped that *C. ledgeriana*, the best-yielding species, did not grow well in India. But the British needed to produce quinine mainly to protect British troops in India, and they conceded to the Dutch its commercial development for a global market. Consequently, Brockway concluded, this economic division of labor meant that "the two empires cooperated when market forces favored cooperation, and in the scientific sphere the cross fertilization of plants and ideas was helpful to the development of both."[55]

Similarly, Kew's willingness to facilitate exchanges of rubber seeds between the major botanic gardens in Britain's south and southeast Asian colonies and Buitenzorg in the late nineteenth century might seem mystifying at first glance, given the level of effort devoted to the colonial acquisition of rubber. At a time without a consensus on the best varieties or species of rubber for plantation agriculture, however, British-Dutch collaboration benefited both parties.[56]

Kew also engaged in more open forms of sharing in the 1880s and 1890s when it published all of its findings on agave-derived Mexican sisal. In this case, Mexican sisal was not amenable to monopolization, and therefore British botany could gain greater benefits by highlighting scientific openness and universalism.[57] For the Dutch in the late nineteenth and early twentieth centuries, Buitenzorg also increasingly traded on the advantages of internationalism. The garden's lavish facilities for foreign visitors did more than simply satisfy director Melchior Treub's desire for botanical company and intellectual stimulation in a place distant from scientific life in the metropole. As Andrew Goss observes, the Foreigner's Laboratory also served as an avenue for cultural diplomacy that showcased the virtues of Dutch colonial science. Treub, according to Goss, became "the fashioner of a beautiful image of international-level scientific research being done in the Indies" and promulgated an image of enlightened imperialism through "an enduring fantasy that the colony was on the road to civilization."[58]

Scientific exchange constitutes a historical conundrum precisely because knowledge production relies upon communities' accumulated experience and ideas, yet also grants instrumental and material advantages that encourage secrecy. An imperial age that offered European powers novel resources from places yet unknown to each other provided ample opportunities to try to maintain monopolies, particularly within a context of frequent European warfare from the late seventeenth to the early nineteenth centuries. Nonetheless, knowledge crossed national and imperial borders, not because ideas "want" to circulate but because human circumstances conspired to facilitate exchanges, whether on the part of elite communities or under direct state sponsorship. Such impulses gained more regularized, institutionalized, and state-sponsored forms with the "long peace" of the period after 1815, continued efforts by imperial powers to stabilize a global order based on imperial advantage, and the internationalist ideology that came of age during the post-Vienna epoch.

A New Internationalism: Science in an International and Imperial Age, 1815–1914

Historical discussions of the rise of a novel age of internationalism, in which intellectuals, government officials, and reformers saw themselves as living in a newly modern and international age, usually start in the mid-nineteenth century. According to this line of interpretation, the decade of the 1850s launched an era of international congresses, thanks to a technological revolution in steam-powered transportation and telegraphic communications that eased travel and cross-border discussions. Not only could intellectual, scientific, and political elites now access each other's writings more easily or, when necessary, communicate on short notice, but unlike the Republic of Letters, increasingly they could expect to meet face to face with some regularity, rub shoulders and hobnob in person, and thereby coordinate and collaborate on initiatives of mutual interest. Such possibilities stimulated a zeal for interchange among increasingly professionalized and transnational epistemic communities, spurred by the need for standard setting to facilitate commercial relations, along with a growing sense that certain types of problems—particularly as related to sanitation and public health—were global in nature and required cooperative solutions between states. From the pursuit of such activities arose a self-consciously modern, internationalist ethos.[59]

Although the new epoch of conferencing and its infrastructural foundations certainly mattered, the emphasis on the turning point of the 1850s underestimates the significance of earlier geopolitical developments, namely, the Congress of Vienna as the precondition for large-scale, interstate collaborations on projects of measurement in the 1830s and 1840s. The relative stability and peace on the European continent from 1815 until the final collapse of the balance of power system established at Vienna and its dissolution into the order-shattering violence of World War I contrasted sharply with the bouts of frequent military conflict between empires from the late seventeenth century until the end of the Napoleonic Wars. England, for example, spent roughly half of the years between 1688 and 1763 embroiled in various wars involving different configurations of other European powers, particularly conflicts that included the French on the opposing side. England and France went to war again in 1778 when the French allied themselves with the American revolutionaries, and yet again in 1793, which saw the beginning of more than two decades of on-and-off conflict. Although the Crimean War of 1853–1856 and the Franco-Prussian War of 1870–1871 interrupted the post-Vienna period, as did multiple other outbreaks of smaller-scale interstate violence, nonetheless, the long peace proved remarkably stable compared to the era that preceded it.[60]

Respect for balance of power also allowed empires to expand without undue interference from each other, and to consolidate power through collaborative arrangements meant to legitimate imperial borders. At the 1884–1885 Berlin Conference, for example, European colonizers settled on a negotiated partitioning of the African continent, which essentially used diplomacy to contain interimperial divisions that otherwise threatened to undermine colonization altogether.[61] Similarly, Theodore Roosevelt's so-called Roosevelt Corollary of 1904–1905 famously called for "an international police power," in which different imperial powers would oversee political order and enforcement of financial obligations in their respective spheres of influence for the supposed benefit of all. As part of this system of interimperial management, Roosevelt imagined the United States' combination of economic power and gunboat diplomacy would discipline Caribbean governments and compel them to fulfill debt obligations in what the United States considered a financially responsible manner. Such an arrangement, Roosevelt argued, would protect not just US creditors' interests, but those of all foreign investors.[62]

From this standpoint, internationalism and cooperative international institutions developed not as a simple expression of uncomplicated cosmopolitan values, but rather, even as they reflected and strengthened boosters' faith in universalist possibilities, they also simultaneously provided means for empires to contain their tendencies toward conflict and thereby sustain a global order friendly to imperial aspirations.[63] Tellingly, anticolonial movements found their strongest footing as the delicate collusive arrangements among imperial powers collapsed, first in World War I, and then in World War II. Anticolonial activists' own internationalism and the networks that they built among themselves, the Russian Revolution and Soviet communism's ideological and organizational support for anticolonial movements, and the global wars that facilitated colonialism's own slow-burning self-destruction ultimately brought down the edifice of imperial internationalism.[64] In the century prior to World War I, however, scientific internationalism evolved within a broader apparatus of interimperial collaboration and diplomacy of empire that sustained high imperialism in the first place.

Indeed, the very word "international" came into common usage by the early 1820s, in a linguistic innovation that perhaps signaled a new era of possibility for interstate relations in Europe after 1815.[65] A decade later, cooperative ventures in the geosciences during the 1830s—and not the usual historiographical departure point of the first International Sanitary Conference in 1851 and the new era of international congresses—made manifest a novel scientific internationalism mediated by states and their interests. As Michael S. Reidy and Helen M. Rozwadowski observe, in accordance with "the larger geopolitical ambitions of maritime nations," perceptions of the ocean shifted markedly during the nineteenth century, toward "a reconception of the ocean as a physical and intellectual space full of imperial and commercial significance."[66] Nationalist ambitions abounded in the science that accompanied this profound transformation, but they took place within a setting of dramatically reduced interstate violence between imperial powers. Consequently, post-Vienna stability reinvigorated state-sponsored exploration and naval collection of magnetic data needed for improved navigation and better determination of longitude at sea. As a result, Britain and other countries mounted new expeditions in the 1820s, and in the 1830s they established cooperative efforts to coordinate observations and share data. The major international collaborations on geomagnetism began with the Magnetische Verein of 1834–1838 that Humboldt helped to organize, which started out by creating a network of observatories across northern Europe and Russia. The British followed with their own Magnetic Crusade, launched in 1838. Although the Magnetic Crusade involved an ambitious intra-imperial program to build new observatories and collect measurements throughout the British Empire, it also enlisted international partners through the governments of Russia, Prussia, Bavaria, Austria, Spain, and Belgium, as well as university-based collaborators in the United States. This organizational achievement, according to John Cawood's research from the 1970s, arguably mattered more than the actual magnetic data: "The foundation of this global network and the demonstration that such large-scale operations could be organized and carried through were probably the most important consequences of the Magnetic Crusade."[67]

Other joint projects of measurement and data gathering also promised mutual benefits in an era of reduced international tensions. Reidy has written at length about the British Admiralty's 1835 project on tidal measurements, which brought together nine countries and their respective colonial possessions in order to coordinate timed data gathering in hundreds of locales. Hopes for safer navigation encouraged participation in this enterprise and other collaborative endeavors, such as the joint efforts of Great Britain, the United States, and transatlantic cable companies in the 1860s to use telegraphy to increase the accuracy of longitude determination.[68] As in the case of the Spanish Empire's new openness to sharing its cosmographic findings at the end of the sixteenth century, in the nineteenth century, the British and US zeal for data collection, chart making, and publication of charts also provided a means of demonstrating national prowess and projecting power over the ocean. For both countries, according to Reidy and Rozwadowski, "the ocean was constructed as a space amenable to control by any nation that could master its surface and use its resources effectively."[69] In nineteenth-century ocean science, information exchange and international cooperation thus served clearly understood national interests.

As this last point indicates, scientific exchanges in the geosciences and oceanography did not connote the absence of national rivalries but rather their expression as suited to nineteenth-century geopolitical conditions. With the founding of the first peace societies in Britain and the United States in 1816, hopes for peace as a long-term if not permanent condition gained institutional form. Starting in the 1840s, Cobdenite free-trade ideology also promised that low tariff barriers would bring not just economic prosperity but an interdependence among nations that would guarantee global peace. By the second half of the nineteenth century, the London-based Cobden Club and its motto of "Free Trade, Peace, and Goodwill among Nations" reached across the Atlantic to the United States, where Cobdenite ideals strongly shaped the Democratic Party's approach to international trade. The club also boasted members in the far reaches of the British Empire, in western Europe, and in Japan.[70] The era of the "long peace" and the conceptual structures that it generated favored new collaborative possibilities between states, navies, and scientists, with that last assembly a newly coined and nascent professional category, thanks to William Whewell's 1833 neologism. At the same time, none of these groups could take peace for granted. For example, the contingency plans that the Admiralty dictated in its original orders to James Clark Ross, head of the Antarctic Expedition of 1839–1843, sternly warned that in case of war, he was not to respond aggressively to any provocations: "In the event of England being involved in hostilities with any other powers during your absence, you are clearly to understand that you are not to commit any hostile act whatsoever, the expedition under your command being fitted out for the sole purpose of scientific discovery."[71] This directive suggests both a perception of the tenuousness of international stability, as well as a recognition that the new scientific possibilities of the 1830s depended on peaceful relations between nations.

The more familiar internationalism of international scientific conferences and new, permanent international committees followed with the rapidly mushrooming growth of international gatherings in the second half of the nineteenth century, along with an expanded array of cooperative efforts. As Valeska Huber observes, "A list of international meetings on all scales and agendas contrasts twenty-four entries up to 1851 (and only one before 1815) against 1,390 between 1851 and 1899."[72] Science featured prominently in this era of burgeoning international engagement. In her 1992 study of Nobel Prizes and the twin roles of nationalism and internationalism in science, Elisabeth Crawford describes the flurry of international scientific activities that burst onto the scene from the 1860s to the start of World War I. Newly launched scientific congresses in multiple disciplines proliferated rapidly in the 1860s and 1870s, and by the final two decades of the nineteenth century, "hardly a year went by without at least one meeting of scientists in one field or another." Moreover, such meetings were no longer one-off occasions but institutionalized and convened regularly. International scientific committees began to meet in order to negotiate common nomenclature and other forms of standardization necessary to make knowledge flow, and such efforts prompted permanent institutions: "the International Committee on Atomic Weights (1897), the International Commission of Photometry (1900), the International Committee for the Publication of Annual Tables of Constants (1909), and the International Radium Standard Commission (1910), to mention only a few." Where states had compelling commercial interests in facilitating international business and trade, international scientific projects of standardization also proceeded with urgency and official oversight, as in the International Congress on Metric Standards of 1875 or the multiple meetings of the International Electrical Congress that convened in the 1880s and 1890s to establish electrical standards.[73] Tropical medicine and tropical agriculture, as fields that embodied common imperial interests, in which information exchange promised to solidify colonial projects more generally, flourished in an era of imperial internationalism.[74] The rise of international public health, starting with the first International Sanitary Conference and culminating in the early post–World War I period in the League of Nations' reporting apparatus to monitor the spread of infectious disease outbreaks, signified as well the new sense of interdependence that self-identified internationalists associated with modernity and that fueled their political commitments in the early decades of the twentieth century.[75]

Scholars have frequently viewed the new era of international congresses and scientific unions in terms of scientists' self-organization, an evolving international civil society, and a new consciousness among reformers and intellectuals about the interdependence of modern life and the progressive possibilities of modernity itself. These trends came together as an explicitly named "internationalism," which emerged in the 1860s from earlier pacifist and cosmopolitan currents.[76] At the same time, however, congresses also served as a vehicle for states to realize their mutual interests, through the technical administrative functions that so many new organizations served. As Martin H. Geyer and Johannes Paulmann pointed out in an important edited volume in 2001, historians have tended to focus disproportionate attention on moral reform groups, the organized peace movement, or the humanitarian functions of the Red Cross, but the organizations attached to these activities were not really representative of the internationalist currents of the late nineteenth and early twentieth centuries. Rather, scholars needed to look at less glamorous activities, such as those tied to standards and trade: "the Bureau of Measures and Weights . . . with its specialist character, seems more typical of the majority of such organizations."[77]

The Smithsonian Institution and its role in facilitating exchanges of scientific publications bring this nineteenth-century world of international bodies and their administrative functions to life and highlight the elaborate political and international machinery and grand expectations behind the otherwise seemingly mundane practice of the circulation of published government documents. Official governmental support for such exchanges also underscores the place of the state in validating and mediating the flow of scientific information. In 1875, the International Congress of Geographical Sciences met in Paris and initiated the formation of a new commission, the International Bureau of Exchanges, to facilitate the interchange of national governments' official scientific documents. Notably, the sciences of state power and territoriality, namely, the natural and human sciences tied to navigation, surveying, and the occupation and oversight of territory, dominated the Smithsonian's official account of the new body and the publications of interest. The commission's work concerned publications in "first, astronomy, geodesy, cartography, geography, topography, geology, mineralogy, botany, anthropology, hygiene, zoology, entomology, explorations and travels, history, archaeology, linguistics, numismatics, &c.; and, secondly, statistical information of all kinds."[78] This mission signaled that the International Bureau of Exchanges constituted not just an exercise in international comity but an enterprise tied to the projection of state power.

The United States entered the new convention with a decades-long history of government-sponsored document exchanges mediated by the Smithsonian Institution. In its official account, the Smithsonian identified the eighteenth-century Republic of Letters and its transatlantic correspondence networks between scientific societies as the origin point of its own exchange system, followed by the exertions in the 1830s and 1840s of Alexandre Vattemare, a renowned French stage performer turned philanthropist and *Kulturträger*, who took it upon himself to develop a system by which libraries could trade duplicate copies. The sharing of such documents, Vattemare argued, created value from materials that otherwise languished in storage: "precious waste, which the savant, only with regret, buried in the dust of forgetfulness."[79]

Cultural diplomacy and government support proved vital to Vattemare's efforts. During roughly the same time period as Alexis de Tocqueville's famous travels, chronicled in *Democracy in America*, Vattemare also took to the road in the United States, glad-handed with state-level and federal officials (as well as city governments and learned elites), and negotiated agreements such as a $3,000 appropriation from the state government of Louisiana to gather publications for dispatch to France. Throughout the 1840s, he energetically arranged shipments of books and documents from the eastern seaboard of the United States to Paris, and he convinced the secretary of the Treasury to allow duty-free entry of French publications sent in return. Meanwhile, in 1840, a joint congressional resolution authorized the exchange of duplicates and the printing of fifty additional copies of congressional documents specifically for purposes of foreign exchange, and in 1848, Congress's Library Committee appointed Vattemare its agent to oversee US exchanges with France. Results apparently did not measure up to promises, however, and after five years, Congress revoked Vattemare's status after its librarian expressed repeated unhappiness about the substandard quality and disorganization of materials received. Although the Smithsonian publicly incorporated Vattemare into the historical lineage of its own exchange network, in less visible communications Joseph Henry, the Smithsonian's founding secretary and part of the leading elite of American science, belittled the Frenchman's project. As Henry put it in an 1864 letter to the assistant secretary of state, Vattemare was engaged in "an irresponsible personal enterprize," which the Smithsonian had rightfully replaced.[80]

Henry's disparaging attack on Vattemare evinced his own ambitious effort to make the Smithsonian the key American organization for overseeing international document exchanges between individual scientists or learned societies around the world and thereby fulfill the quasi-governmental body's mandate to promote "the increase and diffusion of knowledge."[81] Starting in 1848 with the distribution of its own publications, the Smithsonian acted as a central clearinghouse by assembling a network of consuls "and other responsible individuals"—mainly in European centers, but also in Egypt and scattered parts of Asia and the Americas—to receive and distribute American scientific materials and gather foreign publications in return. Formalized exchange required not just relationships among scientists, but governmental authority, in the form of arrangements for duty-free entry of documents and collaboration with both US and foreign consular officials to receive, sort, and forward printed materials. Governments also sometimes lent a hand in arranging with shipping companies for free carriage or reduced rates, although the Smithsonian managed most of its freight issues on its own.[82]

For the US government, the Smithsonian's exchange service provided opportunities to showcase American scientific prowess and to acquire worthwhile information in return. Congressional resolutions to dispatch volumes from the account of the Wilkes Expedition of 1838–1842 to the governments of Russia and Ecuador in 1849 and authorization for the State Department's purchase of a hundred copies each of John James Audubon's *Birds of America* and *Viviparous Quadrupeds of America* "for exchange with foreign governments for valuable works" in 1856 attested to the diplomatic significance associated with the circulation of knowledge through prominent publications. Hence the ready collaboration of the Coast Survey, the US Patent Office, the commissioner of Indian affairs, and other federal agencies in providing the Smithsonian with copies of reports to trade with foreign governments.[83]

Congress further regularized the document exchange network in 1867, with a resolution that called for every federal agency to submit fifty copies of its publications to Congress's librarian for foreign distribution by the Smithsonian.[84] By 1870, the contents of Smithsonian-mediated shipments were reaching some two thousand recipients in twenty-six different countries, including six hundred in Germany alone.[85] With the Smithsonian system thus well established by the time of the 1875 International Congress of Geographical Sciences, Henry and his successor, Spencer F. Baird, convinced government officials to designate his organization the official exchange agent for the new international bureau, rather than setting up a new office within the Department of State. The Smithsonian eventually earned a congressional appropriation as well, which subsidized the costs of a program that had grown to consume about $10,000 a year, or a quarter of the institution's budget.[86]

The commitment to document exchange reflected both internationalist ideals and perceived national needs. An 1879 essay in the *International Review* adeptly captured the multiple purposes at work: "the [Smithsonian] Institution has not only served to increase and diffuse knowledge, but it has enhanced the reputation of our own country abroad, and has been largely instrumental in aiding to promote a kindly and sympathetic feeling between the New and the Old World . . . the tide of communication flows steadily, freely, and pleasantly."[87] Writing in a similar vein, US secretary of state Frederick T. Frelinghuysen informed President Chester A. Arthur in April 1882 that US participation in the new international document-sharing system promised not just "harmonious relations with the like international exchange bureaus in other countries," but also the prospect to "greatly enlarge the beneficial results obtained under the present system of private enterprise."[88] Sentiment and pragmatism neatly aligned to make information exchange a shared project of scientists and state.

These observations testified to how, at one level, the reciprocal flow of printed material grew out of nineteenth-century hopes that interchange could promote cultural understanding and encourage peaceful international relationships, in accordance with the century's dreams of universal peace and unified humanity. Such discourses were common in the nascent peace movement and its related offshoots within Cobdenite free-trade circles, and they permeated scientific discussions as well.[89] Joseph Henry, writing in the Smithsonian Institution's annual report for 1852, emphasized that the distribution of Smithsonian publications belonged to "a general system of international communication," which helped to generate a universalist culture "of literature and science common to all." Such a culture promoted not just knowledge, but harmonious human relations, by creating "a community of interest and of relations of the highest importance to the advancement of knowledge and of kindly feeling among men."[90] Similar notions pervaded the elite world of nineteenth-century diplomacy. As a Colombian official informed the US secretary of state in November 1869, "The Colombian Government considers the exchange of their respective literary and scientific productions as an effective means of developing the civilization and wealth of nations, of drawing their mutual relations closer, and of rendering the same more fraternal."[91] In 1877, Portuguese commissioners praised the new international exchange bureau in like-minded fashion. The sharing of documents, they declared expansively, "ought to create indissoluble bonds of union between the different groups of the human family—bonds which cannot fail to be most profitable to the great cause of civilization."[92] Interchange as an expression of universalist values and aspirations for a peaceful international order thus established firm footing in the intertwined circles of science and diplomacy by the last third of the nineteenth century.

Such lofty sentiments might not have gained traction if not for the economic and cultural advantages also claimed by advocates of international document sharing, as in Congress's eager purchase of Audubon's volumes for distribution abroad. Benjamin Peirce, part of the scientific elite of the self-dubbed Lazzaroni, who pushed to ally science with American nation-building, praised the Smithsonian's exchange system to Representative Nathaniel P. Banks Jr. in 1855 for its showcasing of American cultural prowess: it "has given the world the opportunity to witness what triumphs American genius can achieve in literature and science."[93] Exchange also had direct economic implications. The state government of Kentucky commissioned James K. Patterson, president of the Agricultural and Mechanical College of Kentucky University, to attend the 1875 geographical congress. In his report on the proceedings, Patterson emphasized the "practical utility" of the geographical sciences. He highlighted first and foremost Kentucky's need to develop its natural resources, particularly coal and iron. Geological surveys, if distributed abroad, could attract much-needed foreign capital. International meetings themselves also offered opportunities for the state to sell its wares and to attract migrants for agricultural labor. Patterson acknowledged the risks of foreign investment and the need to make certain that the development of Kentucky's mining and mineral wealth would benefit state residents and not simply line foreign pockets. His analysis, however, left no doubt about the advantages that would accrue from distributing knowledge about the state and its economic potential.[94]

The history of the Smithsonian's document exchange network and its relationship to the International Bureau of Exchanges underscores the combination of scientific and state interests that structured information flows and made the much-vaunted internationalism of science as much a product of government as it was of science. This history also reinforces insights elsewhere in the literature about the internationalism created by nineteenth-century epistemic communities, their ideological commitments, and the interplay between science and state that brought institutionalized forms of collaboration and government into being. Martin Geyer has shown how problems of standardization in weights and measures proved a constant preoccupation in the international statistical congresses of the 1850s and 1860s, as well as the international expositions staged as pageants of imperial prowess. As in the case of international document exchanges, practical discussions of standards stood side by side with lofty discourses about civilization and international harmony, and they bound together national objectives and visions of international concord. In the case of the Metre Treaty of 1875, an "epistemic community of experts" helped to push the negotiations forward, and in the process, the treaty itself also aided the consolidation of this transnational community of scientific experts, including the scientist-civil servants who staffed the newly formed International Bureau of Weights and Measures, itself housed on ground outside Paris defined as international territory. The treaty and its associated bureaucratic systems, Geyer concludes, simultaneously served international science and the aims of nation-states: "The adoption of the metric system gave scientists the opportunity to join an international epistemic community of like-minded colleagues, yet still left ample room for national styles of precision and control."[95]

Geyer's emphasis on scientists' dual role as experts and as civil servants, as well as the US government's involvement in the Smithsonian's document exchange efforts, also call attention to one of Madeleine Herren's key interventions in the literature, namely her point about the critical role of *formal* government representation in the age of conferencing. Scientific congresses did not simply connote international science as a form of voluntary association across borders. *States* and their representatives, in particular civil servants, had significant interests and roles to play in such gatherings. As Herren writes:

> Officials from the postal ministries took part in the congresses of the Universal Postal Union. Likewise the telegraph offices sent experts to international meetings, as did the government's offices of health, insurance, and labour. Official

> state delegates represented central meteorological offices, geodetic commissions, and national observatories. International contact justified the national expansion of administrations for specific subjects and facilitated the process of developing norms and standardization. Internationally active civil servants who came from these offices constituted a significant group of internationalists which has hitherto been virtually disregarded.[96]

Through such collaborations emerged administrative, legal, and organizational apparatuses for dealing with a wide range of needs generated by interstate relations, some scientific or technical in nature and others not. International document exchanges were just one component of an expansive array of activities and concerns, including international postal delivery, telegraphy, railways, protection of intellectual property, public health and sanitation, agreement on the prime meridian and other geodetic matters, interdiction of the international drug trade, and attention to labor standards. These diverse fields of endeavor constituted just some of the subject matter that a combination of diplomacy and internationalist aspiration made into objects of regulation by international unions and other organizations between the 1860s and the start of World War I.[97] More recently, Glenda Sluga has also stressed the administrative functions carried out by new organizations that, by the early twentieth century, "regulated the movement of people, things, and diseases across the borders of member political states," and that "were attributed a simultaneously political and social influence on internationalism by virtue of their administrative internationality."[98] The scientific internationalism of the late nineteenth and early twentieth centuries thus evolved not just as a result of progressive and egalitarian commitments to the universality of knowledge and its service to the common good, or as a product of transnational epistemic communities and their professionalization, but also as a pragmatic realm of interstate practice, designed for an era of increasingly globalized and interdependent commercial relations.

Conclusion

Over the past several decades, historical scholarship has steadily moved away from the mid-twentieth-century confidence in the inherently international character of science and its roots in an uncomplicated universalism. This brief survey of international science and cross-border movements of knowledge—or their obstruction—from settings ranging from sixteenth-century Spain to the international system of the late nineteenth century has attempted to explore the interplay between state power, geopolitical currents, and international scientific relations. Attention to the changing social and political contexts that have shaped scientific pursuits across long periods of time underscores the constant processes of adjustment and balance between different parties and among competing priorities that have always defined scientific inquiry. States and epistemic communities together navigated—with varying degrees of common purpose and conflict—the push and pull of secrecy and openness, nationalism and internationalism, and imperialism and egalitarianism.

Moreover, these binary categories that formed the social and political terrain in which knowledge moved are not themselves irreconcilable opposites. Given the function of international relations itself as a forum in which states pursue national interests, along with the historical emergence of modern nation-states in parallel with the consciously articulated internationalism of the late nineteenth and early twentieth centuries, this codependency of nationalism and internationalism makes sense.[99] More generally, the fuzziness of these categories and their ability to encompass wide-ranging and contradictory meanings require one to contend with multiple nationalisms, internationalisms, and modernities, which further complicates any effort to get a handle on the nature of internationalism in science.

Few historians today would be discomfited by the idea that the historic ebbs and flows of science, state power, and the pathways through which knowledge moves, with their ever-changing currents and eddies, defy any attempt to identify essentialisms in the nature of science. Science is as national and international as states, societies, and epistemic communities require it to be, through their mutual negotiations and antagonistic dealings, which take place within geopolitical settings that themselves contribute to choices about secrecy and openness. That conclusion might be banal, were it not for all of the attendant risks of such negotiations. In a twenty-first-century world where American scientists face an increasingly complex regulatory environment over their dealings with foreigners, especially from China, and where large fines and long jail sentences loom as all-too-real possibilities, the relative configurations of political power and scientific authority weigh heavily.[100] Scientists at the state and federal levels in the United States, and also at the federal level in Canada, have also reported official prohibitions and pressures against discussion of climate change and related topics in recent years.[101] In addition, allegations early in 2020 that China's government had downplayed the threat from a novel coronavirus strain detected in Wuhan suggest the high stakes surrounding information and its control in an era of a now-global pandemic.[102] If science were inherently international, open, and free, complacency would be rewarded. But life has never been so simple, and when knowledge moves, it is sometimes only through hard-fought political battles and choices.

Acknowledgments

This chapter is dedicated in memory of Walter LaFeber, a legendary scholar, teacher, and mentor, who left behind an indelible legacy at Cornell University and beyond. He is deeply missed.

Notes

1. George Sarton, "Histoire de la Science," *Isis* 1, no. 1 (March 1913): 3–46, at 43–44. On the progressive and cosmopolitan internationalist imagination, see Glenda Sluga, *Internationalism in an Age of Nationalism* (Philadelphia: University of Pennsylvania Press, 2013), Introduction, esp. 2 and 7. Sarton's ideals reflected not just the internationalist moment of the turn of the twentieth century but also the cultural and political project of certain scientists and intellectuals to assign science a particularly distinctive role in promoting peace and internationalism. See Geert Somsen, "'Holland's Calling': Dutch Scientists' Self-Fashioning as International Mediators," in *Neutrality in Twentieth-Century Europe: Intersections of Science, Culture, and Politics after the First World War*, ed. Rebecka Lettevall, Geert Somsen, and Sven Widmalm (New York: Routledge, 2012), 45–64. In the same volume, Brigitte Schroeder-Gudehus also highlights the early twentieth-century origins of scientific internationalism while at the same time asking readers to interrogate with a more critical eye this "master narrative" of scientific identity. Schroeder-Gudehus, "Probing the Master Narrative of Scientific Internationalism: Nationals and Neutrals in the 1920s," in Lettevall et al., *Neutrality in Twentieth-Century Europe*, 19–42.

2. Robert K. Merton, "A Note on Science and Democracy," *Journal of Legal and Political Sociology* 1 (1942): 115–126, at 118–124.

3. On the Manhattan Project generation and American scientists' discourse of internationalism after World War II, see Jessica Wang, *American Science in an Age of Anxiety: Scientists, Anticommunism, and the Cold War* (Chapel Hill: University of North Carolina Press, 1999), 17–18, 22, 122–123, and 134–135.

4. Paul Forman, "Scientific Internationalism and the Weimar Physicists: The Ideology and Its Manipulation in Germany after World War I," *Isis* 64, no. 2 (June 1973): 151–180; Brigitte Schroeder-Gudehus, "Challenge to Transnational Loyalties: International Scientific Organizations after the First World War," *Science Studies* 3 (1973): 93–118; and Schroeder-Gudehus, *Les Scientifiques et la Paix: La Communauté Scientifique Internationale au Cours des Années 20* (Montréal: Les Presses de l'Université de Montréal, 1978). See also Jean-Jacques Salomon, "The *Internationale* of Science," *Science Studies* 1, no. 1 (1971): 23–42, at 30.

5. See, for example, Brigitte Schroeder-Gudehus, "Nationalism and Internationalism," in *Companion to the History of Modern Science*, ed. R. C. Olby, G. N. Cantor, J. R. R. Christie, and M. J. S. Hodge (London: Routledge, 1990), 909–919; Elisabeth Crawford, *Nationalism and Internationalism in Science, 1880–1939: Four Studies of the Nobel Population* (Cambridge: Cambridge University Press, 1992), chap. 2; Geert J. Somsen, "A History of Universalism: Conceptions of the Internationality of Science from the Enlightenment to the Cold War," *Minerva* 46, no. 3 (September 2008): 361–379, at 364–366; and Lettevall et al., *Neutrality in Twentieth-Century Europe*, chaps. 2–8.

6. Antonio Lafuente and Nuria Valverde, "Linnaean Botany and Spanish Imperial Biopolitics," in *Colonial Botany: Science, Commerce, and Politics in the Early Modern World*, ed. Londa Schiebinger and Claudia Swan (Philadelphia: University of Pennsylvania Press, 2005), 134–147, at 134.

7. William Eamon, *Science and the Secrets of Nature: Books of Secrets in Medieval and Early Modern Culture* (Princeton, NJ: Princeton University Press, 1994).

8. Ralph Bauer, "A New Book of Secrets: Occult Philosophy and Local Knowledge in the Sixteenth-Century Atlantic," in *Science and Empire in the Atlantic World*, ed. James Delbourgo and Nicholas Dew (New York: Routledge, 2008), 99–126, at 102–104 and 109–110, quotation at 103.

9. Joyce E. Chaplin, "Knowing the Ocean: Benjamin Franklin and the Circulation of Atlantic Knowledge," in Delbourgo and Dew, *Science and Empire in the Atlantic World*, 73–96, at 73, 75, and 89–90.

10. Alison Sandman, "Navigation, Cartography, and Secrecy in the Early Modern Spanish Atlantic," in Delbourgo and Dew, *Science and Empire in the Atlantic World*, 31–51, at 33–36, quotations at 31 and 33–34, respectively.

11. María M. Portuondo, *Secret Science: Spanish Cosmography and the New World* (Chicago: University of Chicago Press, 2009).

12. Portuondo, *Secret Science*, 60–62. On the territorial and political claims attached to the gathering and creation of official information, see Benedict Anderson's account of maps, surveying, and censuses in *Imagined Communities: Reflections on the Origin and Spread of Nationalism*, rev. ed. (London: Verso, 1991), chap. 10. In this context, consider also James C. Scott's observations about standard measurements, cadastral maps, and standardized surnames for official registers as projects of legibility by which states seek to extend their power: Scott, *Seeing Like a State: How Certain Schemes to Improve the Human Condition Have Failed* (New Haven, CT: Yale University Press, 1998), 25–52 and 64–71.

13. Portuondo, *Secret Science*, 94–95, quotation at 95. See also 119–120, on the systematization of information handling at the Council of Indies, so that authorized persons could access regularly updated findings about the Indies rather than allowing the work of exploration to languish for years in the hands of individuals engaged in life-long missions of scholarship.

14. Portuondo, *Secret Science*, 66 and 79–84.

15. Drake's depredations and their impact are discussed in Portuondo, *Secret Science*, 103, 195–196. On the general information controls governing publication and censorship, see 104–107. The regulatory environment surrounding the cosmographer-chronicler and the notary at the Council of Indies is addressed on 120–131. For a sense of the complex organization and functions of the Council of Indies, see also the chart on 127.

16. On the sensitivity surrounding the new maps and rutters for the Strait of Magellan, consult Portuondo, *Secret Science*, 96–99. On the information controls surrounding the summary of Velasco's *Geografía*, see 193–196.

17. On this point, see Sandman, "Navigation, Cartography, and Secrecy," 37–39 and 45–46.

18. Londa Schiebinger and Claudia Swan, "Introduction," in Schiebinger and Swan, *Colonial Botany*, 1–16, at 3. In a separate essay, Daniela Bleichmar echoes this sentiment by describing eighteenth-century botany in the Atlantic world as "both big business and big science"; Bleichmar, "Atlantic Competitions: Botany in the Eighteenth-Century Spanish Empire," in Delbourgo and Dew, *Science and Empire in the Atlantic World*, 225–252, at 226.

19. Londa Schiebinger, *Plants and Empire: Colonial Bioprospecting in the Atlantic World* (Cambridge, MA: Harvard University Press, 2004), 10; Daniela Bleichmar, "Atlantic Competitions: Botany in the Eighteenth-Century Spanish Empire," in Delbourgo and Dew, *Science and Empire in the Atlantic World*, 225–252, at 225; Schiebinger and Swan, "Introduction," 13; Richard Drayton, *Nature's Government: Science, Imperial Britain, and the "Improvement" of the World* (New Haven, CT: Yale University Press, 2000), chap. 3, esp. 50, 65, and 72–77.

20. A growing literature documents the centrality of the transfer of indigenous knowledge to European natural history and its subsequent erasure in discovery narratives. For a now classic account, see Judith Carney's study of rice culture in west Africa and the technology transfers that brought cultivation and processing techniques to the Americas, including South Carolina, via slavery: Carney, *Black Rice: The African Origins of Rice Cultivation in the Americas* (Cambridge, MA: Harvard University Press, 2001). More generally, studies of early modern science now regularly emphasize how European explorers relied upon indigenous informants or locals from diasporic populations to make sense of the botanical environments in which they found themselves. See, for example, Antonio Barrera, "Local Herbs, Global Medicines: Commerce, Knowledge, and Commodities in Spanish America," in *Merchants and Marvels: Commerce, Science and Art in Early Modern Europe*, ed. Pamela H. Smith and Paula Findlen (New York: Routledge, 2002), 163–181, at 167; and Schiebinger, *Plants and Empire*, 75–82. For a recent overview of studies of indigenous knowledge systems and natural history, along with the author's own account of the interplay between Māori and Linnaean taxonomies, see Geoff Bil, "Indexing the Indigenous: Plants, Peoples, and Empires in the Long Nineteenth Century" (PhD diss., University of British Columbia, 2018). Meanwhile, another

vein of scholarship has emphasized the importance of natural history investigation by aspiring colonials, whose presence has also sometimes been erased in accounts of metropolitan science. See, for example, Jorge Cañizares-Esguerra's account of how creole naturalists in Spanish America contributed to Alexander von Humboldt's ideas about biogeography: Cañizares-Esguerra, "How Derivative Was Humboldt? Microcosmic Nature Narratives in Early Modern Spanish America and the (Other) Origins of Humboldt's Ecological Sensibilities," in Schiebinger and Swan, *Colonial Botany*, 148–165. In the era of the early republic,US naturalists also forged rich correspondence networks in order to try to make their reputations in European metropolitan circles. On this point, consult Susan Scott Parrish, *American Curiosity: Cultures of Natural History in the Colonial British Atlantic World* (Chapel Hill: University of North Carolina Press, 2006), chap. 3; and Victoria Johnson, *American Eden: David Hosack, Botany, and Medicine in the Garden of the Early Republic* (New York: W. W. Norton, 2018). Neil Safier has also noted the effacement of both indigenous and nonindigenous but local informants in La Condamine's accounts of his South American travels: Safier, *Measuring the New World: Enlightenment Science and South America* (Chicago: University of Chicago Press, 2008), 252 and 255. More broadly, scholars have increasingly replaced diffusionist accounts of science as a European invention with an emphasis on how cultural encounters in contact zones around the world generated scientific knowledge in the early modern period. As Kapil Raj has observed, "the contact zone was a site for the production of certified knowledges which would not have come into being but for the intercultural encounter between South Asian and European intellectual and material practices that took place here [in South Asia]"; Raj, *Relocating Modern Science: Circulation and the Construction of Knowledge in South Asia and Europe, 1650–1900* (Houndmills, UK: Palgrave Macmillan, 2007; originally published Ranikhet, India: Permanent Black, 2006), 13.

21. Julie Berger Hochstrasser has described the genocidal violence of the Dutch East India Company and the massacre of the Bandanese people in early seventeenth-century Batavia as part of the spice trade: Hochstrasser, "The Conquest of Spice and the Dutch Colonial Imaginary: Seen and Unseen in the Visual Culture of Trade," in Schiebinger and Swan, *Colonial Botany*, 169–186, at 174–175. The global history of slavery is too vast and complex to even begin to cite here. "The Bibliography of Slavery and World Slaving," hosted by the University of Virginia at http://www2 vcdh.virginia.edu/bib/, does not appear (as of May 2021) to have been updated for the past decade or so, but it still provides a useful starting point for navigating this prominent field of scholarship.

22. On the British theft of tea and the political economy surrounding tea cultivation in British India, consult Courtney Fullilove, *The Profit of the Earth: The Global Seeds of American Agriculture* (Chicago: University of Chicago Press, 2017), chap. 3. The story of Wickham and the transformation of rubber into a colonial plantation commodity is well told in Joe Jackson, *The Thief at the End of the World: Rubber,*

Power, and the Seeds of Empire (New York: Viking Penguin, 2008). Kew's cinchona gathering mission is discussed in Lucile H. Brockway, *Science and Colonial Expansion: The Role of the British Royal Botanic Gardens* (1979; repr., New Haven, CT: Yale University Press, 2002), 114–116. The Dutch government also attempted to acquire cinchona illegally, as in the case of a warship dispatched to pick up J. K. Hasskarl and his ill-gotten booty of Peruvian cinchona plants in 1854, in an effort to establish cinchona in Dutch Java; Andrew Goss, *The Floracrats: State-Sponsored Science and the Failure of the Enlightenment in Indonesia* (Madison: University of Wisconsin Press, 2011), 37.

23. My understanding of the US banana empire derives largely from Jason M. Colby, *The Business of Empire: United Fruit, Race, and U.S. Expansion in Central America* (Ithaca, NY: Cornell University Press, 2011); and John Soluri, *Banana Cultures: Agriculture, Consumption, and Environmental Change in Honduras and the United States* (Austin: University of Texas Press, 2005). The human costs of rubber in the Amazon are documented in Brockway, *Science and Colonial Expansion*, 148–154.

24. For an example of the combination of reward and threat used to extract information from indigenous informants, see Londa Schiebinger, "Prospecting for Drugs: European Naturalists in the West Indies," in Schiebinger and Swan, *Colonial Botany*, 119–133, at 129–130.

25. Bleichmar, "Atlantic Competitions," 226.

26. Brockway, *Science and Colonial Expansion*, 35–37, quotation at 37.

27. As Bleichmar observes, Spanish officials sought to reduce reliance on imported botanicals because they "enriched other European nations at the expense of Spanish pockets": Bleichmar, "Atlantic Competitions," 229. Londa Schiebinger also cites French lamentations about the economic losses that came from reliance on Spanish sources for Peruvian bark (cinchona) and other coveted medicinals: Schiebinger, "Prospecting for Drugs," 133. Although liberal economics began to overtake mercantilist impulses in the mid-nineteenth century, the latter never entirely disappeared. As Marc-William Palen has argued in the US context, the Republican politics of the open door in the late nineteenth century should not be understood as a form of free-trade ideology but as a nationalist project of preferential access to foreign markets: see Palen, *The "Conspiracy" of Free Trade: The Anglo-American Struggle over Empire and Economic Globalisation, 1846–1896* (Cambridge: Cambridge University Press, 2016). I thank Kate Epstein for recommending Palen's study to me, and for introducing me more generally to the literature on the history of British free-trade liberalism. An example from the history of tropical agriculture in early twentieth-century Hawai'i suggests the relevance of Palen's point. Jared G. Smith, head of the US Department of Agriculture's Hawaii Agricultural Experiment Station, imagined that if farmers could grow vanilla or chili peppers in Hawai'i, the United States would no longer need to rely on imports. His position could be read as a historical echo of mercantilist instincts. Jared G. Smith, *Annual Report of the*

Hawaii Agricultural Experiment Station for 1903, in US Department of Agriculture, *Annual Report of the Office of Experiment Stations for the Year Ended June 30, 1903* (Washington, DC: Government Printing Office, 1904), 402–404.

28. Andrea Wulf, *The Invention of Nature: Alexander von Humboldt's New World* (New York: Vintage Books, 2015), 193–194, 203–209, and 213 (on Humboldt's failed efforts to secure permission from the East India Company to undertake an expedition in India); and 50, 53–54 (on the Spanish Crown's assent to Humboldt's South American voyage of 1799–1804). Mary Louise Pratt stresses the reform politics that facilitated Humboldt's access to Spanish America, but she nonetheless characterizes Charles' IV's assent to the expedition as "a diplomatic coup": Pratt, *Imperial Eyes: Travel Writing and Transculturation*, 2nd ed. (London: Routledge, 2008), 112–114, quotation at 114.

29. On trading companies' role in restricting information, consider Londa Schiebinger's observation that "the great trading companies of the early modern period also guarded their investments through scrupulously protected monopolies." Their strategies included efforts to discourage scholars from publication, as well as physical violence against smugglers who attempted black market transactions; Schiebinger, "Prospecting for Drugs," 130–131, quotation at 130.

30. Schiebinger and Swan, "Introduction," 1–2.

31. Drayton, *Nature's Government*, 80. A generation earlier, in the mid-1740s, French naturalist Joseph de Jussieu also lamented that the fruits of his Amazonian travels would likely be intercepted by the British or swallowed by the oceans, due to the War of the Austrian Succession. Neil Safier, "Fruitless Botany: Joseph de Jussieu's South American Odyssey," in Delbourgo and Dew, *Science and Empire in the Atlantic World*, 203–224, at 212–213.

32. Charles W. Peale to Étienne Geoffroy Saint-Hilaire, June 28, 1804, in *The Selected Papers of Charles Willson Peale and His Family*, vol. 2, ed. Lillian B. Miller (New Haven, CT: Yale University Press, 1988), 727.

33. Johnson, *American Eden*, 208, 217. Under the policy of impressment, the British Navy claimed a right to force into naval service any sailor on a foreign ship found to be a deserter. Impressment frequently entrapped naturalized American citizens, since the British government did not recognize their new status. Diplomatic tensions over impressment led to Jefferson's ill-fated embargo policy. George C. Herring, *From Colony to Superpower: U.S. Foreign Relations since 1776* (New York: Oxford University Press, 2008), 116–119.

34. Quoted in Johnson, *American Eden*, 259.

35. These and other examples are discussed in Brockway, *Science and Colonial Expansion*, 57.

36. Portuondo, *Secret Science*, 97.

37. For a brief summary on this point, see Somsen, "A History of Universalism," 363. The historical genealogy of scientific openness also had earlier roots in the Baconian attack on esotericism and its insistence on the power of demonstration as

the source of legitimate knowledge, most notably in the gentlemanly public theater of Britain's Royal Society, where Robert Boyle showcased the new experimental philosophy. See Eamon, *Science and the Secrets of Nature*, chap. 10; and Steven Shapin and Simon Schaffer, *Leviathan and the Air-Pump: Hobbes, Boyle, and the Experimental Life* (Princeton, NJ: Princeton University Press, 1985).

38. Johnson, *American Eden*, chaps. 2 and 3 (on Hosack's British sojourn), 269 (on his induction into the Royal Society), and 285–286 (on his election to the Horticultural Society of London).

39. Johnson, *American Eden*, chaps. 8 and 9 (on Hosack and the world of transatlantic botanical exchange around 1804–1806), 3 (on the number of species at the height of Hosack's Elgin Garden), 178 (for the quotation), and 179, 208, and 235 (on specific plant species in Hosack's greenhouse).

40. Júnia Ferreira Furtado, "Tropical Empiricism: Making Medical Knowledge in Colonial Brazil," in Delbourgo and Dew, *Science and Empire in the Atlantic World*, 127–151; Parrish, *American Curiosity*, chap. 3.

41. Goss, *The Floracrats*, 8, 29–30.

42. See Drayton, *Nature's Government*; Jim Endersby, *Imperial Nature: Joseph Hooker and the Practices of Victorian Science* (Chicago: University of Chicago Press, 2010); and, on Singapore, Timothy P. Barnard, *Nature's Colony: Empire, Nation and Environment in the Singapore Botanical Gardens* (Singapore: National University of Singapore Press, 2016), quotations at 8 and 180, respectively. I have elaborated on my reading of Drayton's and Endersby's work in Jessica Wang, "Plants, Insects, and the Biological Management of American Empire: Tropical Agriculture in Early Twentieth-Century Hawai'i," in "Empires of Knowledge," ed. Axel Jansen, John Krige, and Jessica Wang, special issue, *History and Technology* 35, no. 3 (2019): 203–236, at 219–220.

43. Parrish, *American Curiosity*, 134–135; and Andrew J. Lewis, "Gathering for the Republic: Botany in Early Republic America," in Schiebinger and Swan, *Colonial Botany*, 66–80.

44. Staffan Müller-Wille, "Walnuts at Hudson Bay, Coral Reefs in Gotland: The Colonialism of Linnaean Botany," in Schiebinger and Swan, *Colonial Botany*, 34–48, at 39–40 and 46–47. King George's War (1744–1748) was the North American theater of the War of the Austrian Succession (1740–1748). The conflict was part of a long era of recurrent confrontations from 1688 onward between England and France over their respective colonial holdings, until the French ceded their colonies on the North American continent at the end of the Seven Years' War in 1763.

45. Moritz von Brescius, *German Science in the Age of Empire: Enterprise, Opportunity and the Schlagintweit Brothers* (Cambridge: Cambridge University Press, 2018), 9.

46. Lorraine Daston, "The Ideal and Reality of the Republic of Letters in the Enlightenment," *Science in Context* 4, no. 2 (1991): 367–386, at 375–376 and 378. For the quotation, see *Histoire de la République des Lettres et Arts en France. Année 1780* (Amsterdam, 1781), 76; I have quoted a little more of the passage than in Daston's

essay. Daston mentions the granting of safe passage to Alexander Guy Pingré, who was part of an expedition headed to an island off Madagascar for observations. Another French savant, Guillaume de Gentil, encountered diplomatic misfortune when he was refused permission to land at Pondicherry, then under attack by the British, his travel papers notwithstanding. David Coward, "Transit of Venus: A Tale of Two Expeditions," *The Conversation*, June 3, 2012, http://theconversation.com/transit-of-venus-a-tale-of-two-expeditions-7246.

47. John Cawood, "Terrestrial Magnetism and the Development of International Collaboration in the Early Nineteenth Century," *Annals of Science* 34 (1977): 551–587, at 566.

48. Daston, "The Ideal and Reality of the Republic of Letters," 381.

49. Antonio Barrera-Osorio, "Empiricism in the Spanish Atlantic World," in Delbourgo and Dew, *Science and Empire in the Atlantic World*, 177–202, at 182–184.

50. Sandman, "Navigation, Cartography, and Secrecy," 37–40 and 45–46.

51. Cameron B. Strang, *Frontiers of Science: Imperialism and Natural Knowledge in the Gulf South Borderlands, 1500–1850* (Williamsburg, VA: Omohundro Institute of Early American History and Culture; and Chapel Hill: University of North Carolina Press, 2018), Introduction, chaps. 1–4.

52. Goss, *The Floracrats*, 51–56.

53. Portuondo, *Secret Science*, quotations at 259 and 260.

54. Portuondo, *Secret Science*, chap. 7, quotation at 298. On cartographical openness as a means of claiming territory, see also Sandman, "Navigation, Cartography, and Secrecy," 32–33.

55. Brockway, *Science and Colonial Expansion*, 115–124. The British and the Dutch also eventually formed a cartel in order to fix world quinine prices (8). On the earlier cinchona monopoly, the challenges of cinchona cultivation and Spanish conceptions of cinchona bark as essentially a commodity with a fixed supply rather than a renewable resource, see Matthew James Crawford, *The Andean Wonder Drug: Cinchona Bark and Imperial Science in the Spanish Atlantic, 1630–1800* (Pittsburgh: University of Pittsburgh Press, 2016), 74–76.

56. Brockway, *Science and Colonial Expansion*, 158.

57. Brockway, *Science and Colonial Expansion*, chap. 8, esp. 168. The Germans reaped the greatest benefit from Kew's generosity when they used information from the *Kew Bulletin* to establish their own sisal plantations in German East Africa. All of this colonial effort, however, ultimately came under British control, after the Germans lost their African colonies in the post–World War I settlement; Brockway, *Science and Colonial Expansion*, 178–80.

58. Goss, *The Floracrats*, 61. On Buitenzorg as an international and interimperial center of tropical agriculture, see also Robert-Jan Wille, "The Coproduction of Station Morphology and Agricultural Management in the Tropics: Transformations in Botany at the Botanical Garden at Buitenzorg, Java 1880–1914," in *New Perspectives on the History of Life Sciences and Agriculture*, ed. Denise Phillips

and Sharon Kingsland (Cham, Switzerland: Springer, 2015), 253–275; and Wang, "Plants, Insects," 220–222 and 224. On civilizational ideology, its formal status within international law, and its significance for justifying and maintaining imperial hierarchies, consult Gerrit W. Gong, *The Standard of "Civilization" in International Society* (Oxford: Clarendon Press, 1984).

59. On the distinction between the rise of a *conscious* internationalism in the mid-nineteenth century versus a more basic background reality of international and cross-border relationships in past eras, see Martin H. Geyer and Johannes Paulmann, "Introduction: The Mechanics of Internationalism," in *The Mechanics of Internationalism: Culture, Society, and Politics from the 1840s to the First World War*, ed. Martin H. Geyer and Johannes Paulmann (Oxford: Oxford University Press, 2001), 1–25, at 2. Glenda Sluga has also emphasized the need to understand a new discourse about life in a modern international age that grew increasingly visible and persuasive by the early twentieth century: Sluga, *Internationalism in an Age of Nationalism*, 12–13.

60. On the significance of the "long peace," Eric Hobsbawm provided a moving account of the disjuncture between expectations of peace as a way of life on the European continent before 1914 and the shattering consequences of a new age of total war that started with World War I: Hobsbawm, *The Age of Extremes: A History of the World, 1914–1991* (New York: Pantheon Books, 1994), 22–24. Some critics have pointed to the all-too-real conflicts that took place between 1815 and 1914 and questioned the "long peace" framework. See, for example, Sheldon Anderson, "Metternich, Bismarck, and the Myth of the 'Long Peace,' 1815–1914," *Peace and Change* 32, no. 3 (2007): 301–328. I thank Hannah Exley for alerting me to Anderson's essay. Anderson is certainly correct to point to the interstate and colonial violence of the century prior to 1914 and its human costs. Such conflicts, however, did not have the systemic reach of two world wars, and in that sense the "long peace," as a thesis about relative stability and a functional modus operandi in great power relationships, still holds.

61. Charles S. Maier, *Once within Borders: Territories of Power, Wealth, and Belonging since 1500* (Cambridge, MA: Belknap Press of Harvard University Press, 2016), 222–224.

62. On discourses of great power (i.e., interimperial) collaboration, the Roosevelt Corollary, and US appeals for "an international police power," see David Healy, *Drive to Hegemony: The United States in the Caribbean, 1898–1917* (Madison: University of Wisconsin Press, 1988), 106–108, 121. For Roosevelt's own words, see Theodore Roosevelt, "Fourth Annual Message to Congress," December 6, 1904, the American Presidency Project, https://www.presidency.ucsb.edu/node/206208. For the broader historical context of dollar diplomacy and U.S. imperial power, consult Emily S. Rosenberg, *Financial Missionaries to the World: The Politics and Culture of Dollar Diplomacy 1900–1930* (Durham, NC: Duke University Press, 2003).

63. Mark Mazower coined the term "imperial internationalism" to describe this aspect of internationalism and international organization in the late nineteenth and early twentieth centuries: Mazower, *No Enchanted Palace: The End of Empire and the Ideological Origins of the United Nations* (Princeton, NJ: Princeton University Press, 2009), 18, 23, 30–34.

64. On the transnational networks that anticolonial activists developed in the early twentieth century, see Erez Manela, *Wilsonian Moment: Self-Determination and the International Origins of Anticolonial Nationalism* (Oxford: Oxford University Press, 2007). Evidence of a repressive apparatus of interimperial police collaboration that attempted to undermine nationalist movements comes from Anne L. Foster, *Projections of Power: The United States and Europe in Colonial Southeast Asia, 1919–1941* (Durham, NC: Duke University Press, 2010), chap. 1; and Seema Sohi, *Echoes of Mutiny: Race, Surveillance, and Indian Anticolonialism in North America* (Oxford: Oxford University Press, 2014), esp. chap. 3. Vietnamese nationalism is the case that I know best. Ho Chi Minh's turn to the communist movement after the United States ignored his appeal to the Paris Peace Conference is well known. For other evidence of transnational elements in Vietnamese nationalist circles and Vietnam's own significance in anticolonial nationalisms elsewhere in the early twentieth century, see Christopher Goscha, *The Penguin History of Modern Vietnam* (New York: Penguin, 2017), 97–101 (on Phan Bội Châu and the Đông Du [Go East] movement); Agathe Larcher-Goscha (trans. Kareem James Abu-Zeid), "Bùi Quang Chiêu in Calcutta (1928): The Broken Mirror of Vietnamese and Indian Nationalism," *Journal of Vietnamese Studies* 9, no. 4 (2014): 67–114; and Youn Dae-Yeong, "The Loss of Vietnam: Korean Views of Vietnam in the Late Nineteenth and Early Twentieth Centuries," *Journal of Vietnamese Studies* 9, no. 1 (2014): 62–95. Diplomatic arrangements, such as the Franco-Japanese Treaty of 1907 or the Four-Power Treaty of 1921, attempted to stabilize interimperial relations by requiring signatories to respect each other's territorial claims in Asia. Japan discarded these arrangements when it launched its invasions of colonial territories in Southeast Asia, starting with Indochina in 1940. For several years, the Japanese ruled through the existing Vichy-based colonial government, but on March 9, 1945, it formally wrested power away from the French. The wartime rise of the Việt Minh, the nationalist movement led by Ho Chi Minh that succeeded in taking power and declaring Vietnam's independence on September 2, 1945, was predicated in part on the weakening of colonial power created by World War II. As Mark Bradley has observed of Vietnam's August Revolution of 1945, "Just five years before, neither the communists nor any other anti-colonial political movement appeared to have a chance at overthrowing French rule"; Bradley, *Vietnam at War* (Oxford: Oxford University Press, 2009), 38. This is not to dismiss the agency and political ingenuity of the Việt Minh, but there was also no lack of ingenuity among Vietnamese nationalists in the 1930s. Interimperial warfare defined the difference between a French colonial state that had the power to maintain its ability to rule, if only through massive repression, in the 1930s and a

French colonial state that lost power completely during World War II and was unable to reclaim it after eight years of warfare with the Việt Minh between 1946 and 1954.

65. As Mark Mazower has explained, the term was coined in the late eighteenth century by Jeremy Bentham to distinguish between the rule of law within a state as a relative given and the more problematic challenge of imagining legal regimes that might potentially govern relations between states: Mazower, *Governing the World: The History of an Idea* (New York: Penguin, 2012), 19–21.

66. Michael S. Reidy and Helen M. Rozwadowski, "The Spaces in Between: Science, Ocean, Empire," *Isis* 105, no. 2 (2014): 338–351, at 339.

67. John Cawood, "The Magnetic Crusade: Science and Politics in Early Victorian Britain," *Isis* 70, no. 4 (1979): 492–518, quotation at 516. Regarding the reach of the Magnetische Verein, Cawood observes, "There were 18 stations in 1835 covering Northern Europe and Russia. By 1850 there were 50 stations in Europe, Asia, Africa, North America and the South Seas"; Cawood, "Terrestrial Magnetism and the Development of International Collaboration," 552n1.

68. Michael S. Reidy, *Tides of History: Ocean Science and Her Majesty's Navy* (Chicago: University of Chicago Press, 2008), chap. 5; Benjamin Apthorp Gould, *The Transatlantic Longitude as Determined by the Coast Survey Expedition of 1866* (Washington, DC: Smithsonian Institution, 1869).

69. Reidy and Razwadowski, "The Spaces in Between," 344–346 and, for the quotation, 348.

70. Mazower, *Governing the World*, 32, 38–48; Palen, *The "Conspiracy" of Free Trade*, chap. 3, esp. 60, 64–65, and 81.

71. Edward J. Larson, "Public Science for a Global Empire: The British Quest for the South Magnetic Pole," *Isis* 102, no. 1 (March 2011): 34–59, on 35.

72. Valeska Huber, "The Unification of the Globe by Disease? The International Sanitary Conferences on Cholera, 1851–1894," *The Historical Journal* 49, no. 2 (2006): 453–476, at 458.

73. Crawford, *Nationalism and Internationalism in Science*, 38–43.

74. On tropical medicine and interimperial networks of exchange, see Deborah Neill, *Networks in Tropical Medicine: Internationalism, Colonialism, and the Rise of a Medical Specialty, 1890–1930* (Stanford, CA: Stanford University Press, 2012); and Mari K. Webel, "Trypanosomiasis, Tropical Medicine, and the Practices of Inter-Colonial Research at Lake Victoria, 1902–07," in "Empires of Knowledge," ed. Axel Jansen, John Krige, and Jessica Wang, special issue, *History and Technology* 35, no. 3 (2019): 266–292. My own work has begun to broach similar themes in the history of tropical agriculture and the US empire in the early twentieth century: Wang, "Plants, Insects, and the Biological Management of American Empire," 203–236.

75. On international public health and the information network established by the League of Nations, see Heidi J. S. Tworek, "Communicable Disease: Information, Health, and Globalization in the Interwar Period," *American Historical Review* 124, no. 3 (2019): 813–842. On the intertwined history of ideas about interdependence,

modernity, and internationalism, see, for example, Frank Ninkovich, *Modernity and Power: A History of the Domino Theory in the Twentieth Century* (Chicago: University of Chicago Press, 1994), chap. 2.

76. Madeleine Herren, "Governmental Internationalism and the Beginning of a New World Order in the Late Nineteenth Century," in Geyer and Paulmann, *The Mechanics of Internationalism*, 121–144, at 124.

77. Martin H. Geyer and Johannes Paulmann, "Introduction: The Mechanics of Internationalism," in Geyer and Paulmann, *The Mechanics of Internationalism*, 1–25, at 12.

78. George H. Boehmer, *History of the Smithsonian Exchanges: From the Smithsonian Report for 1881* (Washington, DC: Smithsonian Institution, 1882), 75.

79. *Report on International Exchange*, 47th Cong., 1st sess., Ex. Doc. No. 172 (1882), quotation at 37; Nancy Elizabeth Gwinn, "The Origins and Development of International Publication Exchange in Nineteenth-Century America" (PhD diss., George Washington University, 1996), 95–103.

80. *Report on International Exchange*, 38–40; Gwinn, "The Origins and Development of International Publication Exchange," 104–121, 140–149, 155–156, quotation at 155. The boast of Smithsonian assistant secretary Spencer F. Baird to George Perkins Marsh in 1852 similarly highlighted the superiority of the Smithsonian's system to Vattemare's: "Our scale of operations is on a vastly larger scale than Vattemare's (The Gigantic Humbug) and is carried on in a strictly business-like way"; quoted in Gwinn, "The Origins and Development of International Publication Exchange," 223.

81. Gwinn, "The Origins and Development of International Publication Exchange," 14; Robert V. Bruce, *The Launching of Modern American Science, 1846–1876* (1987; repr., Ithaca, NY: Cornell University Press, 1988), 187 (for the quotation), 193, and 195.

82. *Report on International Exchange*, 42–47. Duty-free entry in the Smithsonian exchange system rested on a history of Congressional exemptions for printed materials imported for educational purposes that dated back to the early national period. Other countries did not necessarily have similar policies, however. For example, Smithsonian representatives negotiated for duty-free entry to the United Kingdom of American documents shipped to the Royal Society, which handled the British end of the exchange. Gwinn, "The Origins and Development of International Publication Exchange," 234–235, 238–240. As Gwinn notes, book dealers, missionaries, and other contacts, and not just consular officials, also participated as agents on behalf of the Smithsonian, but consuls feature prominently in her account (see 287–289), and in the Smithsonian's own official reports.

83. Boehmer, *History of the Smithsonian Exchanges*, 43–44, quotation at 44; Gwinn, "The Origins and Development of International Publication Exchange," 322–323.

84. Gwinn, "The Origins and Development of International Publication Exchange," 327–337, discusses both the 1867 resolution and the complexities associated with making the mandate fully operational.

85. Gwinn, "The Origins and Development of International Publication Exchange," 268.

86. Gwinn, "The Origins and Development of International Publication Exchange," 340–350.

87. Henry W. Elliott, "The Smithsonian Institution," *The International Review* 7 (1879): 688–700, at 695.

88. Frederick T. Frelinghuysen to Chester A. Arthur, 11 April 1882, reproduced in Boehmer, *History of the Smithsonian Exchanges*, 107–108, quotation at 108.

89. Mazower, *Governing the World*, 24–26 and 31–48; David Nicholls, "Richard Cobden and the International Peace Congress Movement, 1848–1853," *Journal of British Studies* 30, no. 4 (1991): 351–376. For examples of Cobdenite discourses that equated free trade with international peace, see "Free Trade and Peace," *The League*, March 7, 1846, 402; and Joseph Barker, *Blessings of Free Trade*, speech of July 27, 1846.

90. Quoted in *Report on International Exchange*, 44.

91. Anto. M. Pradillo to the Secretary of State [Hamilton Fish], November 1869, reproduced in Boehmer, *History of the Smithsonian Exchanges*, 61.

92. Marquis de Souza Holstein and José Julio Rodriguez to the president of the Belgian Commission, March 1, 1877, reproduced in Boehmer, *History of the Smithsonian Exchanges*, 78–79, quotation at 78. Free-trade metaphors did not often appear overtly in such discourses, but one senses resonances between free-trade ideology and calls for free trade in publications and the ideas contained therein. Similar discourses about international understanding and friendly feeling between nations permeated both Cobdenite discussions and encomiums in praise of document exchanges. The concern for duty-free entry of materials arguably reflected not just a practical need to contain costs but the context of free-trade ideology. Tellingly, in 1852 Edward Sabine, vice president and treasurer of Britain's Royal Society, praised "the liberal example which has been set by the United States" when he announced that Britain, with the Royal Society as intermediary, would enter into a document exchange relationship with the Smithsonian. Edward Sabine to Joseph Henry, March 19, 1852, reproduced in *Report on International Exchange*, 43.

93. Quoted in Gwinn, "The Origins and Development of International Publication Exchange," 260.

94. *Report of Prof. James K. Patterson, Ph.D.[,] Commissioner of Kentucky to the International Congress of Geographical Sciences, Held at Paris, France, August 1st to 13th, 1875* (Frankfort, KY, 1876), 6–7, 12–14, and 18, quotation at 6. The Agricultural and Mechanical College separated from Kentucky University in 1878, and it eventually evolved into the University of Kentucky, with the name change confirmed in 1916.

95. Martin H. Geyer, "One Language for the World: The Metric System, International Coinage, Gold Standard, and the Rise of Internationalism 1850–1900," in Geyer and Paulmann, *The Mechanics of Internationalism*, 55–92, on 63, and 86–88

(quotations at 86 and 88, respectively). By contrast, in the case of railway governance in nineteenth-century Europe, Johan Schot, Hans Buiter, and Irene Anastasiadou have attributed the successful establishment of cross-border governance mechanisms to a transnational epistemic community of engineers and professional managers who developed ways to overcome states' intense national rivalries. The authors' own narrative, however, also suggests the constant presence of government officials in the work of the Verein Deutscher Eisenbahn-Verwaltungen and nation-states' strong interest in achieving a system of international cooperation to facilitate cross-border rail transport. Schot, Buiter, and Anastasiadou, "The Dynamics of Transnational Railway Governance in Europe during the Long Nineteenth Century," *History and Technology* 27 (September 2011): 265–289.

96. Herren, "Governmental Internationalism," 128. Valeska Huber has echoed this point with a discussion of diplomats' roles at the International Sanitary Conferences from the 1850s to the 1890s, including the intentional exclusion of scientists from the second conference in 1859 as a means of facilitating efficient diplomatic negotiations. See Huber, "The Unification of the Globe by Disease?," esp. 459–461.

97. Herren, "Governmental Internationalism," 132 and 136.

98. Sluga, *Internationalism in an Age of Nationalism*, 15.

99. Hence Glenda Sluga's characterization of nationalism and internationalism as "twinned liberal ideologies"; Sluga, *Internationalism in the Age of Nationalism*, 3. On the related phenomenon of international relationships' centrality in supporting and shaping nationalist independence movements in the early decades of the twentieth century, see Sohi, *Echoes of Mutiny*; Manela, *Wilsonian Moment*; Michael Goebel, *Anti-Imperial Metropolis: Interwar Paris and the Seeds of Third World Nationalism* (Cambridge: Cambridge University Press, 2015); Minkah Makalani, *In the Cause of Freedom: Radical Black Internationalism from Harlem to London, 1917–1939* (Chapel Hill: University of North Carolina Press, 2011); and Adom Getachew, *Worldmaking after Empire: The Rise and Fall of Self-Determination* (Princeton, NJ: Princeton University Press, 2019), Introduction, chaps. 1–2.

100. The arrest of the prominent Harvard chemist Charles Lieber, who stands accused of lying about the nature of his relationship to China's Thousand Talents Program and other legal violations, provides one of the more recent examples of the US government's efforts to prosecute scientists because of cross-border scientific ties with Chinese institutions. Ellen Barry, "U.S. Accuses Harvard Scientist of Concealing Chinese Funding," *New York Times*, January 28, 2020, https://www.nytimes.com/2020/01/28/us/charles-lieber-harvard.html. On ongoing US-China tensions over control of scientific information and their history, see John Krige, "Scholars or Spies? U.S.-China Tension in Academic Collaboration," *China Currents* 19, no. 3 (2020), https://www.chinacenter.net/2020/china_currents/19-3/scholars-or-spies-u-s-china-tension-in-academic-collaboration/; John Krige, "National Security and Academia: Regulating the International Circulation of Knowledge," *Bulletin of the Atomic Scientists* 70, no. 2 (2014): 42–52; and Jessica Wang, "A State of Rumor:

Low Knowledge, Nuclear Fear, and the Scientist as Security Risk," *Journal of Policy History* 28, no. 3 (2016): 406–446, esp. 424–435.

101. Steven Newton, "Florida Bans the Term 'Climate Change,'" *HuffPost*, March 12, 2015, https://www.huffpost.com/entry/florida-bans-the-term-climate-change_b_6852726; Lindsay Abrams, "World's Researchers to Canada: Stop Censoring Science!," Salon.com, October 21, 2014, https://www.salon.com/test/2014/10/21/worlds_researchers_to_canada_stop_censoring_science/.

102. Chris Buckley, "Chinese Doctor, Silenced after Warning of Outbreak, Dies from Coronavirus," *New York Times*, February 6, 2020, https://www.nytimes.com/2020/02/06/world/asia/chinese-doctor-Li-Wenliang-coronavirus.html.

Part I

Regulating Transnational Knowledge Flows

Chapter Two

Harnessing Invention

The British Admiralty and the Political Economy of Knowledge in the World War I Era

KATHERINE C. EPSTEIN

For many historians of science and technology, the US national-security state and the nuclear-secrecy regime of the early Cold War have functioned as ideal types—as representations par excellence of what a national-security state and secrecy look like. But they were not unprecedented, nor are they necessarily representative. After all, the global hegemony of the United States and the national-security practices necessary to sustain it are relatively recent developments.

At the dawn of the twentieth century, the most powerful polity in the world was the British Empire. British power rested on the Royal Navy, which controlled access to the global commons. The Royal Navy's preeminence, in turn, rested on money and the maintenance of technological supremacy. Naval technology was to Edwardian Britain what nuclear technology was to the Cold War United States: the most important kind of technology for national (or imperial) security. As with nuclear technology in the postwar United States, therefore, the production and control of knowledge about naval technology were vital to Edwardian Britain.

Then as later, producing and controlling this knowledge posed a knotty challenge. On the one hand, the British Admiralty wished to incentivize invention, or the production of new naval-technological knowledge. Within a domestic political economy ideologically committed to liberal property norms, this meant respecting the commercial interests of inventors and compensating them for the use of their knowledge—not only when they were contractors, but even when they were naval officers or civilian employees of the Admiralty. On the other hand, the Admiralty had to prevent knowledge from falling into the wrong hands within an international political economy populated by knowledge-hungry rising powers and profit-seeking market actors—even if it meant harming the financial interests of inventors. What the Admiralty wished to keep secret, foreign governments wanted to learn and defense contractors wanted to monetize abroad as well as at home.

Thus, while stimulating invention, the Admiralty sought to regulate knowledge flows across two borders: an internal border between the public and private sectors, and an external border between Britain and foreign nations. The internal border mattered because knowledge could flow much more easily to foreign nations from the private sector than from the public sector. Constructing and policing these borders meant managing a "triad" of conduits, to borrow Mario Daniels's phrase, through which knowledge could move: it could be encoded in information (like manuals), carried by people (like inventors), and embodied in "things" (like weapons).[1] Alternatively, we might say that knowledge can be textual, human, and mechanical. Since these three forms of knowledge were potentially interconvertible, all three required management. The act of managing had to balance security interests with liberal property norms.

Britain developed two foundational legal tools—one might say legal technologies—for controlling knowledge flows across these borders decades in advance of the United States. The first was the establishment of secret patents as a permanent, peacetime legal category through the Patents for Inventions (Munitions of War) Bill of 1859.[2] The United States did not pass equivalent legislation until the Invention Secrecy Act of 1951, relying on temporary ad hoc legislation during the two world wars. Literally a contradiction in terms (the word "patent" comes from a Latin root meaning "open"), secret patents sought to reconcile the security state's interest in secrecy with inventors' interest in property rights. Inventors retained ownership of secret patents, with the Crown acquiring control through assignment. The patent law passed in 1883, which effectively reversed the famous 1865 ruling in *Feather v. Regina* that the Crown could use patents without compensating the owners, confirmed that patents had the same effect against the Crown as against private subjects.[3] The Crown could not compel assignment of patents; consistent with liberal property norms, the default setting, so to speak, was to require the Crown to negotiate with inventor-owners over the terms of assignment. That said, the law did give the Crown the right of nonconsensual, though not uncompensated, use (as distinct from assignment), by permitting it to settle terms with the patentee after use of the patent or to have the Treasury settle terms if the patentee refused to permit use.[4] Secret patents also enabled the Crown to mimic inventors' strategic use of patents to establish priority and protect against future demands for royalty payments.[5]

Britain's second novel legal tool for controlling knowledge flows was the Official Secrets Act of 1889. The United States did not pass equivalent legislation until the National Defense Secrets Act of 1911, the precursor to the better-known Espionage Act of 1917, which remains in force today as part of Title 18 of the US Code. The Official Secrets Act functioned as a trump card over property rights if patentees showed too much interest in foreign sales: the British government used it as improvised export-control legislation to prevent transnational flows of sensitive knowledge. If secret patents squared illiberal security norms with liberal property norms, then the Official Secrets Act squashed the latter with the former.

This chapter describes the British government's, and especially the Admiralty's, use of secret patents and the Official Secrets Act to regulate transnational flows of knowledge produced by domestic inventors decades before such practices were institutionalized by the United States in World War II. It examines three cases: the fire-control (gunnery-targeting) system of Arthur Pollen, the torpedo propulsion system of Sydney Hardcastle, and the gyro-technologies of Sir James Henderson.[6] These cases make for a comparison about as close to laboratory conditions as is possible in history. All three technologies were among the most secret of their day. As an American naval observer remarked in a 1912 report, "From a confidential standpoint, torpedo subjects stand next only to fire control, in the British Service" (and both incorporated gyroscopes).[7] All three inventors made their key breakthroughs during roughly the same time period—1901 to 1917—and thus dealt with the same naval officials. The independent variable was the status of the inventors, which structured the knowledge-regulation options available to the Admiralty. Pollen was an independent civilian contractor (an outsider), Hardcastle a naval officer (an insider), and Henderson a civilian Admiralty employee (a hybrid insider-outsider). The Admiralty faced different legal environments in trying to control the knowledge of inventors of different statuses. In all three cases, the Admiralty prioritized security over the inventors' intellectual property rights—though perhaps because Pollen's outsider status made his knowledge the most difficult to control, the Admiralty treated him more harshly than it did Hardcastle or Henderson.

Pollen's Fire-Control System

In 1900, Pollen was serving as the managing director of the Linotype Company, the premier supplier of mechanical printing equipment to Britain's newspapers, when he became intrigued by the problem of aiming naval guns at moving targets. He spent the next dozen years developing a system that would enable navies to aim their guns accurately. In this quest, he benefited from his earlier legal training and from his management role at Linotype, which manufactured patented equipment under license from an American firm and thus afforded him extensive day-to-day experience dealing with intellectual property in the commercial world. Linotype also familiarized him with complex machinery and gave him access to talented designers.

Pollen's fire-control system consisted of a definition of a new problem and a set of ideas and instruments to solve it. When he began, naval gunnery had entered a period of ferment. After centuries of firing at ranges below a thousand yards, navies had begun experimenting with ways to increase ranges by thousands of yards, such that shell trajectories became parabolic and the target moved along its course while the projectiles were in flight toward it. Pollen quickly perceived that these conditions created new challenges. Accuracy now required canceling the effects of relative motion—one's own and the target's course and speed—in order to predict where the target would be at the moment a shell landed on it. By 1905, Pollen had defined the parameters of the problem. Predicting where the target would be when a shell landed on it required precise knowledge of the enemy's range (distance) and bearing (angle), and accurate bearings required gyroscopic correction of the effects of yaw on one's own ship.[8] Changes in range and bearing over a period of time could be expressed as rates of change, which could be integrated continuously to predict where the target would be when a shell landed on it. Critically, Pollen realized that the range and bearing rates varied instantaneously, such that taking meaned averages over a period of time would be inaccurate, and he realized that the rates varied with each other, such that knowledge of both rates was necessary. Also by 1905, although he did not complete the process of transforming his ideas into material form until 1912, Pollen had identified the three-part instrumental means of solving this new fire-control problem. The first part consisted of devices to observe enemy range and bearing, namely a rangefinder (an optical device for measuring ranges not superseded by radar until World War II) and a gyro-stabilized device for taking yaw-free bearings. The second part consisted of a graphical device known as a "plotter" or "plotting table" for instantly recording observed data. The third part consisted of a highly sophisticated mechanical analog computer, known as a "clock" (or "rangekeeper"), for integrating the observed data so as to predict the target's future location.

Neither Pollen's definition of the problem nor his solution was obvious to expert observers at the time. Although aspects of his invention incorporated well-known trigonometric techniques and existing technology (like rangefinders), both he and the Admiralty's leading fire-control expert, Edward Harding, agreed that he had done something novel and important. For both men, the essential value of Pollen's new knowledge was not its reduction to particular material form but its systematic quality—the thoroughness with which Pollen had identified the relevant variables—and his insistence that these variables had to be known with an unprecedented degree of accuracy. As he put it to the Admiralty:

> What I have to sell is not instruments but a system, the embodiment of certain laws of gunnery which I have been the first to codify. My instruments as at present designed, will no doubt soon be superseded by others that embody the essential features of my system more completely. The monopoly of instruments is only incidental. It is the knowledge of the system which they make workable, and the exclusive knowledge of it, to which high, possibly supreme, value attaches.[9]

Harding agreed, writing of Pollen's system that "its national value cannot be regarded only from the point of view of ingenious instruments for obtaining even a great advance in Artillery technique but must be looked upon as an essential link in a far reaching chain of development."[10] Harding thought there was little chance that foreign powers would independently discover Pollen's system because "the fundamental conditions of the problem are unknown outside the service and have been known inside only comparatively recently and to a limited extent only." Since the problem was unknown, the solution was "unsought for."[11] Thus, the question of independent discovery "really depends on the state of mind and habits of thought of the possible discoverers."[12] Like Pollen, Harding believed that the material flowed from the mental, or intangible, and it was Pollen's mental knowledge, not its reduction to material form, that defined the system and safeguarded its secrecy. In other words, to return to Daniels's "triad," the essential knowledge—indeed the essential technology—was more human than mechanical: it was more Pollen's brain than his instruments. On this logic, the intangible and the material formed a continuum and could not be neatly separated from each other.

This epistemology had significant legal and commercial implications. Pollen, who was not indifferent to financial considerations but also had a powerful sense of patriotism, wanted the Royal Navy (which he described himself as "almost abjectly" admiring) to have a monopoly on his system and to keep it secret.[13] However, his understanding of the nature of his invention and of the law meant that a monopoly contract could not define the invention in terms of patents. He did not conceptualize his intellectual property in terms of patents, secret or otherwise, or look to the patent system—with which his Linotype job familiarized him—for its protection. Over the past century, British patent law had come to objectify inventions: to detach them from the environment that produced them and to imagine them as closed things, represented in the texts of patents.[14] This involved simultaneously a process of dematerialization, in that the law increasingly identified the representation with the abstract essence of an invention rather than its material embodiment, and a process of rematerialization, in that the representation—the printed patent—was itself a reduction of the invention to material form. For Pollen, who understood his invention as inseparable from the creative process that had produced it, the representation of inventions-as-objects in patents was impoverished. It placed too much emphasis on the material representation and not enough on the mental labor behind the real invention. Moreover, this material-representational turn did not fully eradicate an older and, from Pollen's perspective, equally problematic materialistic bias in the patent system.[15] From the eighteenth century on, widespread suspicion of monopoly grants impelled defenders of patents to define a boundary between natural principles, which comprised unpatentable knowledge that should be open to all, and applications of principles, which resulted from mental labor and could thus be patented. The former was the realm of discovery and the latter the realm of invention.[16] Pollen saw this division as artificial and cramped, believing that his "discovery" of principles (or "laws") was an essential part of the process of "inventing" his instruments. An excessively materialistic definition of intellectual property exposed him to commercial risk: he could not adequately protect the money he had invested in developing his invention, which he regarded as far more than a set of material objects. Writing to a fellow inventor who had published a sketch of an invention "totally unlike" something that Pollen wanted to use but "undoubtedly embod[ying] its principle," Pollen explained that he wished to pay the inventor for the use of his invention, regardless of whether it was covered by patents, "as I am rather sensitive to the rights of inventors, and do not think them adequately protected by the patent laws of any country."[17]

Conversely, an excessively mentalistic definition of intellectual property exposed the Admiralty to commercial risk. Normally, the Admiralty did not purchase knowledge until it had been reduced to material form and the Admiralty knew what it was getting. As an official explained to Pollen, "inventors usually have to perfect their inventions at their own expense, or else grant very favorable terms to others who will find the money for them."[18] Buying intangible rather than mechanically embedded knowledge—or the inventor as much as the invention, so to speak—required a radical shift in the Admiralty's contracting procedures and a new tolerance for risk. Instead of buying a finished instrument, it would have to invest in the development of a system that would materialize only at some future date. Along the way, Admiralty officials feared that the navy would communicate service knowledge to Pollen in order to help him develop his system and then "be almost obliged to accept his terms so as to keep that knowledge from others."[19] The final instrumental form could mix up the Admiralty's knowledge with Pollen's. In short, the Admiralty feared that this new kind of contract amounted to a license for blackmail.

Nevertheless, Pollen and the Admiralty negotiated a mutually satisfactory system-development contract between 1906 and 1908. The contract permitted Pollen to take out patents and assign them to the Admiralty for secrecy; despite his skepticism of patents, Pollen nevertheless availed himself of this option under the contract and eventually held more than a dozen secret patents. The contract also obliged Pollen to keep his system secret (and thus to forgo foreign sales) and gave the Admiralty a monopoly over his invention. It further obliged him to share his evolving knowledge with the Admiralty by communicating all improvements to his system. From Pollen's perspective, these obligations created a reciprocal one for the Admiralty to protect his interests, including by treating his invention as a mixed mental-material system rather than as a mere collection of instruments or patents.[20] The contract was renewed in somewhat modified form in April 1910, with the monopoly and secrecy provisions extended.[21] Twice during the life of this agreement—in 1908 and 1910—the Admiralty decided to let it lapse and permit Pollen to sell abroad, believing that it had a cheaper and equally effective alternative to his system; on both occasions, Pollen and his supporters in the service lobbied the Admiralty to continue the agreement, and he took less money than he believed he had earned under the contract in order to preserve the monopoly and secrecy of the system for the Royal Navy.[22]

Although Pollen kept his side of the bargain, the Admiralty did not protect his interests. A naval officer named Frederic Dreyer became familiar with Pollen's system in the course of his official duties and began working on a rival system, which came to differ from Pollen's in certain respects yet unmistakably bore Pollen's influence. Whereas Pollen preferred to plot target course and speed from range and bearing observations ("course plotting"), for instance, Dreyer preferred to plot range and bearing observations from which he determined target course and speed ("rate plotting"). Although the method was different, Pollen had experimented with it before Dreyer, and indeed Dreyer began to experiment with rate plotting only and immediately after the admiral overseeing the trials of Pollen's system in early 1908 saw Pollen do it. Like his plotting, Dreyer's clock design illustrated both Pollen's influence and his own misunderstanding. Initially, Dreyer's clock did not mechanically generate the continuously varying range and bearing rates, instead relying on manual adjustment. Only after becoming familiar with Pollen's clock, which did mechanically generate the changing rates, did Dreyer develop a clock that did so—albeit one missing certain important features of Pollen's, such as a variable-speed drive that minimized friction and slippage. But because Dreyer did not fully understand the theory behind Pollen's clock, his mimicry remained imperfect. To borrow language from John Krige's chapter in this volume (chapter 5), it might be said that Dreyer got the "know-how" but not the "know-why" of Pollen's system.

Dreyer's imitation of Pollen revealed the limitations of the patent system for protecting intellectual property and those of patent infringement as a paradigm for understanding technological piracy. Dreyer may well not have infringed Pollen's patents, because the Admiralty actively helped him to formulate his patent applications in such a way as to avoid doing so—operating on precisely the reductionist definition of Pollen's invention that his interests dictated it eschew.[23] Plagiarism, a term used to describe literary rather than technological piracy, better captures what Dreyer did.[24] Dreyer's instruments may have differed from Pollen's, but he owed the ideas that had enabled him to produce them almost entirely to Pollen. It was Pollen's knowledge, not Dreyer's, that was embodied in Dreyer's instruments; the material and the mental were on an inseparable continuum.

While seeking to break this continuum in order to evade Pollen's patents, the Admiralty sought to restore the continuum in order to preserve the secrecy of Pollen's invention without paying for it. In 1912, the Admiralty decided to end its relationship with Pollen and to adopt Dreyer's bowdlerized system instead. Chafing at Pollen's prices and attracted by the cheapness of Dreyer's system, Admiralty officials believed, or rationalized, that Pollen was exploiting them: they feared that extending monopoly arrangements would be "the signal for Mr. Pollen to demand a continually higher price for this secrecy and monopoly as he had already done on two previous occasions."[25] For Pollen, the end of monopoly necessarily implied the end of secrecy: if the Admiralty would no longer purchase his system, then it was obliged to let him sell it abroad. He had formed his own firm, the Argo Company, to develop his system; unlike in 1908 and 1910, when he had forgone his freedom, he now had investors and employees to whom he felt responsible.[26]

The Admiralty checked him at every turn. First, it tried relying on intellectual property rights, asserting that Pollen's patents infringed Dreyer's. When that failed, it fell back on secrecy, arguing that Pollen had gotten the knowledge embodied in his patents from the Royal Navy—through sea trials and communications with officers—and that they disclosed service secrets. Specifically, the Admiralty objected to the mention of dials for setting the clock by range and bearing rates in one of Pollen's secret patents as the matter that must be kept secret. While dials might seem an unlikely subject for secrecy, the Admiralty insisted that they would reveal the Royal Navy's use of rate plotting, which is what it really wanted to keep secret. If "you were offering your Clock to a foreign purchaser, and putting your patent [in question] before him, and if he should ask the use of the two dials referred to, you would be debarred from answering his question," the Admiralty informed Pollen. "He would, however, speedily draw his own conclusions as to the alternative method of using the Clock, and as to your reasons for silence."[27] Here, the Admiralty was arguing that instruments (dials) could not be separated from the idea (rate plotting) that produced them, and thus ideas could be reverse-engineered from instruments. While the logic of patentability dictated a break between the material and the intangible, the logic of secrecy dictated a continuum between them.

The Admiralty coupled its argument about secrecy with a threat: if Pollen revealed the knowledge embodied in the dials to a foreign power, he would violate the Official Secrets Act—a criminal offense carrying a mandatory prison sentence of up to seven years.[28] Sometimes the Admiralty linked its deployment of this act to its property claims, acknowledging that it would be obliged to pay Pollen for secrecy if the knowledge it wished to keep secret embodied "the inventor's own unaided ideas alone."[29] Pollen's legal advisers, who included a former attorney general, agreed that applying the Official Secrets Act to knowledge that Pollen had not acquired from the Admiralty would be "taking ground outside the law."[30] At other times, however, the Admiralty took precisely that illegal ground, telling Pollen that the act applied to the knowledge in question "whether or not your own invention."[31] Here, the Admiralty severed its deployment of the act from its property claims and insisted that the act could cover even knowledge that Pollen had invented himself. Particularly in the latter case, the deployment of the Official Secrets Act marked a shift from a strategy of knowledge control based on the Admiralty's alleged property rights to one based on security; in Pollen's words, the Admiralty was mixing questions of "ownership" with questions of "policy" and "administrative necessity."[32] In effect, the Admiralty was using the Official Secrets Act as export-control legislation. If patents served as the legal technology for converting knowledge into property, then the Official Secrets Act served as the legal technology for converting intellectual property into national-security knowledge. By ensuring secrecy, deployment of the act preserved the Admiralty's monopoly over Pollen's knowledge without paying him for it.

When Pollen refused to succumb, the Admiralty turned to the last arrow in its quiver: smearing Pollen's name. It decided not to follow through on its threat of prosecution, thereby depriving Pollen of his day in court, since sovereign immunity effectively precluded him from initiating a suit against the Crown.[33] Rather than prosecute him, the Admiralty issued a navy-wide order forbidding communications with him and struck him from the list of contractors eligible to bid on navy contracts, on the grounds that he could not be trusted with official secrets. It took these steps certainly in order to punish him—Admiralty officials considered that Pollen's action in selling his system abroad "was most reprehensible and rendered it necessary to consider punitive measures"—and probably also in order to injure his commercial prospects.[34] Although other navies placed orders with him anyway, defaming Pollen was the opposite of a good-housekeeping seal of approval, signaling to the market that he could not be trusted.

Pollen spent the next fifteen years seeking the money and credit he believed he deserved for his influence on the Dreyer-Admiralty fire-control system. In the 1920s, after exhausting efforts to resolve his grievances against the Admiralty privately, he decided to initiate legal proceedings before the Royal Commission on Awards to Inventors (RCAI) in the 1920s. Pollen persuaded the tribunal that he had invented rate plotting and a rate clock before Dreyer, that he had broadly influenced the development of the Dreyer system, and that Dreyer's plagiarism of his clock was so severe as to be tantamount to patent infringement, entitling Pollen to a large award of £30,000. This sum included damages—which in British patent law could not be awarded if the infringement was innocent.[35]

Hardcastle's Superheater

Hardcastle was an engineer officer in the Royal Navy when he began working on torpedo propulsion outside his official duties. The modern torpedo had been invented in the 1860s by a British engineer living in the Austro-Hungarian Empire named Robert Whitehead. Early versions relied on unheated compressed air stored in a flask to power their engines. At the turn of the century, engineers began working on devices known as "superheaters" to heat the compressed air through a combustion reaction so that it would do more work per volume. Hardcastle was the first to invent a working third-generation superheater, which heated the air outside (as opposed to inside) the flask and used steam (as opposed to air) as the working fluid of the engine. He began working on superheaters in 1905 and had a prototype ready for production in 1908. It enabled torpedoes set for short range to run roughly twice as fast as, and those set for long range to travel five times farther than, the Navy's last pre-Hardcastle torpedoes.

Given its high importance, the Admiralty wanted to keep Hardcastle's invention secret. Hardcastle's status as a "service inventor"—that is, as a serving naval officer—gave the Admiralty powers to control his knowledge that it lacked when dealing with outside inventors. While the Patents and Designs Act of 1883 entitled all inventors to compensation for use of their patents by the Crown, Section 415 of the *King's Regulations and Admiralty Instructions for the Government of His Majesty's Naval Service* required service inventors, unlike outside inventors, to seek their departments' permission before applying for patents, thereby compelling them to reveal their knowledge. Since their knowledge (like Pollen's) was the chip being bargained over, this requirement potentially reduced their negotiating power. It also exposed service inventors to theft and made them reluctant to reveal their ideas too fully. As an interdepartmental committee charged with investigating the status of inventors in government service acknowledged, the requirement to pass an invention through a long channel of communication in order to obtain patent protection "is apt to arouse the suspicion of the inventor that the nature of his invention may be divulged before he has obtained protection."[36] However, holding information back could damage service inventors' ability to prove that their inventions predated Admiralty assistance in developing their ideas and therefore lower the amount of any award. Moreover, the *King's Regulations* forbade them from appealing the terms offered by their departments to the Treasury.[37] These provisions gave the Admiralty more power to control Hardcastle's knowledge than it had to control the knowledge of outside inventors. As Hardcastle later put it, "whatever terms were embodied in the [patent] assignment had to be accepted with good grace by the inventor if serving in H.M. Service, or incur Their Lordships displeasure, a proceeding which no junior Officer with any regard for his future prospects in the Service dare to risk."[38]

Service culture worked alongside the letter of employment contracts to influence officers' behavior. Although Hardcastle seems to have had no qualms about seeking compensation for his invention, others did. One service inventor putting forward a request for an award expressed his "regret at being obliged, by financial embarrassment, to occupy their Lordships attention upon a mercenary subject."[39] Debased profit-seeking behavior was for outside contractors; officers preferred to see themselves as acting from nobler, public-minded motives.

While Hardcastle's status as a service inventor meant that he did not have the same freedom of action vis-à-vis the Admiralty enjoyed by outside inventors, this service culture gave him a key advantage over the Pollens of the world. He wore the uniform; they did not. The uniform conferred legitimacy that outside "mercenary" inventors had to generate by other means; hence, in part, Pollen's attempt to draw on the prestige of science by speaking of the "laws" of gunnery. Naval officers did not have to fight what Pollen referred to as the "fraudulent contractor theory,"[40] nor did Admiralty officials instinctively suspect them of blackmail.

In 1908, Hardcastle and the Admiralty reached an agreement about his invention. In return for Hardcastle assigning his secret patents to the Admiralty, the Admiralty made him a three-part award. First, it gave him £5,000, judging this the sum that "an outside firm would have been likely to have given for the crude invention before the details had been worked out and the ultimate practical success obtained."[41] In line with court rulings about intellectual property in the private-sector employer-employee context, the Admiralty felt obliged to pay Hardcastle for what he had invented on his own but not for what he had invented after being reassigned to work on developing his invention with public support. In recommending a monetary sum, the Admiralty also factored in a second part of the award, namely, a promise that Hardcastle would be eligible for early promotion, indicating that it did not regard £5,000 as adequate in and of itself. In 1912, approximately three years early, Hardcastle was duly promoted to Engineer Commander, giving him a higher salary and pension. The third part of the award, which must be inferred from Hardcastle's subsequent pursuit of foreign patent rights and from knowledge of Admiralty practice in other cases, was an intimation that he would be permitted to exploit his patents commercially once the Admiralty assigned them back to him, both sides having preserved their secrecy and thus their commercial value.

The Admiralty made this award not out of generosity but to control Hardcastle's knowledge. It wanted to keep him in the service and prevent him from bolting to the private sector, as several other officers had recently done. Officials agreed that the relevant metric for judging Hardcastle's award was not his naval salary but what he could command for his invention and services outside the government—either as a private-sector employee or by selling his invention on the open market. "You must remember that the inventor may be lured away from the Government service for his brains," an official at the Navy's torpedo factory reminded an interdepartmental committee investigating service inventors in 1905, "and then the Government will have to pay a very much higher price for his inventions."[42] This was not an abstract concern for the Admiralty: for example, two of its first three heads of submarine development had left the service for the private sector, and the third had tried. When officers left the navy for private industry, the Admiralty lost control of the knowledge they carried—the human part of the knowledge triad could reveal the textual and mechanical parts of the triad. As Pollen's case showed, controlling the knowledge of defense contractors could be tricky.

World War I upended the agreement between Hardcastle and the Admiralty. Compensation in the form of early promotion had been predicated on the assumption of stability and a normal career path, while the possibility of commercial exploitation depended on preserving the secrecy of the invention. During the war, however, the Royal Navy divulged Hardcastle's superheater to allied navies (including the United States), calculating that British security now demanded not secrecy but sharing of knowledge. The result was that he could no longer obtain foreign patents for his invention and exploit them commercially for profit. Moreover, the Admiralty had kept him on shore for seven years ensuring that all torpedoes equipped with his invention were fit for service, and his lack of sea service made him ineligible under the rules for further promotion. The Admiralty did its best to secure higher pay for Hardcastle without the promotion by seeking an exception to the usual pay scale, which the Treasury resisted because exceptions caused bureaucratic complications. Although it relented under pressure, Hardcastle still believed that his stalled promotion had significantly reduced his pension and thus nullified much of the 1908 award. Moreover, the Admiralty could do nothing to put the genie of secrecy back in the bottle. It weakly requested the allies with whom it had shared Hardcastle's invention to compensate him for his inability to take out patents in their countries; unsurprisingly, they declined to do so. Accordingly, Hardcastle decided to seek redress with the RCAI, but it recommended against a further award.

Henderson's Gyro-Technologies

A professor at the Royal Naval College, Henderson was the Royal Navy's leading expert on gyroscopes in the World War I era. First used to maintain torpedoes' course, gyroscopes rapidly found multiple navigational, stabilization, and guidance applications in naval technology after the turn of the century. Henderson invented numerous gyro-technologies, the most important of which, from the Admiralty's perspective, was an improvement to a part of the Navy's fire-control system known as the director. Invented by the naval officer Percy Scott before the war, the director, placed high in a ship to maximize visibility, functioned as a master gunsight for the individual guns. Henderson improved the director by stabilizing it relative to a gyroscopic artificial horizon, an invention known as Gyro Director Training (GDT). The first trials in early 1916 immediately demonstrated its importance, and the Royal Navy found it invaluable in the low-visibility conditions of the North Sea during the war.[43]

Henderson occupied a hybrid insider-outsider position within the Royal Navy. On the one hand, his professorship made him an Admiralty employee. On the other hand, he had obtained his education that enabled him to make his inventions independent of the Admiralty. University-trained in both Britain and Germany, Henderson had studied with leading scientific lights like Lord Kelvin, Hermann von Helmholtz, and Max Planck before taking a position as head of the scientific department at the Scottish optical firm Barr & Stroud, which supplied rangefinders to the Royal Navy.[44] Interested in both teaching and industry, he left Barr & Stroud first for a professorship at the University of Glasgow and then for the Royal Naval College in 1905, where he remained through World War I.[45] As he never tired of reminding Admiralty officials (because he had to), his professorship did not come with a pension, unlike an officer's commission or civil-service position.[46] In addition, his employment contract with the Admiralty explicitly permitted him to perform outside work as a consultant, and he carried on his inventive work outside his duties as a professor—meaning that the Admiralty did not own the products of his inventive mind.[47] The consulting provision was almost certainly necessary to induce him to take the job, and he maintained a partnership contract with Barr & Stroud after joining the Royal Naval College.[48]

Keen to control Henderson's knowledge, the Admiralty directed him to take out secret patents on his inventions and assign them in accordance with Section 415 of the *King's Regulations*, as though he were an insider. In 1907, he complied with instructions from the Admiralty, which feared that Pollen (who was then working on gyroscopically stabilized rangefinder mountings) might invent something similar, to take out his first secret patent on what would become GDT.[49] Henderson's willingness to do so saved the Admiralty from having to negotiate terms for the assignment of the patent. By 1911, however, he had begun to develop a sense of grievance at the Admiralty's treatment of his secret patent. As he was preparing to apply for his second patent on GDT, he found that several patents taken out by others in 1910 infringed his original 1907 secret patent on GDT, and he believed that the Patent Office never would have issued them had his 1907 patent not been secret. In point of fact, the Patent Office did not search secret patents for anticipations, because the single examiner assigned to secret patents for security reasons did not have time to do so[50]—a point that irritated Admiralty officials as much as it did Henderson when they suspected that contractors were reinventing inventions covered by Admiralty secret patents.[51] From Henderson's perspective, the Admiralty, as assignee, was failing to defend his secret patent from infringement. "Although I have assigned away all rights in my patent, I am nevertheless still interested in the development of the invention, and from an inventor's point of view the situation is not a satisfactory one," he wrote to the Admiralty. "It would be much more so if My Lords could see their way to publish all the secret patents on this subject."[52] His unhappiness increased when the Admiralty declined to publish any secret patents, rejected his argument that the 1910 patents infringed his, refused to let him see relevant secret patents so that he could judge whether they infringed his own, and informed him that it doubted the validity of his 1907 patent.[53]

In 1915, Henderson's simmering dissatisfaction boiled over. He complained to the Admiralty, "My Lords have throughout compelled me to sign the same agreements as a Naval Officer whom they have educated and trained, and who invents something in the course of his duties; whereas all my inventions were outside of my duties and I had to pay dearly for the education and experience which have enabled me to make them."[54] Henderson also resented his loss of a professional reputation due to secrecy.[55] Expressing a similar sense of the continuum between the mental and material aspects of invention as Pollen, but through the particular discourse privileging theoretical over applied science in which his university training had socialized him, Henderson described his inventions as "applying the scientific principles which I learned at the feet of Kelvin and von Helmholtz, to the solution of problems having a practical application in the Service."[56] In a subsection entitled "Invention and Discovery," Henderson wrote:

> It is very difficult to draw a sharp line between invention and discovery, but to my mind the latter is greater than the former. A new mechanism for doing something which is already being done is called an invention, but I would not call it a discovery. Where an applied science like Naval Gunnery has developed by repeated invention until it seems to have reached the limit of perfection possible with the means available, and a new step is suddenly made which opens up a new line of advance giving rise to many inventions, such a step I would call a discovery.

On top of his own path-breaking discoveries, Henderson had directed many students who wrote important scientific papers, but the Admiralty had forbidden publication and robbed him of the reflected glory. "Let My Lords consider for a moment how the scientific reputation of, say, Sir J. J. Thomson or any of our great experimentalists would be depreciated," he lamented, "if deprived of all the work of the schools of research which they directed."[57]

Admiralty officials acknowledged that Henderson's arguments contained merit. Because it was vital to keep Henderson's knowledge secret, the director of contracts explained, the Admiralty had indeed required him to comply with the *King's Regulations*, thereby "depriving him of the fruits of his own brain-work and efforts." This official also admitted that secrecy had prevented Henderson from publishing papers, "which could have enhanced his scientific reputation and given him a standing with the Royal Society."[58] Temporarily putting aside the question of a monetary award, the Admiralty arranged to send a letter to the Royal Society supporting Henderson's candidacy for a fellowship.[59] Officials also agreed to put pressure on the recalcitrant Treasury to improve Henderson's emoluments as professor.[60]

The Royal Society letter pleased but did not pacify Henderson, who wanted financial security as well as an enhanced reputation. Knowing that the Admiralty wanted to control his knowledge, he now threatened to stop assigning his patents in accordance with the *King's Regulations* unless the Admiralty increased his salary and made him a satisfactory monetary award.[61] While decrying his "querulous attitude," Admiralty officials admitted that the *King's Regulations* did not apply to him the same as to naval officers and decided to recommend his case to the Admiralty Awards Council, which issued awards to inventors.[62] At the same time, however, they refused to let Henderson exploit GDT commercially in Allied nations or to agree to merely temporary secrecy of his patents, insisting that secrecy must be permanent.[63] In October 1916, the Admiralty Awards Council suggested an award of £5,000 to Henderson, on the condition that he assign all past and future patents to the Admiralty.[64]

Henderson had no intention of acceding to this demand. Not only was the Admiralty proposing to pay him far less than he thought his past patents merited, it was also requiring him to turn over all future inventions no matter how much he might make from commercially exploiting them. From his perspective, the Admiralty was attempting to buy ownership of his inventive mind, regardless of what he might do in the future, for the price of a set of existing inventions; put differently, it was trying to buy human knowledge, still in mental form but pregnant with applied potential, for the lower cost of mechanical knowledge already reduced to material form. In so doing, the Admiralty was asking him to give up all his negotiating leverage—control of his knowledge—for an inadequate sum. Again reminding the Admiralty of his outsider status, Henderson claimed that he had the right to be treated "as though I were wholly unconnected with the Service." At the same time, valuing his insider status and worried like Pollen about charges of unpatriotism if he lost it, he asked the Admiralty not to place him "in the invidious position of choosing between resigning and claiming my full rights, and thereby making it possible for anyone to say that in a time of danger to the country I had vacated my post for the purpose of personal gain."[65] Admiralty officials again backed down. Blaming the Treasury for refusing to place him on a pensionable salary and accepting that he deserved to be treated as an outside inventor, they agreed to refer his case back to the Admiralty Awards Council.[66] Upon reconsideration, the council increased its recommendation to £7,500, still on the condition that Henderson assign all past and present future patents to the Admiralty.[67]

Henderson still balked. He could not accept the condition that he assign all future patents to the Admiralty without "a voice in the assessment of the sum" he should receive for them. But he indicated a willingness to lower his pecuniary expectations if the Admiralty agreed to let him exploit GDT commercially in Allied nations; otherwise, he would expect an additional £10,000—tax-free—to offset the loss.[68] The Admiralty duly agreed to let him sell GDT abroad subject to its approval. This concession enabled him, unlike Hardcastle, to protect his foreign commercial position now that the Admiralty had decided that wartime security favored sharing knowledge rather than keeping it secret. Officials showed mounting frustration, however, over his unwillingness to assign all his patents and their consequent inability to control his knowledge. In a reversal of previous policy, one official complained that his refusal constituted a breach of the *King's Regulations*.[69] Both sides dug in: Henderson refused to assign his future patent rights, and the Admiralty refused to drop its demand that he do so.[70]

As the stalemate dragged on, Henderson's leverage grew. For one thing, the leakage of his knowledge of GDT to Allied nations authorized by the Admiralty threatened to turn into a flood. Henderson and his invention greatly impressed the US Navy.[71] According to him, the US Navy went so far as to ask "if I thought there was any possibility of them acquiring my services even temporarily to direct the work of the department of Fire Control Design." Henderson promptly informed the Admiralty of this tempting offer to sell his knowledge abroad, warning that "I cannot carry on much longer in my present state of uncertainty and impecuniosity."[72] In addition to this negotiating windfall, Henderson also benefited from the Admiralty's growing desperation to secure his services to help design a new fire-control system to replace Dreyer's, which, for reasons predicted by Pollen, had failed to cope with combat conditions during the war.[73] On legal advice, Henderson refused to sign the agreement governing the disposition of patents arising from the work for fear of prejudicing his ongoing battle with the Admiralty over control of his patents.[74] Without his signature, the Admiralty could not put him on the committee designing the new system.[75] The committee president implored the Admiralty to settle its differences with Henderson, as he "has a very much greater knowledge of the Director and of Fire Control requirements generally, than any other Gyro expert within the knowledge of this Committee."[76]

In an extraordinary minute, the director of contracts mused that any reconsideration of the award should factor in the extent to which the Official Secrets Act would have prevented the exploitation of Henderson's inventions abroad, even if the Admiralty had not required assignment of his patents in accordance with the *King's Regulations*. The director of contracts was contemplating, in other words, that the Admiralty could have used the Official Secrets Act to prevent exports of Henderson's knowledge to the private sector or abroad—just as it had done with Pollen.[77] Similarly, the secretary of the Admiralty suggested that the Official Secrets Act "is adequate to bind Dr. Henderson to keep secret all his present knowledge concerning naval gunnery requirements" regardless of the wording of any agreement with him.[78] If the Admiralty could not control inventors' knowledge with their consent, the Official Secrets Act provided a means to do so without their consent—and without the obligation to negotiate with or compensate them as dictated by patent law.

Even so, the Admiralty worked to improve its offer to Henderson. He received a knighthood, and the Admiralty agreed to reassign a number of patents to him for commercial exploitation.[79] With "the development of fire control at a standstill" due to the committee's inability to consult Henderson, the Admiralty official responsible for procurement worked out a deal to enhance the inventor's reputation by appointing him as "Expert Adviser to the Admiralty on the Application of Gyroscopes for Armament Purposes" and giving him a five-year deal worth £3,000 a year (nonpensionable) in return for Henderson agreeing to let the Admiralty control his knowledge by signing over any patents it wanted.[80] The Naval Staff concurred on the condition that Henderson's new title not indicate to foreign powers the importance attached by the Royal Navy to gyroscopes' potential in weapons—this itself was valuable knowledge.[81] Henderson agreed to sign the agreements covering the work of the fire-control committee in April 1920[82] and the five-year deal later that year,[83] making him "Adviser on Gyroscopic Equipment" (but not "for Armament Purposes") and pledging him to keep secret for ten years all knowledge acquired in the course of his work for the Navy.

The Treasury, as it had with Hardcastle, now complicated matters. Charged not with controlling inventors' knowledge but with safeguarding the public purse, it balked at the Admiralty's suggestion to pay Henderson for assignment of his patents, on the grounds that the *King's Regulations* obliged him to assign them without payment.[84] At the Admiralty's prodding, however, the Treasury consented to allow Henderson bring his claims before the RCAI.[85] Henderson and the Admiralty agreed to let the RCAI decide any award for both his secret and published patents,[86] and he finally assigned a number of patents to the Admiralty in late 1920 and early 1921.[87] When he went before the RCAI, he received additional awards totaling £18,000 for his various inventions.[88] This was on top of the knighthood and the £15,000 he earned from his five-year deal, plus the right to control his knowledge and sell his inventions after ten years. All told, he got at least four times what the Admiralty had tried to buy him for.

Conclusion

The cases of Pollen, Hardcastle, and Henderson all illustrate the challenges faced by the state, in the shape of the Admiralty, as it tried to control knowledge in a domestic and international political economy with a robust private sector and liberal property norms. In each case, the Admiralty had to deal with either the reality or the specter of knowledge controlled by independent contractors who might try to sell to foreign governments. It faced different legal environments depending on the status of the inventor.

With Pollen, a contractor, the Admiralty fought to wrest control of his knowledge, misjudging him as dispensable and Dreyer as an equivalent substitute. In the first instance, it had Dreyer take out secret patents covering his ostensibly independent invention. When intellectual property arguments failed to deter Pollen from maintaining secrecy without payment, the Admiralty fell back on intimidation. When this too failed, the Admiralty retaliated by striking Pollen from its list of approved contractors and forbidding communications with him, warning shots to anyone considering doing business with him.

With Hardcastle, whose insider position both helped and hindered him, the control afforded by the *King's Regulations* rendered the Admiralty's treatment of Pollen unnecessary, but its power was no less real for being hidden within the terms of Hardcastle's employment contract rather than wielded nakedly through the Official Secrets Act. The fear of driving Hardcastle, whose human knowledge was convertible into the textual and mechanical parts of the knowledge triad, out of the Navy and into the private sector incentivized the Admiralty not to wield its power arbitrarily, but instead to compensate him for submitting to it. When the Admiralty's calculations shifted during the war to favor sharing, rather than denying, Hardcastle's knowledge to foreign governments, however, it unilaterally abrogated its agreement with him by destroying his overseas commercial prospects. As in Pollen's case, the Admiralty treated secrecy as a one-way street: after enjoining it on an inventor, who maintained it, the Admiralty shared the inventor's knowledge with others (a rival inventor in Pollen's case, foreign governments in Hardcastle's) to his detriment.

Henderson alone avoided damage to his interests. Skillfully exploiting his hybrid insider-outsider position to avoid accusations of profiteering while demanding to be compensated like a contractor, Henderson protected his intellectual property—his chief leverage—by refusing to assign it to the Admiralty. He also taunted the Admiralty with his other leverage: the threat of leaving its employment. Judging him indispensable and increasingly desperate to control his knowledge, ironically thanks to the need to replace Dreyer's failed system, the Admiralty gave way. Tellingly, in so doing, it contemplated deploying the Official Secrets Act against Henderson. This suggests that use of the act against Pollen was not isolated: the idea got into the Admiralty's bloodstream.

Because the law offered more robust protection to the property rights of contractors—or, put differently, since the law gave the Admiralty more power over serving officers—the Admiralty struggled to control contractors' knowledge without breaking the law. Its employment contract with Hardcastle was a classic case of how contracts, notwithstanding liberal ideology, could encode and legitimize power imbalances between ostensibly equal parties. Its contracts with Pollen and Henderson did not give it the same level of control over their knowledge; hence its turn to the intrinsically less liberal Official Secrets Act. By its own admission, using this act to preserve the secrecy of independently discovered knowledge was illegal—amounting, in effect, to arbitrary expropriation. Yet the Admiralty told Pollen that the act applied to knowledge "whether or not" he had invented it, and it contemplated using the act against Henderson despite acknowledging that it had no ownership claim to his inventions.

Unsurprisingly, those on the receiving end of this behavior were left unhappy. Though the behavior was bullying, it did not stem from malice or conspiracy. Rather, it reflected the difficulty of pursuing the illiberal goal of national security in a liberal society, which was less willing to tolerate coercion outside the confines of contracts than within them. National security did not always prove compatible with the security of property, and in such cases, the former trumped the latter.

All parties to these fights used law as a technology to control techno-knowledge—so much so that we might think of law as the uber-technology of weapons technology. As Mario Biagioli and Marius Buning write, law has a "constructive feature": it does not merely represent "the objects that the technosciences confront it with," but actually constructs those objects itself.[89] Contracts and patents made knowledge into intellectual property. The Official Secrets Act made it into national-security information. Secret patents made it into both at the same time. In each case, law created new borders, not only around the object of knowledge but also around its container. Contracts, patents, and the Official Secrets Act all had to define the borders of the knowledge that was to be property or a secret. In particular, they had to locate those borders within a mental and/or material realm. Was the object of the contract the inventor's brain or something he had invented? Was the property a system or its mechanical instantiation? Was the secret a dial or the idea revealed by the dial? By the same token, law created borders limiting the circulation of knowledge. Patents in theory destroyed such borders—they threw open to all the knowledge they contained—but secret patents and the Official Secrets Act re-erected them. These borders should not be reified: they were constructed through a historical process and contested between actors inside the Admiralty, between the Admiralty and the Treasury, and between the state and outside inventors.

Finally, this chapter emphasizes that the determination to control the transnational flow of sensitive knowledge to maintain military superiority predates by many decades the institutionalization of such measures by the US national-security state. Indeed, the United States looked to Britain for inspiration when it wanted to create its own version of the Official Secrets Act and secret patents.[90] Just as the borders constructed in Britain before World War I should not be reified, so the US national-security state and nuclear-secrecy regime during and after World War II must not be black-boxed. There are long lines of continuity between the regulatory strategies adopted to protect British imperial power in the early twentieth century and those institutionalized by the US national-security state and nuclear-secrecy regime during and after World War II.

Notes

1. Mario Daniels, "Controlling Knowledge, Controlling People: Travel Restrictions of U.S. Scientists and National Security," *Diplomatic History* 43, no. 1 (January 2019): 57–82, at 58–59.

2. See T. H. O'Dell, *Inventions and Official Secrecy: A History of Secret Patents in the United Kingdom* (Oxford: Clarendon Press, 1994).

3. Patents, Designs, and Trade Marks Act of 1883, 46 & 47 Vict. c. 57, sec. 27.

4. Patents, Designs, and Trade Marks Act of 1883, 46 & 47 Vict. c. 57, sec. 27.

5. Minutes by Black, n.d. but November 17, 1913, and Treasury Solicitor, February 6, 1914, on CP Patents 1981/1913, ADM 1/8464/181, the National Archives, Kew, London, UK (hereafter TNA).

6. "Fire-control systems" are so called because they seek to control the gunfire of ships. For full treatment of the Pollen case, see Jon Tetsuro Sumida, *In Defence of Naval Supremacy: Finance, Technology, and British Naval Policy, 1889–1914* (Boston: Unwin Hyman, 1989). For Hardcastle, see Katherine C. Epstein, "Intellectual Property and National Security: The Case of the Hardcastle Superheater, 1905–1927," *History and Technology* 34, no. 2 (2018): 126–156. To my knowledge, the Henderson case has not been written up.

7. Babcock to Twining, September 1, 1912, B73-315, Naval Torpedo Station records, Newport, RI, USA.

8. Yaw is the rotation of a ship around a vertical axis.

9. Pollen to Tweedmouth, August 27, 1906, PLLN 5/3/1, Churchill Archives Centre, University of Cambridge (hereafter CAC).

10. Edward Harding, "Memorandum upon the Professional and Financial Value of the A.C. System," September 4, 1906, fol. 41, T 173/91, Pt. 7, TNA (hereafter Harding report).

11. Harding report, fol. 44.

12. Harding report, fol. 46.

13. Arthur Pollen, "The *Jupiter* Letters: Extracts from Letters Addressed to Various Correspondents in the Royal Navy Principally from HMS *Jupiter*" (May 1906), in *The Pollen Papers: The Privately Circulated Printed Works of Arthur Hungerford Pollen*, ed. Jon Tetsuro Sumida (London: George Allen and Unwin for the Naval Records Society, 1984), 73.

14. Brad Sherman and Lionel Bently, *The Making of Modern Intellectual Property Law: The British Experience, 1760–1911* (Cambridge: Cambridge University Press, 1999), 173–82.

15. Sherman and Bently, *The Making of Modern Intellectual Property Law*, 199–204.

16. Sherman and Bently, *The Making of Modern Intellectual Property Law*, 28–35, 44–47.

17. Pollen to Fiske, February 17, 1911, PLLN 5/4/4, CAC.

18. "Pollen's Aim-Correcting Apparatus. Notes of a Meeting held at the Admiralty in the Board Room on 9th August 1906," CP 13313/1906, PLLN 5/2/2, CAC.

19. "Pollen's Aim-Correcting Apparatus," CP 13313/1906, PLLN 5/2/2, CAC.

20. Admiralty to Pollen, September 21, 1906, PLLN 5/2/2, CAC; contract of February 18, 1908, Docket "Mr. A. H. Pollen. Original agreement for the 'Pollen Aim Corrector System, 1908," ADM 1/7991, TNA. For background, see Pollen, brief for counsel, ca. July–November 1924, pt. II, pp. 32–38, PLLN 8/2, CAC.

21. Admiralty to Argo, April 29, 1910, and enclosure, CP 14084/93S, PLLN 5/2/6, CAC.

22. For 1908, see Pollen to Tweedmouth, March 10, 1908, and enclosed memorandum, PLLN 5/3/1, CAC, and the correspondence between Pollen and the Admiralty for April–June 1908, PLLN 5/2/4, CAC. For 1910, see Admiralty to Pollen, April 11, 1910, PLLN 5/2/6, CAC; Pollen to Spender, April 12 and 13, 1910, PLLN 5/4/14, CAC; Pollen to McKenna, April 13 and 19, 1910, PLLN 5/3/2, CAC; Pollen to Admiralty, April 29, 1910, PLLN 5/2/6, CAC; and Pollen to Battenberg, May 9, 1910, PLLN 5/3/8, CAC.

23. Minutes on CP Patents 582/1910 and 600/196S/1910, ADM 1/8131, TNA.

24. See Mario Biagioli, "Recycling Texts or Stealing Time? Plagiarism, Authorship, and Credit in Science," *International Journal of Cultural Property* 19, no. 3 (2012): 453–476.

25. Contract and Purchase Department, *Pollen Aim Correction System. General Grounds of Admiralty Policy and Historical Record of Business Negotiations*, typescript addendum, para. 198, Admiralty Library, Portsmouth, UK.

26. Pollen to Peirse, September 16, 1912, PLLN 5/3/8, CAC; Pollen to Craig Waller, 24 December 1912, PLLN 5/3/9, CAC.

27. Admiralty to Pollen, February 21, 1913, PLLN 5/2/8, CAC.

28. Admiralty to Pollen, February 21, 1913, PLLN 5/2/8, CAC; Official Secrets Act of 1911, 1 & 2 Geo. 5 c. 28, section 1.

29. Admiralty to Coward et al., April 30, 1913, CP 16787, PLLN 5/2/8, CAC.

30. Pollen to Admiralty, July 14, 1913, PLLN 5/2/8, CAC.

31. Admiralty to Pollen, July 21, 1913, CP 22454, PLLN 5/2/8, CAC.

32. Pollen to Admiralty, July 14, 1913, PLLN 5/2/8, CAC.

33. Sovereign immunity is a legal doctrine with long and complex historical roots that may be briefly summarized as the immunity of a sovereign government to suit without its consent (that is, without a waiver of its sovereign immunity for a particular type of suit).

34. Contract and Purchase Department, *Pollen Aim Correction System*, typescript addendum, para. 221.

35. Transcript of hearing, August 1, 1925, pp. 110–111, Pt. 14, and August 7, 1925, pp. 17 and 24, Pt. 19, T 173/547, TNA; transcript of hearing, August 6, 1925, pp. 54, 97–98, 100, Pt. 18, T 173/547, TNA; RCAI recommendation, October 30, 1925, T 173/90, TNA; Tindal Robertson to Tomlin, October 20, 1925, Pt. 8, T 173/91, TNA.

36. "Report of the Inter-Departmental Committee Appointed to Consider the Regulations as to the Taking out of Patents by Officers and Subordinates in Government Employment, with Appendices, 1905–06," April 30, 1906, p. 5, WO 32/5080, TNA.

37. See, e.g., Section 415 of *The King's Regulations and Admiralty Instructions for the Government of His Majesty's Naval Service* of 1906 and 1913.

38. Hardcastle to Robertson, April 15, 1926, T 173/257, TNA. "Their Lordships" was a reference to the Lords Commissioner of the Admiralty—the formal name for those comprising the Board of Admiralty when it was "in commission" rather than held by a single individual (the Lord High Admiral).

39. Dumaresq to Admiralty, September 7, 1912, CP Patents 1366/1912, ADM 1/8330, TNA.

40. Pollen to Thomas, March 1, 1910, PLLN 5/2/6, CAC.

41. Admiralty Awards Council, Report 26, "Award to Engineer Lieutenant S. U. Hardcastle," November 3, 1908, ADM 245/1, TNA.

42. Testimony of Colonel H. C. L. Holden (Superintendent of RGF), October 25, 1905, Appendix VII, "Report of the Inter-Departmental Committee Appointed to Consider the Regulations as to the Taking out of Patents by Officers and Subordinates in Government Employment, with Appendices, 1905–06," WO 32/5080, TNA.

43. Minutes by Singer, February 8, 1916, CP Patents 2805/1915, and for DNO, March 19, 1916, CP Patents 3583/1916, ADM 1/8590/111, TNA.

44. "Our Weekly Biography: Professor James Blacklock Henderson," *Page's Weekly* 7, no. 57 (October 13, 1905): 823.

45. Barr to Henderson, March 16, 1898, fols. 33–36, UGD 295/4/744, Barr & Stroud (hereafter B&S) mss., University of Glasgow Archive Services.

46. See, e.g., Henderson to President, Royal Naval College (hereafter RNC), "Report of Ten Years Scientific Work for the Admiralty," March 10, 1915, CP Patents 2805/1915, ADM 1/8590/111, TNA.

47. Henderson to Jackson, February 28, 1917, CP Patents 4856/1917, TNA.

48. Barr & Stroud to Henderson, March 20, 1907, fols. 423–424, UGD 295/4/744, B&S mss.

49. Minute by Jellicoe, June 10, 1907, CP Patents 46/1907, ADM 1/7936, TNA.

50. Minute by Black, March 24, 1909, CP 12451/1909, ADM 1/8046, TNA.

51. Minute by Tudor, April 2, 1914, CP Patents 2246/1914, ADM 1/8464/181, TNA.

52. Henderson to President RNC, October 27, 1911, CP Patents 996/1911, ADM 1/8222, TNA.

53. Henderson to President RNC, November 27, 1911, and minutes thereon, CP Patents 1031/1911, ADM 1/8222, TNA; Admiralty to President RNC, March 21, 1912, CP Patents 1031/1911/79S, TNA; Henderson to President RNC, June 25, 1912, CP Patents 1298/1912, TNA.

54. Henderson to President RNC, "Report of Ten Years Scientific Work for the Admiralty."

55. Black to Greene, January 20, 1916, ADM 1/8590/111, TNA.

56. Henderson to President RNC, "Report of Ten Years Scientific Work for the Admiralty."

57. Henderson to Controller, December 24, 1919, CP Patents 8573/1920, ADM 1/8590/111, TNA.

58. Minute by Black, March 26, 1915, CP Patents 2805/1915, ADM 1/8590/111, TNA.

59. Admiralty to President, Royal Society, February 17, 1916, CP Patents 2805/15/S17, ADM 1/8590/111, TNA.

60. Minute by Ewing, June 23, 1915, CP Patents 2805/1915, ADM 1/8590/111, TNA.

61. Henderson to President RNC, February 22, 1916, CP Patents 3583/1916, ADM 1/8590/111, TNA.

62. Minutes on CP Patents 3583/1916, ADM 1/8590/111, TNA.

63. Minutes quoted in Admiralty Awards Council, "Consideration of the services of Prof. J. B. Henderson in respect of his inventions relating to gyro-compasses and director gear," October 13, 1916, CP Patents 3583/1916, ADM 1/8590/111, TNA.

64. Admiralty Awards Council, "Consideration of the services of Prof. J. B. Henderson in respect of his inventions relating to gyro-compasses and director gear," October 13, 1916, CP Patents 3583/1916, ADM 1/8590/111, TNA; Greene to President RNC, January 14, 1917, CP Patents 3583/3099, ADM 1/8590/111, TNA.

65. Henderson to Jackson, February 28, 1917, CP Patents 4856/1917, ADM 1/8590/111, TNA.

66. Minutes on Henderson to Jackson, February 28, 1917, CP Patents 4856/1917, ADM 1/8590/111, TNA.

67. Admiralty Awards Council report, September 4, 1917, and Admiralty to President RNC, September 30, 1917, CP Patents 4856/1917, ADM 1/8590/111, TNA.

68. Henderson to President RNC, November 29, 1917, CP Patents 5768/1917, ADM 1/8590/111, TNA.

69. Minutes on Henderson to President RNC, November 29, 1917, CP Patents 5768/1917, ADM 1/8590/111, TNA.

70. Henderson to President RNC, June 4, 1918, CP Patents 6470/1918, ADM 1/8590/111, TNA; minutes thereon, and Admiralty to President RNC, July 19, 1918, CP Patents 6470/1918, ADM 1/8590/111, TNA.

71. McCormick to Craven, July 25, 1919, BuOrd 37711/10, RG74/E25A-I/B2995, National Archives and Records Administration, Washington, DC (hereafter NARA I).

72. Henderson to Controller, December 24, 1919, CP Patents 8573/1920, ADM 1/8590/111, TNA.

73. See *Reports of the Grand Fleet Dreyer Table Committee, 1918–1919*, ADM 186/241, TNA.

74. Henderson to Drury-Lowe, September 4, 1919, G11058/1919, ADM 1/8590/111, TNA.

75. See minutes on Henderson to Drury-Lowe, September 4, 1919, G11058/1919, ADM 1/8590/111, TNA.

76. Minute by Drury-Lowe, October 30, 1919, G11058/1919, ADM 1/8590/111, TNA.

77. Minute by Jenkins, n.d. but ca. January 19, 1920, CP Patents 8573/1920, ADM 1/8590/111, TNA.

78. Minute by Murray, May 25, 1920, CP Patents 8748/1920, ADM 1/8590/111, TNA.

79. Minutes on CP Patents 8700/1920, ADM 1/8590/111, TNA.

80. Minute by Field, April 20, 1920, CP Patents 8748/1920, ADM 1/8590/111, TNA.

81. Minute by Chatfield, April 21, 1920, CP Patents 8748/1920, ADM 1/8590/111, TNA.

82. See "Extract from Agreement between the Commissioners for Executing the Office of Lord High Admiral of the United Kingdom of Great Britain and Ireland (hereinafter called the Admiralty) of the one part, and Professor James B. Henderson of R.N. College, Greenwich, of the other part," and "Conditions as to patents arising in connection with Experimental or other Work undertaken by firms or persons in Collaboration with Government Departments," enclosed with minute by Drury-Lowe, May 7, 1920, G11058/1919, ADM 1/8590/111, TNA.

83. Multiple sources, including Henderson's IEEE obituary, say that Henderson served as "Adviser" from 1920 to 1925. While Henderson signed a provisional version of the five-year deal in April, the exact date of the final agreement is unknown, and it may have been back-dated, for which see Bristows, Cook & Carpmael to Admiralty, August 12, 1920, CP Patents 9217/1920, ADM 1/8590/111, TNA.

84. Admiralty to Treasury, February 28, 1920, CP Patents 8701/20881, ADM 1/8590/111, TNA; Treasury to Admiralty, April 7, 1920, CP Patents 8862/1920, ADM 1/8590/111, TNA; Admiralty to Treasury, May 19, 1920, CP Patents 8862/22106, ADM 1/8590/111, TNA; and Treasury to Admiralty, June 17, 1920, CP Patents 9038/1920, ADM 1/8590/111, TNA.

85. Treasury to Admiralty, November 3, 1920, CP Patents 9038/1920, ADM 1/8590/111, TNA.

86. Admiralty to Henderson, August 5, 1920, CP Patents 9038/23076, ADM 1/8590/111, TNA.

87. Henderson, "Assignment of Certain Inventions for Improvements in Sighting Devices, Fire Control Apparatus for Naval Guns and Bomb Dropping Gear," December 31, 1920, and "Assignment of Letters Patent for Certain Inventions for Improvements to Sighting Devices, Fire Control Apparatus for Naval Gun and Bomb Dropping Gear," March 1, 1921, CP Patents 8700/1920, ADM 1/8590/111, TNA.

88. RCAI report, October 21, 1924, Cmd. 2275, and October 18, 1937, Cmd. 5594.

89. Mario Biagioli and Marius Buning, "Technologies of the Law / Law as a Technology," *History of Science* 57, no. 1 (2019): 3–17, at 17.

90. See, e.g., "HR 26656" (National Defense Secrets Act of 1911), RG233/B404, NARA I; and the correspondence between the British Comptroller-General of Patents and the US Commissioner of Patents in RG241/EA1-1038/B142/F1-132, National Archives and Records Administration, College Park, MD.

Chapter Three

Culture Diplomacy

Penicillin and the Problem of Anglo-American Knowledge Sharing in World War II

MICHAEL A. FALCONE

Mass-produced penicillin was one of the most significant research and development (R&D) achievements of World War II. From the moment of its first successful use in March 1942, the antibiotic made surgeries less risky; it allowed a huge reduction in amputations; it freed precious time and resources for field hospitals, doctors, and nurses; and it restored countless troop hours previously lost to treatment of sexually transmitted diseases and other maladies. At least 100,000 soldiers were treated with penicillin in the European theater, bringing surgical infection under control for the first time in the history of warfare.[1] As a 1944 cover story in *TIME* proclaimed, penicillin seemed to be a miraculous cure that might even "save more lives than war can spend."[2] Beyond this therapeutic success, the antibiotic also looked like a heartening triumph for international scientific cooperation. Penicillin was a British discovery, but when war broke out it was hastily brought to the United States to take advantage of superior American production capacity. Together, the two allies seemed to set in motion a pharmaceutical revolution, united in selfless devotion to a common cause, just as they had been for so many other critical war technologies. After VJ Day, US science administrator Vannevar Bush proclaimed that "American and British scientists have worked so closely together that it will be utterly impossible, and a matter of no vital interest, to attempt to assign many explicit accomplishments to one or the other."[3]

But, of course, technological nationalism doesn't work that way. The question of who controlled penicillin in the postwar world turned out to be a matter of *great* interest. Penicillin was to spawn a multibillion-dollar antibiotics industry and prove a vital medical resource amid the rubble of a conflict-ravaged, reconstructing postwar world. It would thus bestow enormous leverage on the state that could lay global claim to its therapeutic gifts, its ultramodern sublimity, its embedded discourses of developmental altruism, and its commercial profits. Penicillin became, in short, an intensely *diplomatic* problem. British influence over its future began slipping away from nearly the moment the first secret vials arrived on American shores, while the convergence of geostrategic and commercial benefits embodied by the drug helped the United States to consolidate its technological power as it emerged as a global hegemon. In this way, the fungus cultures that birthed the modern antibiotics industry became a site of intensive struggle among state and corporate actors within and across national borders, as clashing interests sought to delimit and control this assemblage of highly valuable knowledge for themselves.

But penicillin was not just the *object* of these struggles, it was also a catalyst for them. Before World War II, the American state had shown an uneven interest in intellectual property issues. As this chapter details, however, the state's desire and capacity to control knowledge flows began to crystallize during the war years, thanks in part to the timely acquisition of technologies like penicillin. The drug quickly took its place in the thinking of American science and industry leaders as a national strategic asset and war technology, one that should fall under the rightful authority of the country that produced most of it: the United States itself. At the same time, members of the British diplomatic staff in Washington sensed that their claim to joint oversight of penicillin production was eroding as the war went on, and they fought doggedly to preserve the United Kingdom's say in controlling the drug's global distribution. By doing so, they hoped to conserve Britain's modernist reputation, its commercial profits, and its self-appointed role as the able guardian of its empire.

Both sides recognized that postwar global power would require a mix of military-strategic command and strength in world markets, and that is the reason penicillin became such a flashpoint—it represented both. Relatively new to geopolitical primacy, however, the Americans had more work to do to ensure that public and private interests would operate in alignment. With dominance over new technologies like antibiotics potentially there for the taking, American officials sought to link strategy and commerce by building material and legal infrastructures of state technological control. On one hand, this meant using public resources to mobilize industrial research, development, and production on an unprecedented scale—the taxpayer-funded "Arsenal of Democracy" that diffused costs and risks but concentrated profits, leveraging existing capitalist infrastructures. On the other hand, pursuit of postwar technological power also meant changing the rules of the game itself—modifying the legal structures governing intellectual property, institutional processes, and technical know-how across both public-private and Anglo-American divides. The institutional changes necessary to manipulate these legal frameworks might have seemed like dreary bureaucratic minutiae to many Americans at the time—and, indeed, to many historians since—but their effects would prove momentous for US corporations and deleterious for British hegemony.[4]

Thanks to its therapeutic and economic importance, penicillin has drawn considerable attention from historians. Numerous first-rate works have documented controversies over the antibiotic's priority of discovery; the public-private partnerships (and enmities) of its development; the discourses, reception, media attention, and meaning making behind the drug's long public life; and the birth of homegrown industries in places like France, the Netherlands, India, and China.[5] But for all that they deftly explain about one of the signal technologies of the twentieth century, these works have had little to say about the invention's role in the projection of US power on a global scale, or in the development of knowledge-control systems in the American context.

By contrast, this chapter examines both the new intellectual property machinery that technologies like penicillin helped to inspire in the United States and the diplomatic frictions that that machinery engendered, particularly with respect to the United Kingdom. By highlighting the transnational ramifications of corporate-friendly US war mobilization, as well as the perspectives of oft-overlooked diplomats and administrators plying the knowledge divide between global powers, this chapter spotlights a dynamic that has been all too easily overlooked before now: the ways that transnational technology flows led the US government to construct a knowledge regime that served the interests of both American capitalism and American hegemony. The buildup of this regime marks one understudied precondition for America's ascent to global power—and the transition from Pax Britannica to Pax Americana.

More specifically, the diplomatic furor over penicillin demonstrates how the American state created conditions in which corporations could not only function with minimum resistance but could do so on the backs of both US taxpayers and foreign partners. During World War II, the US government created mechanisms—patent powers, secrecy provisions, and the strategic manipulation of international commercial channels—to selectively enforce the permeability of the national border along a plane of technical knowledge. Sometimes the barrier was fluid and ideas circulated rapidly, but at other times—particularly when advantages to US commercial and strategic interests converged—the barrier was nearly impermeable. British collaborators often found that they could not get back enough of what they put in and could not capitalize on R&D that ripened on the other side of the knowledge wall. The effects of this selective permeability reverberated not only in the commercial interests of the two countries after the war but the strategic ones as well.

Using the transatlantic development of penicillin as its case study, then, this chapter demonstrates one way in which World War II marked an important pivot for the rise of the United States to world power, since it spurred the creation of a new regime in which *knowledge* moved to the center of American state builders' strategies of diplomacy and influence.[6] Far from the rosy portrait of collaboration hailed by Bush, the history of penicillin tells us much about the new mechanisms of national technological control that were built by the United States during the 1940s. It also lays bare the forcefulness of British officials' reciprocal attempts to maintain their own technological authority, as they forecast what it would take to preserve a hegemonic posture in the postcolonial world. Predictably, these simmering transatlantic tensions boiled over at key points during the war. One such case is an overlooked diplomatic dispute from the time of the D-Day invasion in 1944, when American officials attempted to regulate use of penicillin in British Empire and Commonwealth countries just as they did for the rest of the world, prompting outrage from London. Episodes like this conveyed a growing American assertiveness over the command and disposal of shared knowledge and an increased US willingness to restrict the circulation of knowledge for nationalistic ends. In an effort to maximize strategic advantage—as well as in the struggle to coax sometimes-reluctant industrialists to serve government and national interests—the United States built its Arsenal of Democracy in particular ways that had the effect of absorbing knowledge from the outside and then ring-fencing it.

US Pharma's Very Good Day

In 1928, Alexander Fleming, a medical researcher at St. Mary's Hospital, London, chanced to discover that a fungal mold known as *Penicillium rubens* possessed antibiotic properties. Twelve years later, a research team at the University of Oxford directed by Australian pathologist Howard Florey proved the mold's efficacy in curing bacterial infections. Benefiting from the bioassay work of fungal expert Norman Heatley and the purification work of German émigré Ernst Chain, the team devised a method to reproduce penicillin cultures, and in 1940 published a paper in *The Lancet* suggesting its ability to keep infected mice alive.[7]

But if 1940 was a good year for Oxford's rodent population, it was also Britain's "darkest hour," the nadir after the defeat at Dunkirk and before the relative relief of Lend-Lease and the arrival of the United States and Soviet Union as co-combatants. With war raging, the British pharmaceutical industry was too overburdened to provide the huge increase in supply needed to conduct clinical penicillin trials, and indeed, the first human patient died after the small existing stock of the drug ran out.[8] Given these obstacles, the Rockefeller Foundation gave the Oxford group funds to travel to the United States to seek production there. Aware of the security implications of their discovery, and worried in case their vials of penicillin should be stolen or sabotaged, Florey and Heatley smeared samples of the fungus into the linings of their coats for their June 1941 flight to America.[9]

Penicillin was far from the only technology that crossed the Atlantic in the early years of World War II. In fact, an array of strategic technologies—among them radar, jet engines, DDT, microwave generators, and, in nascent form, the atomic bomb—came under joint development by Britain and the United States during the conflict. The United Kingdom in 1940 possessed arguably the world's preeminent military-scientific-industrial complex, but the war stretched the metropole's resources to their limits. Coordinating with American researchers and accessing the seemingly limitless capacity of US factories was thus of vital strategic importance to Britain and its empire.[10] For their part, the still-noncombatant Americans were drawn in by the cutting-edge research breakthroughs and processes the British were offering with essentially no strings attached. Both sides stood to gain strategically from an open technical interchange.[11]

Penicillin arrived nine months after these exchanges began. Professional contacts led Florey and Heatley to the US Department of Agriculture, which maintained a fermentation program at its Northern Regional Research Lab (NRRL) in Peoria, Illinois. Collaborating with NRRL researchers Kenneth Raper, Andrew Moyer, and Charles Thom, the team improved on the Oxford group's prior research by developing a new fermentation method using a broth of corn steep liquor and phenylacetic acid. They also devised a new "submerged culture" process that allowed the penicillin to grow throughout the culture medium rather than only on the surface. Together, these advancements increased the production volume by many orders of magnitude, allowing for trials—if a manufacturer could be found.[12]

But convincing American pharmaceutical firms to join a state-led R&D program was far from a straightforward proposition in 1941. Despite their considerable bluster about patriotic productivity, many American corporations had been hostile to the Roosevelt administration since the early New Deal, a tension that visiting British missions observed and wrote home about in blunt terms.[13] In the case of penicillin, corporate skepticism was evident from the first joint meeting on the drug, held in October 1941 in Washington.

The meeting was convened by A. N. Richards of the Office of Scientific Research and Development (OSRD), the new agency designed to coordinate civilian R&D projects for what promised to be a technology-heavy war. Richards had recently been appointed head of the OSRD's Committee on Medical Research (CMR), and penicillin interested him greatly.[14] Buoyed by the Oxford and Peoria group's confidence that their mold could be mass produced, Richards proposed a cooperative penicillin program among the leading drug firms of Merck, Lilly, Pfizer, Lederle, and Squibb. With the state lacking compulsory powers, however, big businesses possessed the leverage simply to refuse government contracts point-blank if they were unsatisfied with the potential rewards.[15] Merck, for one, considered it unfeasible to produce the kilogram of testing material that the British visitors wanted. The meeting ended with no agreement.[16]

Two months passed. The conundrum for science administrators at the OSRD and other agencies was thus twofold: How could they induce industry to work on priorities deemed by the state to be strategically critical? And how could they convince companies to take part in a collaborative program that would require the sharing of precious technological knowledge not only with other US interests but with British ones as well?

The solution was to sweeten the pot. Ten days after the Japanese attack on Pearl Harbor, Richards called another meeting with pharmaceutical heads in New York. This time he stressed that the Oxford/Peoria team had increased production nearly twentyfold since the previous meeting and that he was eager to turn the advancements over to them. This news was sufficient to stoke the enthusiasm of George Merck and his counterparts, and a collaborative program commenced between the government and firms, under the coordination of Richards's CMR.[17] But although the pharmaceutical companies agreed to join the project, they uniformly refused direct government funding, as they were keen to avoid any administrative Trojan horse that might invite coercion from the state.[18] Without direct subsidy, the government would have to find other ways to exert oversight.

Recognizing that firms would only cooperate if their interests were protected, the CMR and the War Production Board (WPB)—the latter of which took primary responsibility for the program in June 1943—devised a corporation-friendly system that attended to the profit motives of industry first and foremost. Aiming at producing enough supply for a conjectural

D-Day invasion of Europe, the government underwrote large-scale research at university and in-house laboratories for the benefit of private-sector producers. Federal funds pushed research forward at Stanford, Penn State, the Universities of Wisconsin and Minnesota, Cold Spring Harbor (New York), and the National Institutes of Health. The CMR sponsored clinical trials and made decisions on usage of the drug. The NRRL continued its work streamlining fermentation techniques. And the government ensured a steady supply of raw materials: the WPB assigned blanket AA-I priority status to all ingredients and equipment needed for production; the Army Transport Command brought soil samples from all over the world to Peoria for strain isolation work; and the Jar Food Administration oversaw the doubling of the production and rationing of lactose, vital to the corn steep liquor method.[19] At the heart of it all, the Chemical Division of the WPB distributed materials and data, built facilities, and coordinated essentially all facets of the project, harmonizing the efforts of the twenty-one pharmaceutical firms selected to produce the antibiotic.[20] All told, when major production of fermented penicillin was ready to begin in May 1943, firms were reliant on the government for a significant portion of their costs, data, materials, and equipment.

The state also shielded companies from the risks of factory construction by offering 100 percent rapid tax depreciation on new antibiotic plants.[21] Corporations were quick to cash in on this arrangement, to the consternation of the WPB's head of penicillin, Albert Elder. Elder observed companies like Commercial Solvents ordering the finest equipment and commissioning elaborately appointed factories to rival the most magnificent prewar facilities, in some cases driving construction costs to double what had been estimated. Yet the WPB dutifully paid the bills, taking it upon itself to make up for industrial delay and overreach by procuring from the army as much advance equipment as possible, so that firms could be ready to roll their assembly lines on day one.[22]

Just as enticing to corporations were the intellectual property carrots that government officials dangled in front of them. Richards's boss, OSRD head Vannevar Bush, assured company leaders that when he drew up royalty and licensing provisions he would always keep corporate profit margins in mind. His only regulatory aim, he said, was to prevent unreasonable monopolies.[23] Similarly, when the state conducted research and clinical trials, the resulting data and production methods were patented and assigned to the federal government, then turned over to industry for free. The implication of this was that any further improvement made by the companies would be privately patentable and marketable in its entirety, as if it represented new research from scratch. Such terms also left open the possibility of filing patents on production processes overseas.[24] These liberal terms were notably more generous than those imposed by the OSRD on academic researchers: unlike in industry, academics were required to turn all inventions over to the government for public patenting.[25] Merck, for one, was quick to take advantage of the favorable rules for the private sector, and filed its first US penicillin patent on May 15, 1943.[26]

Industrialists lapped up these benefits, and made sure to frame the results as if they were the product of private-sector dynamism. Conscious public relations strategies leveraged the seemingly miraculous properties of penicillin to boast of individual companies' patriotism and inventiveness, and even firms making auxiliary products like refrigerators and machine belts got in on the act.[27] These spirited narratives obscured both the indispensable role of the state and the fact that, even with the assistance they were receiving, firms *still* looked for ways to circumvent the program's terms. For example, although companies were permitted to conduct their research entirely independently, they were held to the simple condition that they keep the government informed of their activities so that vital information could be distributed to other participants as necessary.[28] But firms bristled at the requirement. The WPB's Elder despaired when one industry liaison told him that the government should merely "go from one plant to another collecting honey, but [should] not . . . distribute pollen along the way."[29]

A forceful case was made at the time by certain observers—and by some historians since—that the US government was too much in the thrall of industry during the penicillin program.[30] Among the strongest evidence supporting this critique was the manner in which officials prioritized certain avenues of research that favored the beliefs—and, potentially, the pocketbooks—of industrialists. For there were in fact *two* simultaneous penicillin development programs: mold fermentation (growing the drug from fungus) and chemical synthesis (artificially producing drug molecules from laboratory-induced reactions). Pharmaceutical companies believed that they stood to profit much more from the latter. So too, the critics allege, did the CMR. With the modern drugs trade based overwhelmingly on organic chemistry, it was widely assumed that antibiotics synthesized in the laboratory, rather than ones fermented from complicated and unpredictable fungi, would represent the commercial future.[31] This belief prevailed despite the fact that mold fermentation was proving consistently successful while penicillin synthesis was not.[32]

Indeed, industrial scientists spoke of nonsynthetic, fermented penicillin as a probable commercial "flop," and their treatment of it reflected this. A WPB investigation in November 1943 revealed some firms to be adopting a "lackadaisical attitude" on fermentation, believing that "it would be foolish to waste a lot of time and money on a relatively inefficient method of production when large scale production of a synthetic may be just around the corner."[33] This foot dragging gravely concerned the WPB's Elder, whose task was to obtain medicine for war theaters regardless of method. But while his WPB backed the proven program of fermenting fungal penicillin, Richards's CMR continued backing synthetics, signing contracts with eleven industrial laboratories and thirteen educational institutions and government labs, while negotiating an agreement with British officials for a synthesis exchange with UK companies and agencies. The results failed to meet expectations time and again, yet the CMR forged ahead, underwriting the program to the tune of at least $2.25 million from December 1943 through October 1945.[34] This commitment was ultimately for naught—the CMR had thrown its weight behind industry logic in vain. In fact, true synthesis was not achieved until 1957—based, again, on British research, building on Oxford chemist Dorothy Hodgkin's 1945 modeling of penicillin's molecular structure. Even then, fungus fermentation remained the most efficient production method for the rest of the century.[35]

The WPB's Elder believed that the CMR's blind commitment to the seductive and potentially lucrative synthetic program imposed the largest single impediment to speedy mass production of fermented penicillin. Elder trained his sights in particular on CMR head Richards, who had previously served as a paid consultant for Merck and who regularly spoke in florid terms about the firm's pioneering spirit and sure ability to meet all future supply needs.[36] The CMR's cozy relationship with the private sector was also questioned by Robert Coghill, the head of fermentation at the NRRL Peoria lab, who accused Richards of acting as an industrial gatekeeper, one who allowed outside researchers to liaise with the "closed corporation" of Merck, Pfizer, Squibb, and Lederle only when they had something to give.[37] Whatever the motive, the combined dedication of the CMR and private firms to synthesis fruitlessly took up "the best efforts of probably the largest number of chemists ever concentrated upon a single objective," in the words of Nobel laureate R. B. Woodward.[38]

With the backing of Elder and the WPB, however, mold fermentation sallied forth. With the government spending millions on production capacity, the WPB insisted on retaining more oversight than the CMR.[39] Production of fermented penicillin thus became a hybrid enterprise. On top of the government's contributions—and to avoid state intellectual property controls—Merck, Pfizer, Squibb, and others collectively invested several million dollars, merging public, private, British, and American research to devise a new "deep tank" fermentation method that significantly increased yields. Moreover, Oxford's Heatley joined the Merck staff for a half year, contributing the cup-plate assay technique that became the industry standard.[40]

The government's incentivizing paid off—a vital new strategic technology had arrived as desired. A patient with streptococcal septicemia was successfully treated with drugs made via the new fermentation methods in March 1942, and in September 1943 Pfizer began converting an old ice factory on Marcy Avenue in Brooklyn into the world's first commercial penicillin plant. By D-Day nine months later, total allied supplies had surpassed 100 billion Oxford units, easily meeting the needs of the Normandy invasion. By VJ Day the following year, the number had grown to a remarkable 650 billion units.[41]

Transatlantic Tensions from the Earliest Days

But if the vast therapeutic success of the drug seemed on one side of the Atlantic like a triumph of American public-private industrial might, collaborators from Britain saw a decidedly less rosy picture. In transatlantic R&D collaborations across dozens of technical fields, British procurement officers in Washington frequently complained that US industry was playing a tilted game. Since much vital R&D data was of an unpatented or unpatentable nature, significant industrial know-how fell through the cracks of legal agreements between the United States and Britain, giving much more latitude to industrialists to share knowledge—or not—as they saw fit. Penicillin was a noteworthy such case—technology transfer can scarcely get more unpatentable and informal than fungal cultures smeared into scientists' coat linings—but perceived US industrial secretiveness plagued many other programs, too. British diplomats protested to their American counterparts that UK firms essentially disclosed all their trade secrets while US companies refused to give anything back, citing insufficient patent protection.[42]

This is not to say that the British were simply charitable. Britain's leaders saw their country as a global power whose self-conception as a power derived in part from its technological knowledge, and they leveraged that knowledge in an explicit strategy to secure material benefits from the United States. They desperately needed US resources and production, and discoveries like penicillin gave them bargaining power. Moreover, most British officials were convinced that US leaders would take pains to secure British intellectual property rights. The British embassy, Purchasing Commission, and other missions in Washington pressed their case in good faith at the highest levels, drawing US military secretaries Frank Knox and Henry Stimson directly into negotiations.[43] Soon, delegates began work on a bilateral patent interchange agreement, talks for which were still underway when Florey and Heatley arrived in Peoria.

As for UK private intellectual property holders, they were far from pliant in following their leaders and often recalcitrant. British corporations found themselves in an unenviable bind. Firms whose research was being released to the United States were at the same time barred by the British government from filing for international patents, since their inventions were subject to the Official Secrets Act. Many were thus reluctant, despite the government's authority, to release information into what seemed blatantly like a losing proposition.[44] Their room to maneuver, however, was limited by the fact that the British state's powers over them had been vastly expanded for the war. They also received repeated assurances that US authorities like Knox would safeguard their interests. As a result, by January 1941 all but a handful of UK licensors had agreed to transmit their information to the United States, confident in leaving the nitty-gritty legal wrangling to the governments.[45]

But this confidence turned out to be misplaced. As we have seen, American officials like Bush and Richards were preoccupied with the enormous task of implanting an R&D bureaucracy into a state apparatus that had never had a centralized science structure before, as well as by the need to coax industrial laboratories to cooperate with state-strategic research goals in the absence of effective compulsion powers. In such a context, the expectation that the United States would have the bureaucratic capacity—let alone the will—to safeguard British intellectual property claims was short-sighted. The British, for their part, did not have the luxury of time to ensure fully that their knowledge would be enshrined in protective legal frameworks. Since the purpose of the exchange scheme had been to leverage scientific secrets in return for quickly acquired materiel and cooperation, delays to seek intellectual property assurances would only negate the exchange's importance to fighting the war. Thus, as MIT president Karl Compton, head of radar on the American side, wrote, "We have been inclined, I think quite properly, to consider [patent rights] as of very secondary importance and to follow the policy of not permitting the consideration of patent rights or trade advantages to slow up in any way the prosecution of the defense effort."[46] This was to prove a fateful decision for both sides.

Beyond mere neglect, however, the American government was also actively evolving to exert more control over the knowledge circulating within its borders. The state was initially not well organized for easy government control over public-private breakthroughs like penicillin. It was thus necessary for R&D administrators to transform new and existing bureaucratic instruments into an improvised suite of control mechanisms that could increase state authority over research, production, and dissemination. Taking advantage of the United States' unique financial and productive position in the war, government decision makers used these mechanisms to gain control over what they considered to be endogenous national security knowledge. But while the new bureaucratic frameworks came into being for targeted war purposes, by conceiving of and deploying such new expressions of authority, the state laid the groundwork for more muscular structures of power over knowledge and knowledgeable bodies in the decades to come.

Among the new mechanisms were, first, administrative incursions into patent laws; second, new secrecy provisions; and third, an active manipulation of America's globally dominant commercial production and distribution flows. While these might seem at first blush to be boilerplate administrative innovations, they nevertheless formed a crucial—and largely hidden—aspect of the broader global power transition from Britain to the United States. They thus demonstrate the importance, for foreign policy historians, of devoting careful attention to the structural, procedural, and administrative machineries that lay behind sites of trans-border interactions and hegemonic politics.

Control Mechanism I: Patents

The first of the new mechanisms was in the realm of patents. Intellectual property represented one of the most pressing military-industrial hurdles at the beginning of the war, since, as we have seen, US industrial firms made clear that their participation in the Roosevelt administration's schemes would be contingent on their competitive interests remaining unharmed. Patents also became increasingly urgent as the sheer volume—and cost—of technological materials commanded by the military ballooned to unprecedented levels during the war.

Unlike Britain, which for centuries had reserved a "Crown right" system for the state to utilize subjects' patented inventions, the United States government had no general patent use policy prior to World War II. In fact, from the nineteenth century into the 1930s, the Supreme Court often ruled *against* the government when officials attempted to use or redistribute inventions without clear authorization or consent of the owner.[47] Lacking central guidance, individual departments and agencies constructed their own regulations on matters of ownership, assignment, and license of inventions, navigating the difficult terrain of how patent policies should apply to public employees versus private contractors versus "extramural" third parties.[48]

In August 1940, the OSRD's predecessor agency, the National Defense Research Committee, hammered out a policy delineating its right to command and dispose of the intellectual fruits of domestic R&D. The state would retain the right to a royalty-free license to any invention made under the auspices of a government contract. If a private firm declined to file a patent, the government could file one itself, but would be required to give the firm a license to use it. By the fall of 1941 the OSRD had a team of twelve patent lawyers evaluating contracts and inventions, distributing information to other agencies and the military, and deciding which patents the government would exercise rights over.[49]

How do these expansions of government interest into patent issues relate to diplomacy? In early 1941 President Roosevelt negotiated the Lend-Lease agreement with Britain, permitting the still-neutral United States to send a torrent of aid across the Atlantic via unrestricted transfers of "defense articles."[50] In a matter of months the program was expanded across the world, sending military technology and equipment to dozens of other allies. American companies soon realized that Lend-Lease would mean massive outflows of not only materiel but also, potentially, their own proprietary knowledge. As foreign service officer Elbridge Durbrow reported, US firms were loath to share trade secrets with other countries lest the information be "used to produce goods commercially which will compete with American goods in the world market after the war."[51]

As a resolution, Section 7 of the March 1941 Lend-Lease Act placed the responsibility for safeguarding American intellectual property at very high levels of authority. The legislation required the "Secretary of War, the Secretary of the Navy, and the head of the department or agency . . . in all contracts or agreements for the disposition of any defense article or defense information[, to] fully protect the rights of all citizens of the United States who have patent rights," and to ensure the payment of royalties, war emergency notwithstanding.[52] Responsibility for patent royalties, in fact, reached the *highest* of levels. The bilateral Lend-Lease agreements included language, modeled after Article IV of the original accord with Britain, that mandated that foreign governments cover the costs of intellectual property payments to American firms, "tak[ing] such action or mak[ing] such payment when requested to do so by the President of the United States of America."[53] This clause gave autonomy and flexibility to the White House in conducting Lend-Lease, to be sure, but it is also telling of the government's evolving relationship with high-tech national industry that a specialized legal matter such as patent protection should be vested at such a high level in what was primarily an arms and equipment transfer. By invoking both the president and the cabinet, the global intellectual property concerns behind the transfers were hard to miss. For their part, the British found the language to be ambiguous regarding their own responsibilities. As P. H. Goffey, a patents officer from the Ministry of Aircraft Production, put it, "Whatever Section 7 may mean, we are bound to abide by it."[54]

Soon the American state began to exercise even more muscular control over patents, beyond simply safeguarding the commercial interests of US inventors. For one thing, in the face of exploding wartime costs for manufacturing and intellectual property licenses, on Halloween of 1942 Congress passed the Royalty Adjustment Act. This act allowed agencies and departments to challenge corporate royalties if they deemed them to be unreasonable or excessive, thus opening the way for negotiation with industry over what constituted "fair and just" rates. It also allowed lawful "government use" to apply to contractors and subcontractors working on the government's behalf, thus opening the way for the internal redistribution of knowledge when deemed necessary by the increasingly influential research agencies.[55]

Nowhere were the new intellectual property interests of the state better in evidence than in the failed penicillin synthesis program. Although contracts between government and industry were voluntary—or, rather, precisely *because* they were voluntary—the OSRD was able to vest considerable power over the project's output in the hands of just one individual: Vannevar Bush. Marking a departure from prior US intellectual property paradigms, under the terms of the synthesis agreements, Bush had the power to require contractors to grant licenses at reasonable rates to other companies, whether they were program participants or not; he could determine unilaterally whether or not inventions were attributable to the collective program and if patents should be filed on them; and he had final say on the use and distribution of all patent rights in the United States among the contractors, domestic *or* foreign.[56] Such wide-ranging discretion was highly unusual in the US context, and much more akin to the powers enjoyed by a British war minister than a US civilian administrator. Indeed, by design and agreement, Bush's authority in the synthesis program precisely matched that of his UK counterpart, Sir Edward Mellanby, head of the Medical Research Council in London.[57]

Nevertheless, by dint of his country's stronger material position, Bush's de facto patent powers turned out to exceed those of Mellanby, an asymmetry that would bring significant advantage to the US pharmaceutical industry. When it came time to mete out postwar patent rights for new processes developed during the synthesis program, Bush pressed for the agreement to define fairness in ways that validated US corporate claims about the origins of their techniques and discoveries, to the great consternation of Britain's Imperial Chemical Industries (ICI) and Therapeutic Research Corporation (TRC). Mellanby's Medical Research Council exchanged letters with Bush and his associates for five years after the war trying to hammer out the particulars of the agreement, but they repeatedly reached impasses regarding the definition of "patents," the inventions to be included in the agreement, and which companies and consortia should receive what proportion of the intellectual property spoils. For example, in determining whether knowledge relevant to both the synthesis and fermentation programs would fall under the synthesis agreement, Bush said yes—but only in certain cases, which would be determined by investigation where necessary. This allowed US firms like Lilly and Upjohn to claim that several of their inventions (pertaining to ingredients known as "precursors" or "adjuvants") fell outside the transatlantic agreement and were thus not subject to sharing, license distribution, or any other oversight. Predictably, the British were dissatisfied with these loopholes, and ICI and the TRC vehemently contested Bush's determinations in statements of case all the way through 1957.[58]

The overall contours of the synthesis agreement also reflected the newfound grasping for intellectual property control by the American state, and by Bush in particular. While Mellanby suggested that each side determine the fate of penicillin inventions made by its own nationals, Bush insisted—based partly on the limitations of the US government—that each side instead control the fate of all patents on its soil, regardless of whether the patent holder was British or American. As a legal statement by Britain's TRC argued in 1953, it appeared that American authorities were "particularly anxious to have full control over all American patents whether owned by American or British participants."[59] As we have seen, it was precisely the American state's newfound involvement in the development of strategic materials like penicillin that created such an anxiousness in the first place. Development of technologies and extension of control over them were mutually constituting phenomena during World War II.

Indeed, Bush's negotiations were just the tip of the iceberg. As intellectual property arrangements like the synthesis agreement went forward at the agency level, the US Congress took sweeping measures to create even broader knowledge-control frameworks. An unprecedented legislative amendment in August 1941 (Public Law 239) forbade any US invention from being patented in a foreign country unless express consent was given by the US Patent and Trademark Office (PTO). While this was primarily intended to prevent dissemination of sensitive defense information, it also logically had the effect of regulating access to US companies' licenses for foreign manufacturers without potential government vetting. The consequences for American inventors violating the statute (that is, sharing with foreigners without permission) were severe—two years in prison, a $10,000 fine, and permanent prohibition from filing for future patents, regardless of the invention's potential relevance to the war or the presence of any secrecy orders. When it came to the unregulated export of American intellectual property, lawmakers now meant business.[60]

Notably, the PTO *did* sometimes grant exceptions to this statute, but generally only to allow Americans to file patents in the United Kingdom.[61]

And companies generally requested such waivers only when they perceived that filing abroad would ensure their future competitive advantage over foreign companies.[62] In other words, the government declined to invoke "national defense" conceits if a persuasive case could be made for multinational profits—particularly if they were being earned in Britain, still the United States' primary industrial-technological competitor.

The British were not without agency in the matter of intellectual property, of course, and their missions to Washington spent several years negotiating a general Patent Interchange Agreement, signed in August 1942. But there were several problems with the accord that worked to the advantage of those trying to keep American knowledge in America. First, relentless political difficulties ratifying the deal in the US Congress meant that the pact was subject to continual modification and erosion by a bilateral committee through 1946.[63] Second, while the agreement set precedents for the disposal of royalties and evaluation procedures on *patented* information, it did little to codify the transfer of *unpatented* knowledge like production processes—which remained a highly lucrative asset for US industry.[64]

Finally, the lopsided solution offered by the Patent Interchange Agreement to resolve transatlantic intellectual property disputes was that each government would pay its own nationals' claims for the wartime use of inventions by the other government. That is, the US government would pay to settle the complaints of US patent holders if the British used their inventions, and vice versa. Fortuitously for the United States, many more such claims were made by *foreign* inventors against *US* producers than the other way around. In this way, the British Treasury paying its own domestic inventors for the use of British inventions by Americans ended up serving as an unintended form of Reverse Lend-Lease.[65]

Control Mechanism II: Secrecy

A second—and more direct and robust—power the US government claimed for itself to regulate knowledge flows during World War II was its use of secrecy orders. Such powers in the United States dated to an October 1917 law that allowed the PTO to withhold grants and order that inventions be kept secret for war purposes. But that law lapsed with the signing of the armistice, and secrecy powers lay fallow until 1940. With entry into another war looming, and despite considerable controversy and debate over the harm that it would do to private inventors, Congress renewed the 1917 legislation as Public Law 700 in July.[66] Under the new statutes, inventors were forbidden from disclosing information declared secret unless authorized to do so by the government. They were also forbidden from filing secret patents abroad.[67] The PTO convened a special office with representatives from the military, WPB, OSRD, and others to determine the secrecy status of all new filings, and they exercised these powers liberally. From 1942 to 1944 more than half of all patent applications in war-related fields—radar, electronics, materials science, cryptography, and others—were put under lock and key by the authorities.[68]

These provisions fulfilled both security and practical purposes for the government. Up to that point, departments and agencies outside the PTO did not have access to pending patents. But using the gravitas of wartime secrecy orders, officials from across the federal bureaucracy now had a new incentive at their disposal to coax inventors to cooperate with research agencies. The text of the orders reflected administrators' brazen exploitation of that gravitas. When secrecy letters were sent out to people and companies, the document included a warning that if the inventor wished their creation to be considered for war use—and more directly, if they wished to "preserve [their] rights"—then "it is suggested that you promptly tender this invention to the Government of the United States for its use."[69]

A brief aside will illustrate just how quickly the government consolidated its intellectual property power through secrecy provisions. As Alex Wellerstein has detailed, in April 1942 a member of Frédéric Joliot-Curie's team from the Collège de France filed patents in the United States on various prewar nuclear energy innovations, including designs for a reactor. This was a potentially serious problem. In the eyes of US officials, it would be out of the question to grant nuclear patents to French scientists, since that would threaten US national security. Approval of the patents would also potentially let France break America's intended postwar monopoly on atomic energy. On the other hand, openly contesting the French claims might reveal the extent of the United States's own nuclear knowledge, or even require detailed argumentation in court over the most closely guarded secret in American history. As it was—and despite the fact that the French patents did not even derive from knowledge originating in the United States—Bush and commissioner of patents Conway Coe took advantage of the newly revived US secrecy laws to declare the applications secret, preventing access to them until peacetime. In doing so, they affirmed their own willingness to break prior small-state taboos when it came to the circulation and restriction of knowledge.[70]

These measures were powerful, but the growing warfare state quickly extended an even more muscular hold over intellectual property. The Second War Powers Act of March 1942 included a provision allowing the White House to transmit to any government agency all information "now or hereafter in the possession of the Department of Commerce, or any division or bureau thereof." Hidden behind that simple language was the handy fact that one of the divisions of the Department of Commerce was the Patent Office. The statute thus used national security provisions to permit any interested agency to use any given patent application, even if the inventor declined to tender it to the government.[71]

In the case of penicillin, the enactment of the Second War Powers Act allowed for the restriction of publication on the drug just as the first successful civilian treatment test was being administered.[72] This information blackout applied to all countries outside the Anglo-American exchange. Even the chemical structure of penicillin remained a classified secret through at least 1946, which meant that the four principal variants of the drug received temporary scientific designations (F, G, X, and K in the United States, and I, II, III, and IV in the United Kingdom) in place of accepted chemical nomenclature.[73] Moreover, since the synthesis agreement yoked both countries together when it came to releasing information, mutual consultation would be required before any secret research data on synthesis could be disseminated to outside scientists, even after the exchange period ended.[74]

US authorities had now turned a corner when it came to the state imposition of restrictions on private intellectual property. In 1940, the renewal of secrecy legislation had still been controversial and a sunset clause had been included to allow the orders to lapse after only two years. Over the course of the war, though, opposition to government impositions of secrecy gradually fell away, and administrators' power over domestic inventions grew.

There was to be no turning back. Despite the fact that in August 1945 the incoming patents commissioner, Casper Ooms, issued a "general rescinding order" to cancel restrictions on the more than 11,000 inventions that had been declared secret during the war, the emergency secrecy paradigm behind them remained in place. A total of 799 inventions remained in the dark, mostly in the nuclear field, and more than two thousand *new* secrecy orders were added during the next half-decade. Finally, under pressure from the Department of Defense, in 1951 Congress passed the Invention Secrecy Act, making secrecy powers a permanent fact of US law.[75] The prerogatives cobbled together around technologies like penicillin and atomic bombs in the early part of the war thus became new, fixed mechanisms of state control over intellectual property.

Control Mechanism III: Commercial Muscle, Diplomatic Leverage

The third mechanism the American state built around its new strategic knowledge was the exploitation of overseas commercial channels, engaging in a clear-eyed manipulation of the alignment between national-strategic and commercial interests. The government both helped lubricate the US production machine at home *and* facilitated the distribution of that machine's output overseas, thereby ensuring domination over the global antibiotics market both during and after the war. Having suitably indigenized a technology initially brought to its shores from Oxford, American state actors put antibiotics at the center of efforts to exert influence over postwar European reconstruction; to project US power abroad as an advanced and benevolent technological hegemon; and to funnel countless new customers to private American industry. The result—sometimes through formal action, sometimes through diplomatic neglect—was to isolate Britain not only from much of the development of penicillin but also from the drug's global distribution, a circumstance that set off a diplomatic imbroglio across the Atlantic and inspired bureaucratic restructuring in the corridors of power in London.

At the beginning, penicillin was intended for military use. But as production mushroomed, a natural question arose: Who was next in line? Public awareness of the "wonder drug" grew wildly between 1941 and 1944, and newspapers on both sides of the Atlantic extolled its miraculous properties, running regular stories about patriotic production and ill families in desperate need of cures.[76] With output increasing a thousandfold between 1943 and 1945, and the price per unit dropping from $20 to near $0.60, the question of conversion for civilian use was not a matter of "if" but "when."[77]

But the looming decision to apportion stocks for civilians made British liaison officers nervous, since production was overwhelmingly concentrated in the United States—American factories provided 95 percent of the worldwide supply.[78] Moreover, as stockpiles swelled, US corporations began bristling at their British commercial counterparts' desires to increase production in the United Kingdom. A June 1944 dispatch from Nigel Campbell, the Ministry of Production's head of nonmunition supplies, reported that US firms believed that they alone should "satisfy the needs of the whole world, especially after the war."[79]

These worries came to a head around the time of the D-Day invasion in June 1944. That same month, America's fledgling international distribution agency, the Foreign Economic Administration (FEA), made an explicit break from the country's general policy of distributing medical and other supplies through the Lend-Lease aid program.[80] In a move that coincided neatly with the agency's first allocation of penicillin vials for civilian use, the FEA now decreed that international recipients of US materiel would be obliged, whenever and wherever possible, to utilize commercial channels for all purchases and transfers.[81] Forcing allies away from Lend-Lease was no small matter. By 1944 more than forty countries outside the British Commonwealth qualified for the program, comprising three-fourths of the earth's landmass and three-fifths of its population.[82] Selling penicillin commercially instead of "loaning" it through Lend-Lease thus meant tapping into a latent business market of staggering proportions.

What drove the FEA's new policy? The agency had been created in September 1943 to absorb and centralize more than forty foreign economic programs, including the Board of Economic Warfare, the foreign relief programs of the State Department, and the Lend-Lease Administration. Combining those activities brought the aid impulses of Lend-Lease into friction with the other agencies' mandates to safeguard US commercial interests. As it turned out, the latter often took precedence.

Milo Perkins, the outgoing head of economic warfare, explained the balance of concerns to Congress in 1943. "The whole job," he said, "is complicated by the need to protect United States commercial exporters, just as far as it is physically possible to do so in a war economy." This was vital work, since it was precisely these exporters who would "spearhead United States commercial activities abroad when the war is over."[83] For its part, the FEA's own statement of objectives specified the agency's obligation to promote "mutually advantageous economic relations between the United States and other countries"—a clear reflection of the shifting priority toward trade among allies and away from the unidirectional US beneficence that had prevailed in 1940–1942.[84]

But penicillin wasn't just any commodity, and in taking over its global distribution the FEA was inheriting a high-level diplomatic concern. This was evidenced by the fact that emergency appeals for the drug from friendly and neutral countries had heretofore been handled by the State Department.[85] Soon, however, procedures surrounding the circulation of the American "gift" of penicillin regularized. The FEA and WPB announced that they would first export one billion Oxford units of penicillin to Latin America, then quickly expand the program to European neutrals, the Middle East, and French areas, among others. Following the new FEA policy, all of these transactions were to be made on a commercial basis by US companies shipping the drug to American agents in the field.[86] Recipient governments would be required to submit full reports on usage to the United States every month. Most importantly, the United States required that each country convene a penicillin committee to tightly control the drug's distribution, using a tiered list of usage priorities devised by Dr. Chester Keefer, chairman of the National Research Council's chemotherapy committee. This monitoring was justified, officials said, by the fact that supplies for civilian use were still limited, and that US hospitals were subject to restrictions as well.[87] As part of its blanket program, the FEA did not exempt Britain from the new policy.

The British were incensed. Sir Walter Venning, head of the British Supply Mission in Washington, immediately put the matter before the

Principal Commonwealth Supply Committee, the coordinating body for Commonwealth and empire procurement in the United States. Venning wanted all British missions in Washington to oppose the FEA mandate with a forceful united front, which they promptly did. T. W. Childs, the British Supply Council's general counsel, informed the FEA that Britain's numerous missions "would not tolerate dictation."[88] Within a week the FEA backed down, agreeing that Britain was free to receive penicillin through government channels as it pleased. This was only a temporary concession, however. Two months later, on September 4, the FEA sent another letter to the British Colonies Supply Mission (BCSM), informing it that the mandate for Britain to buy penicillin commercially was now a matter of "settled policy" and would be enforced.[89]

The British were infuriated once again. The Colonies Supply Mission complained that the FEA had "shown the cloven hoof," while the Ministry of Supply scrambled to find ways to increase production in the United Kingdom to reduce reliance on the United States. And as the Ministry of Production saw it, the new US decree "would open export of penicillin to commercial channels while placing technical control of its usage throughout the world in American hands. This is of course quite unacceptable to us." Together with ongoing difficulties British researchers were experiencing in obtaining information on the submerged culture process from American pharmaceutical firms—an area of knowledge critical for expanding production in the United Kingdom—many British diplomats and mission staff perceived the new arrangements as a deliberate effort on the part of US administrators to crowd Britain out of a vital and lucrative new industry.[90] In their view, the United Kingdom should not be indebted to American firms for what they considered to be an essentially British technology.[91]

British diplomatic staff also demonstrated an acute awareness of the emerging commercial and geopolitical implications of penicillin. On June 26, the British Supply Mission cabled Ministry of Supply headquarters in London warning of the harm to Britain that would be caused by a unilateral US program of civilian penicillin exportation. Receipt of the American drug in foreign capitals, said the mission, was creating "a steadily growing widespread impression that [the] U.S. is the only source of supply, and it might be well if steps could be taken to counteract this." As a solution, the cable suggested that a token amount of Britain's diminutive output could be set aside for civilian and empire use, as well as small amounts for third countries. Sending British penicillin overseas would be a diplomatic coup that could "satisfy some of the enquiries which have been received from British Embassies in various parts of the world."[92] This would show that Britain was still a major global player, as well as preserve its prospects as a commercial powerhouse. In the mission's view, the WPB and the FEA could hardly object to a modest export scheme from Britain, since America surely "recognised that it is only fair that the UK should have some share in the development of trade in this drug which has been made available by our joint efforts."[93]

But reputation was not the only thing at stake. Britain also had an empire problem. A week after D-Day, the Ministry of Supply acknowledged that British production would not be sufficient to cover the needs of the colonies or Commonwealth.[94] Although there were seven authorized penicillin manufacturers in the United Kingdom, shortages of materials, know-how, equipment, and labor continued to hamper their output.[95] The deficiency was especially acute for the case of India, whose army would need, by an early estimate, more than a third of all the supply of penicillin available to Britain—or ten times the amount required for the Royal Navy and Royal Air Force combined.[96] If the metropole couldn't meet the need, it would mean a further slippage in Britain's dominance over its territories and associated states—away from London and, in this case, *toward* the American private sector, since the FEA rules would force Britain's entire bloc to buy the life-saving drugs from agents of Merck and other companies. Such a capitulation would have implications beyond penicillin, too. As the BCSM in Washington urgently telegrammed to Undersecretary J. B. Williams of the Colonial Office, failure to resist the FEA on penicillin would open the floodgates for a more general commercialization of vital commodities for Britain and the empire—something that the FEA had threatened directly in its letter on September 4, along with an assurance that it had the full backing of the State Department in doing so.[97]

Unsurprisingly, given the imperial and economic stakes, the BCSM lodged fierce protests with American officials to try to exempt British dependencies from the FEA's commercial directive. On September 5, the FEA agreed to a compromise of sorts. It would allocate 100 vials each of 100,000 Oxford units of penicillin to Gold Coast, Nigeria, Kenya, and Uganda. But Jamaica, Trinidad, and British Guiana would be required to receive their penicillin commercially like everybody else, since they had well-established trading channels with the United States. Although colonial supply officials strongly opposed this division, their attempts to resist by presenting another united diplomatic front quickly fell apart: to their dismay, the missions from India, South Africa, New Zealand, Australia, and the British West Indies—likely concerned more with their own immediate supply needs than the interests of Whitehall or metropolitan drug companies—all readily agreed to the FEA's purchasing paradigm. It was a capitulation that the FEA was quick to point out to the BCSM.[98]

If Britain was hemmed in by the FEA's dominion over the penicillin supply, it was also hamstrung by its own ongoing, desperate need for

Lend-Lease aid—the conditions of which stifled its freedom to export anything, let alone to sell penicillin. Following foreign minister Anthony Eden's informal agreement with the Roosevelt administration in September 1941, Britain had pledged to drive down its export activities to near zero in exchange for a $30 billion Lend-Lease lifeline. It had also agreed that no articles from Lend-Lease would enter commercial channels in Britain or elsewhere, even indirectly.[99] With a wash of raw materials, equipment, and knowledge funneling across the Atlantic in both directions via Lend-Lease, the prospect of Britain exporting a jointly produced antibiotic from its own private sector raised eyebrows. Thus, proposals like those from the Ministry of Production's Nigel Campbell for Britain to divert some of its penicillin production for civilian use while continuing to import American penicillin for the military were, in the words of Lord Moore of the Joint American Secretariat (JAS), politically "dangerous" and liable to put Britain "in a position which would be hard to defend."[100]

Awkwardly, mass penicillin production was also coming online at precisely the time that certain vocal elements in the United States were fighting to curtail the generous terms of Lend-Lease, particularly vis-à-vis goods like penicillin that were perceived as trickling into Britain's postwar commercial plans.[101] Even worse, one of the leaders of this movement was Leo Crowley, head of the FEA.

In the words of historian Alan Dobson, Crowley was "pro-business and anti-British and this soon became apparent." Through 1943 and the first half of 1944, Crowley waged a punitive campaign against Britain. He sidelined UK-friendly administrators at the FEA, and when British officials tried to negotiate permission to increase their gold reserves and loosen limitations on exports, Crowley set up an executive policy committee to *tighten* the restrictions.[102] By mid-June the Ministry of Supply was so vexed by Crowley's crusade that, in a move perhaps akin to cutting off one's nose to spite one's face, it proposed abstaining from acquiring civilian penicillin through Lend-Lease altogether in order to avoid being indebted to the United States and its increasingly coarse restrictions.[103] For its part, the FEA simply responded that Britain wasn't eligible to receive penicillin for noncombatants through Lend-Lease anyway.[104]

The broader issue of Lend-Lease remained strained for the rest of the war. Despite negotiations between Winston Churchill and Franklin Roosevelt at the Quebec Conference in September 1944—and notwithstanding the sympathies of State Department negotiators—Britain's export industry received little reprieve.[105] US domestic political considerations prevented a meaningful relaxation of Lend-Lease terms until VE Day. Thus, while mushrooming American stockpiles allowed penicillin to be sent liberally to foreign governments as of July 1944, as well as to be available in corner drugstores across the United States in March 1945, the drug did not even go on sale to the British public until June 1946.[106] While Britain waited out the war, American pharmaceutical companies enjoyed a year's head start in selling penicillin to the world.

All in all, the United States used the overwhelming dominance of its material wealth and production, and a clear understanding of the dependence of other nations upon it, to control the transnational flow of knowledge. With vast Lend-Lease needs still vital to everyday survival up to and beyond the end of the war, Britain had no choice but to accede to American whims. In mid-September 1944, Britain's JAS—a blanket policy body coordinating all UK supply missions—conceded that while British and imperial self-determination was important in principle, the utter production imbalance and unanticipated muscularity of the US state in wielding it meant that Britain and its colonies had no choice but to accept whatever conditions the FEA imposed. With UK supplies simply outmatched and the FEA digging in, the penicillin case was "[not] one on which to make a last ditch stand." The empire would have to take advantage of procurement from the United States however it could. The Secretariat briefed the Colonial Office, and finally asked the BCSM to stand down, terminating its communication with the FEA on penicillin problems.[107]

As this episode shows, wherever British policymakers turned in terms of divvying up the knowledge spoils of the war they met a new instrument of the American state or a new expression of US productive dominance to block the way, particularly as the end of the conflict drew near. In the case of penicillin, British mission officers made one other attempt to achieve some sort of parity in the intellectual property realm. Henry Self and Robert Sinclair from the Ministry of Production and Combined Production and Resources Board (CPRB) believed that the FEA's hard line was simply the culmination of an ongoing American campaign of knowledge denial to British firms, and that it was past time to make a comprehensive, bilateral study by the two ostensible world penicillin leaders to determine once and for all their mutual requirements, production, usage, and export plans. The CPRB could undertake the task, and both partners could finally have equal seats at the negotiating table. Self and Sinclair insisted that American dictation had become "unacceptable," and the need for high-level joint action was now "urgent and acute."[108]

This proposal was doomed from the beginning, however, because while the basics of penicillin production were known to both sides, delays among British manufacturers were primarily due to shortages in *unpatented* knowledge—the production techniques and proprietary know-how of US firms. Corporations guarded these jealously, and the US government had no authority to disclose or compel them under the framework of the public-private program it had cobbled together earlier in the war.[109] This led some British missions instead to approach US companies directly, as in the case of the Ministry of Supply's director of medical supplies, Frank Warburton, who began delicate negotiations with Squibb and Merck over data sharing with Britain's Glaxo. By July 1944 a tentative accord had also been reached between Britain's Distillers Ltd. and the US Commercial Solvents Corporation.[110]

These backdoor agreements, quietly arrived at amid a summer's worth of tempestuous conflict between diplomatic missions, demonstrated something simple and ironic: Washington could now subsidize, direct, and ring-fence national security knowledge, all while pleading powerlessness at the state level when it came to sharing it. Elegantly tying together the overlapping interests of national security and the commercial marketplace, the very compromises and intellectual property concessions that the US government had made to secure the cooperation of industry at home had now also become key devices for preventing of the outflow of knowledge from American shores.

Modernism and Mammon

As the above episodes have shown, the penicillin question was not simply a matter of uneven research collaboration or commercial competition. Rather, it represented an urgent and strategic foreign-relations matter at the highest level. The ongoing penicillin synthesis negotiations, for example, centered on protracted talks between Bush and Mellanby, but also looped in personalities all the way up to British ambassador Lord Halifax and foreign minister Anthony Eden.[111] What this indicates is that to both sides, the strategic, economic, political, and indeed, nationalistic characteristics of high technologies like penicillin were becoming central policymaking and diplomatic preoccupations. In the postwar world, high-technological industry and domestic control over the knowledge embedded therein would go hand in hand with national strength, in a way that exceeded the paradigm before the war.

The fight over penicillin rights persisted in rankling people on both sides of the Atlantic for years. The OSRD and MRC continued to pick over the bones of the failed synthesis program, clashing over which side's companies would fall into what tier of licensing rights.[112] British journalists and politicians hotly debated what they perceived as the wanton loss of a national strategic asset to a commercial rival. And before the war even ended, members of Parliament (including one who shot a pointed question at soon-to-be prime minister Clement Attlee) questioned why it had been necessary for British antibiotic scientists to seek help from the United States in the first place.[113]

Indeed, believing Americans to be technological thieves was a theme that both right- and left-wing Britons could agree on in the 1940s and 1950s, and the "theft" of penicillin represented one of the most painful abuses of all.[114] In the corporate world, the director of ICI complained that the Americans had "shown a lack of good neighbourliness in claiming the penicillin story as their own achievement."[115] And as late as 1959, Labour Member of Parliament Julian Snow fulminated in the House of Commons that "[Florey and Heatley] went to America and before we knew where we were the drug was patented in the United States and throughout the world by American interests."[116]

In the 1950s these hostilities were exacerbated by a further expansion of US pharmaceutical dominance. Penicillin set off a frenzied search for new antibiotics among biomedical companies, which in turn unleashed a patenting arms race. Between 1950 and 1959 US pharmaceutical giants patented drugs on a hitherto unknown scale in the United Kingdom and Europe, with nearly 3,000 new applications filed by the top six firms alone.[117] Among those were various attempts to patent penicillin processes, which particularly twinged Britons. As might be expected, the patenting bonanza inspired new accusations that US firms like Merck had deceived their transatlantic counterparts by charging royalties on the fruits of the minds of Fleming, Florey, Heatley, and Hodgkin.[118]

Merck's general attorney (and later president) John T. Connor responded to British accusations of malfeasance by insisting that "no British firm has paid one shilling" for royalties, a finding that allowed Americans—and scholars since—to disavow British claims as mere conspiracy-mongering.[119] But the finding that British firms were not paying licenses on American-held patents obfuscated the reality that they were paying for something else. Despite waging legal battles until 1952 to avoid paying a percentage of their net sales to the United States, British firms Distillers, ICI, and the consortium TRC eventually reached agreements with Merck and Pfizer to pay for *unpatented* process knowledge. Glaxo and Distillers took out fifteen-year licenses costing many millions of pounds for this know-how, and in the case of Glaxo, this amounted to the equivalent of 3 percent of the company's net profits. Even if these fees were somewhat modest by commercial standards, for British observers, paying foreigners for penicillin was a matter of principle and primacy, as well as a cause of anger at having been crowded out of a modern boom industry.[120]

Scholars, led by Robert Bud, have identified British accusations of US technological thievery as part of an exaggerated declinist narrative borne of postwar anxieties. Postimperial British cultural discourse directed grievances at an ascendant American superpower as the United Kingdom struggled to redefine Britishness and recast its place in the world at midcentury. The result was a circling of the wagons around exemplars of British technocratic glory—a display of "defiant modernism," in Bud's memorable phrase.[121]

Bud's case is persuasive, but beyond the geopolitics of grievance we must also pay close attention to how intensely *diplomatic* penicillin was—and from an earlier date than scholars have generally articulated. As we have seen, the archives reveal that penicillin was treated by both sides in terms of national pride, strategic power, and commercial strength from the earliest days of the transatlantic collaboration—and well before the final imperial unraveling of the following decades.

To take one example: In its first meeting in October 1942, the Ministry of Supply's General Penicillin Committee discussed how best to censor publication on penicillin to prevent information falling into enemy hands. Yet within a year, the same committee was lamenting that Britain's role in penicillin development was underappreciated by the world, and its members were brainstorming ways to publish pamphlets for mass distribution that would tout British achievements.[122] In the same vein, the priority of discovery of penicillin became a sizzling press controversy in Britain in the middle of the war. This was not just because of conflicting accounts between Fleming and Florey, but because of the early and intense *nationalism* of penicillin discourses. As some historians have noted, Fleming's ultimate triumph in the dispute—and his resulting celebrity status after 1942—may have been in large part due to the seamlessness with which his heroic discovery fit within the providential narratives of national chosenness surrounding Dunkirk, the Battle of Britain, and the Blitz.[123]

It is hardly surprising, then, that penicillin became a national rallying cry in Britain. The antibiotic—as both medication and discourse—was a "never again" case that gnawed at the pride of many Britons. The story of its ostensible "donation" to the United States was repeatedly cited as a dark parable in parliamentary debates until 1948, when the Development of Inventions Act passed into law. That act created a new body, the National Research and Development Corporation, which was intended to bolster state capacity in the arenas of technology and intellectual property. The agency—important through the 1980s in coordinating revenue for such technologies as hovercraft, computing, and carbon-fiber composites—explicitly hoped to remedy what J. C. Cain, its head of applied science, labeled as "Penicillin Syndrome."[124]

British patent law was also overhauled, with a new Patents Act in 1949. Notably, the legislation permitted the patenting of drugs for the first time.[125] Under the new act, inventions in any government sphere of interest would remain with the Crown, rather than the inventor, although the inventor would retain the right to claim a reward should the invention be used commercially. The outgrowth of this policy was to increase pressure on government departments to seek patents and to look for commercial exploitation for inventions made under their auspices, including overseas.[126] As Bud notes, for example, when Oxford researchers in the 1950s determined the structure and efficacy of cephalosporins, they were quick to obtain patents.[127] Britain, it seemed, had learned a lesson from the disastrous diplomacy of penicillin.

A Pharmaceutical Superpower

For their part, the Americans resented what they perceived as British undermining of US contributions to the development of penicillin. In practical war terms, Fleming's mold was unusable without mass production, and it had been in the United States that the submerged culture and deep tank processes had been perfected. Moreover, both the US government and American pharmaceutical firms had invested substantial funds and resources into the project, something that no less an authority than Florey hailed as critical to the successful deployment of the drug before the Normandy landings.[128] For these reasons, American actors believed that it was their right to retain intellectual property rights over penicillin. As Vannevar Bush said in 1952, "the British tendency to call us robbers annoys me considerably."[129] By this time, however, Bush was sitting on the board of directors of none other than Merck & Co.

However annoyed the chairman of Merck might have been, it was his and other US companies that ultimately profited from the state-lubricated technology exchange of the war, and it was his government that enjoyed the strategic and geopolitical results. The production expertise acquired in this period—and the new mechanisms leveraged by the American state to control knowledge over it—meant that before the war had even ended, American firms already held a monopoly on the natural antibiotics market at home and abroad, a sector that for three decades after the war made up 25 percent of the entire global pharmaceutical trade. Moreover, the penicillin methodologies developed during the war were central to the discovery and manufacturing of an enormous range of later antibiotics—as of 2009, β-lactams (penicillins and their spinoffs, carbapenems and cephalosporins) continued to make up 45 percent of the global antibiotic market, generating some $18.9 billion a year. Similarly, all production today of monoclonal antibodies and other bioengineered bacteria like insulin producers depends on fermentation techniques first devised by the Oxford group, the Peoria laboratory, and the industrial firms during the war.[130]

Penicillin allowed US pharma to break into markets from which it had always been excluded by European brands. By July 1945, American drug companies' overseas sales surpassed sales in the United States, a feat never before achieved in the history of US pharma. In fact, penicillin caused such a rush of foreign trade that supplies in hospitals in the United States began to dry up in October 1945, requiring temporary reestablishment of export controls and rationing. This was merely a speedbump, however. An analysis in March 1946 showed that the value of US penicillin exported to foreign countries represented three times the amount of *all* American drug exports to the same countries in 1938.[131]

Before the war, US pharmaceutical companies had paled in size and activity compared to their European counterparts. But control of penicillin knowledge suddenly allowed them not only to profit directly from sales but also to dictate commercial futures among their former rivals. Requests for factories and licenses flooded in. Merck built plants in Francoist Spain. In France, after the failure of a state-run penicillin project, private firms argued that it would be cheaper to buy patents and know-how from the United States than to waste time and money developing domestic deep-fermentation processes. In October 1945 the Ministère de la Santé Publique permitted the industry giant Rhône-Poulenc to purchase a license from Merck. German firms Bayer and Hoeschst—which had been undersupported in penicillin research by the Nazi regime—quickly did the same.[132]

Even the mere specter of US knowledge drove international decisions on penicillin. Companies like Rhône-Poulenc and Bayer sought American licenses in part to head off the encroaching threat of new players in the industry: importers and distributors from outside the pharmaceutical business using get-rich-quick schemes to sell US penicillin or acquire US licenses in Europe. Pressure from these new players obliged governments and firms to play by Merck's rules and obtain licenses of their own in order not to fall behind. And American industry could be choosy: for example, as Daniele Cozzoli has detailed, Merck refused to sell penicillin know-how to the diminutive Danish firm Løven, as the US firm correctly perceived that the Danish one had plans to sell and license the antibiotic elsewhere in Europe.[133]

The concrete benefits of this domination—namely, the propelling of US firms like Merck and Pfizer to global status—are self-evident, but the intangible ramifications for national geopolitical strategy are equally important, giving heft to the ideologies behind America's Cold War aid and development programs, and facilitating the global projection of the country as a biomedical power. Once again, here in the expression of new projections of American influence and self-conception we see a blurring between commercial and geostrategic objectives.

The benefits to US diplomatic power came immediately, particularly because as the war ended the state retained a control over disposal of national technologies that it had never previously enjoyed in peacetime. The concentration of penicillin production centered on the United States; the stipulations the FEA put in place for recipient countries during the war; the strict rationing through 1946; the continued vesting of critical knowledge in American corporate laboratories in the decades that followed; and the ability of the government to order and subsidize distribution for its own foreign policy purposes, all meant that the state could now use privately produced antibiotics as a diplomatic tool. It did so with particular vigor in overseeing the reconstruction of Europe, positioning itself at the head of a broad sphere of influence in the bipolar world order already in formation.

Through the United Nations Relief and Rehabilitation Administration (UNRRA), the United States government began in 1945 to "donate" penicillin factories overseas, under the stipulation that the resulting drugs not be sold commercially or exported to other countries. This was a relief to UNRRA, which had feared that liberated areas would not have access to the life-saving drug if US industry declined to sell there.[134] When UNRRA was replaced by the massive aid program of the Marshall Plan, both the mass shipments of penicillin and the construction of local antibiotics facilities continued apace. Along with them came large-scale propaganda efforts extolling the ways that the United States was lifting Europe's health from "suffering and misery on the one hand [to] health and hope on the other." Boats plied the canals of France showing films about American aid, and administrators even sent troubadours to Sicily to sing songs praising Marshall Plan penicillin.[135]

The "magic saving agent" thus became a powerful symbol of US technological diplomacy. As Mark Merrell, assistant chief of health supplies for the Office of International Trade, boasted, "the foreign visitor of today besides wanting to visit Niagara Falls and T.V.A. has added a tour of penicillin plants to his 'musts.'" One mythical story circulated of a foreign dignitary whose government was overthrown while he was visiting the United States; he went home to be reinstated on the strength that he had brought American penicillin back with him.[136] Robert Coghill, the leader of NRRL Peoria from the Oxford mission days, better encapsulated what penicillin meant for Americans' ideas about themselves as global actors: "The United States shared penicillin with other countries for military and civilian use on the basis of an equitable division of the available supply and at comparable cost, thereby contributing to the improvement of public health throughout the world. Penicillin, thus, has helped to establish a better understanding among nations, which should aid in the establishment of world peace."[137]

Ultimately, the British Ministry of Production's warning in 1944 that US knowledge control "would open export of penicillin to commercial channels while placing technical control of its usage throughout the world in American hands" turned out to be remarkably prescient, for that is precisely what happened.[138] And the United States only augmented its command of global affairs because of it. Because the US government could dictate the terms of other governments' acceptance of its "gifts," it could generate the geopolitical goodwill of providing health and reconstruction aid while simultaneously funneling an enormous new customer base to US industry. And it could send a powerful message of diplomatic clout to foreign capitals: there is a new way to treat myriad previously incurable diseases, and you can have access to it—on American terms.

Conclusion

This chapter has been about penicillin, but it could as easily have been about the jet engine, or sonar, or microwave and radar technology, or, in an extreme example, the atomic bomb—other technologies where transnational knowledge flows, secrecy, military-strategic interest, and commercial considerations washed together for the benefit of the postwar United States. So what else do these episodes tell us?

To begin with, World War II represented the first moment for the United States in which the government successfully mobilized the totality of its potential resources for knowledge creation and, it follows, it was the first time the state could effectively regulate who possessed that knowledge and how it moved. By the latter portion of the conflict, both British and American officials understood the converging importance of commercial and national-strategic interests to the coming postwar order, as well as the importance of technology in maintaining both. Thus, it is important not to discount the probusiness maneuverings of people like Bush and Richards as merely a matter of domestic politics or corporate capture. Rather, we must see the simultaneous buildup of an associationalist, corporate-serving state R&D system and a muscular national security apparatus as of a piece with the broader phenomenon of a resource-endowed hemispheric power expanding to new global ambitions.

Penicillin was a technology that ticked all the boxes in terms of power: military, economic, technological, and ideological. By the end of the war, then, evolving ideas about knowledge, capitalism, and hegemony made penicillin into an urgent diplomatic question. In Britain the answer to this question was not salutary. The early priority disputes between Fleming and Florey—framed so triumphantly as a competition between geniuses *within* the British national research powerhouse—gave way to diplomatic unease, economic anxiety, bureaucratic reorganization, and wounded national pride. Facts on the ground bore out this gloomy outlook. Inadequate know-how on the fermentation process became an acute production bottleneck for British pharmaceutical firms at precisely the moment that the United States moved into position to satisfy the therapeutic demands of India, Kenya, and myriad other places slipping away from the British metropole's grasp. Far from being just a discourse, losing credit for penicillin must have symbolized the loosening bonds of empire itself in the minds of many British mission officers.

To be sure, Britain was not a hapless victim. The self-aggrandizing narrative making of corporations like ICI, the propaganda efforts, the "defiant modernism," all of it closely mirrored the chest thumping of American counterparts.[139] Moreover, the lines between industry and state were no cleaner in Britain than they were in the United States—the wartime Minister of Supply, Sir Andrew Rae Duncan, became a board member at ICI in much the same way that Vannevar Bush did at Merck. Yet however much Britain initially thought it was engaging its exceptional technology complex in a hegemonic game with the United States, by the middle of World War II a stark reality had set in, a reality we can see clearly through the desperate cables of the BCSM and other missions: Britain was simply outmatched. It did not possess the resources or the corporate gigantism of the United States. Neither could it penetrate the intellectual property and national security regime that the American state had built up around ostensibly shared technologies. The calculus had changed. When Florey and Heatley first brought penicillin to American shores in 1941, the US machine of knowledge control barely existed. Its buildup by 1945 ultimately made US industry nearly impregnable to outside competition, and high-tech knowledge was ably integrated into national strategy as an instrument of geopolitical power.

Not that the public-private relationship in the US context was an easy one. In thinking about the midcentury interdependence between the US government and its corporations, historians often assume that national goals and private interests neatly overlapped. But looking more closely at episodes like that of penicillin shows us the ways in which that interdependence had to be actively engineered. In this case, corporations were hard-headed and short-sighted about the methodologies that would be best suited to hastening immediate production of a vital wartime material, to the exasperation of combat-focused bureaucrats. In order to pursue their strategic goals, then, state actors had to create a set of incentives that insulated corporations from more than just foreign rivals—they also had to shield them from market forces and regulatory politics.

Yet Big Pharma marketed penicillin largely as the triumph of maverick corporations, whose singular vision and singular risk-assumption justified the massive concentration of wealth that antibiotics afforded them. The speciousness of that narrative is clear when we reflect that the success of fermented penicillin was *not* their vision—that was a transnational conception, guided by the calculating lucidity of state actors and science administrators on both sides of the Atlantic. And neither was the risk-assumption theirs—that was shouldered by taxpayers, and in any case was far offset by the billions in global profits that swept quickly in behind the crash development program. Transnational penicillin development mutated into multinational corporate empire, and it did so along a geography and chronology that neatly mapped onto the global presence of the United States itself—in occupied Europe and Japan in 1945 and 1946, and worldwide quickly thereafter.

Emphasizing these themes thus helps us to reorient certain scholarly conventions. Historians of technology have often framed their analyses of technological innovation across borders as being matters of national security, capitalist development, or discrete knowledge-production flows and practices, frequently in the context of a borderless globalization. Diplomatic historians follow a similarly divided path—with exceptions, of course, as in discussions of the corporatist postwar state—but they rarely take R&D seriously as an arena of US power.[140] Penicillin thus represents a powerful demonstration of the ways that a focus on capitalism and the state—and especially the state as an actor asserting sovereignty over knowledge—can illuminate both technology and foreign relations scholarship. By closely studying the fraught foreign politics of technology and technological capitalism, we can arrive at a synthetic framework that demands the attention of both diplomatic historians and those who seek to understand the scope and content of American power beyond borders.

Ultimately, by looking at the lengths American administrators were willing to go to safeguard the strategic and commercial primacy of US research and production, we see dawning ideas about what expertise, resources, and powers the state thought it would need to effectively manage America's new place in the world. An analysis of the ways that state builders improvised controls of knowledge for hegemonic purposes helps us understand both *that* they considered knowledge to be a hegemonic commodity and *what* knowledge they considered to have hegemonic properties. It also emphasizes that—unlike after the last war—hegemony itself had become an unambiguous national goal. The frictions between the United States and Britain over controlling those commodities—even when they came in the guise of an unassuming fungus culture—tell us much about the use of knowledge as a tool, or cudgel, in international power politics.

Notes

1. G. A. G. Mitchell, "The Value of Penicillin in Surgery," *British Medical Journal* 1, no. 4488 (1947): 41–45.

2. "Dr. Alexander Fleming," *TIME*, May 15, 1944.

3. Transcript, "Response of Vannevar Bush on Occasion of Presentation of the Medal of the National Institute of Social Sciences," New York, May 23, 1945, box 83, Vannevar Bush Papers, Library of Congress, Washington, DC (hereafter LoC).

4. For more on the role of the transatlantic alliance and matters of science and technology in the conscious American rise to power, see Michael A. Falcone, "The Rocket's Red Glare: Global Power and the Rise of American State Technology, 1940–1960" (PhD diss., Northwestern University, 2019). In it, I argue that the United States lagged significantly behind other powers in pursuing state science and technology for much of the industrial era, and that the technological, bureaucratic, and doctrinal tutelage of the United Kingdom during World War II played an important and overlooked role in coaxing the American state into pursuing what would eventually become known as the military-industrial complex. I also argue that US inclination toward global hegemony was neither "present at the creation" nor a reluctant assumption of responsibility in the aftermath of war, but rather it represented a conscious doctrinal pivot, one informed in large part by the technological changes of the war. The British, desperate and eager to prop up their ally, helped their American partners to erect new institutions to accommodate further state research into technology—institutions that were previously wholly lacking in the United States. To a surprising degree, then, the military-industrial complex that defined the postwar American landscape had a distinctly multinational origin.

5. For a cross-section of this literature, and a sense of its volume from the 1980s to the present, see for example David Wilson, *Penicillin in Perspective* (London: Faber, 1976); David P. Adams, "The Penicillin Mystique and the Popular Press (1935–1950)," *Pharmacy in History* 26, no. 3 (1984): 134–142; Adams, *"The Greatest Good to the Greatest Number": Penicillin Rationing on the American Home Front, 1940–1945* (New York: Peter Lang, 1991); Jonathan Liebenau, "The British Success with Penicillin," *Social Studies of Science* 17, no. 1 (1987): 69–86; Donald J. McGraw, "On Leaving the Mine: Historiographic Resource Exhaustion in Antibiotics History," *Dynamis* 11 (1991): 415–436; Nicolas Rasmussen, "Of 'Small Men,' Big Science and Bigger Business: The Second World War and Biomedical Research in the United States," *Minerva* 40, no. 2 (2002): 115–146; Nasir Tyabji, "Gaining Technical Know-How in an Unequal World: Penicillin Manufacture in Nehru's India," *Technology and Culture* 45, no. 2 (2004): 331–349; María Jesús Santesmases and Christoph Gradmann, "Circulation of Antibiotics: An Introduction," *Dynamis* 31, no. 2 (2011):

293–303; Robert Bud, "Innovators, Deep Fermentation and Antibiotics: Promoting Applied Science before and after the Second World War," *Dynamis* 31, no. 2 (2011): 323–341; Gilbert Shama, "Déjà Vu: The Recycling of Penicillin in Post-Liberation Paris," *Pharmacy in History* 55, no. 1 (2013): 19–27; Shama, "The Role of the Media in Influencing Public Attitudes to Penicillin during World War II," *Dynamis* 35, no. 1 (2015): 131–152; Daniele Cozzoli, "Penicillin and the European Response to Post-war American Hegemony: The Case of Leo-Penicillin," *History and Technology* 30, no. 1–2 (2014): 83–103; Robert Gaynes, "The Discovery of Penicillin: New Insights after More than 75 Years of Clinical Use," *Emerging Infectious Diseases* 23, no. 5 (2017): 849–853, https://www.ncbi.nlm.nih.gov/pmc/articles/PMC5403050/; Mary Augusta Brazelton, "The Production of Penicillin in Wartime China and Sino-American Definitions of 'Normal' Microbiology," *Journal of Modern Chinese History* 13, no. 1 (2019): 102–123.

6. For more on the assumption of global consciousness and hegemonic design, see, for example, John Thompson, *A Sense of Power: The Roots of America's World Role* (Ithaca, NY: Cornell University Press, 2015); Andrew Preston, "Monsters Everywhere: A Genealogy of National Security," *Diplomatic History* 38, no. 3 (June 2014): 477–500; and Stephen Wertheim, *Tomorrow, the World: The Birth of U.S. Global Supremacy in World War II* (Cambridge, MA: Harvard University Press, 2020).

7. E. Chain, H. W. Florey, A. D. Gardner, N. G. Heatley, M. A. Jennings, M. Orr-Ewing, and A. G. Sanders, "Penicillin as a Chemotherapeutic Agent," *The Lancet*, August 24, 1940; Gaynes, "The Discovery of Penicillin."

8. Joan Bennett and King-Thom Chung, "Alexander Fleming and the Discovery of Penicillin," *Advances in Applied Microbiology* 49 (2001): 163–184.

9. The Alexander Fleming Laboratory Museum, London, "The Discovery and Development of Penicillin, 1928–1945," American Chemical Society and Royal Society of Chemistry, November 19, 1999, https://www.acs.org/content/dam/acsorg/education/whatischemistry/landmarks/flemingpenicillin/the-discovery-and-development-of-penicillin-commemorative-booklet.pdf. This story is corroborated in various accounts.

10. For more on Britain's war footing, state-commanded R&D apparatus, and prewar power politics, see David Edgerton, *Warfare State: Britain, 1920–1970* (New York: Cambridge University Press, 2006); Edgerton, *Britain's War Machine: Weapons, Resources and Experts in the Second World War* (New York: Oxford University Press, 2011); David Edgerton and Sally Horrocks, "British Industrial Research and Development before 1945," *Economic History Review* 47, no. 2 (1994): 213–238; Christine MacLeod, *Heroes of Invention: Technology, Liberalism, and British Identity, 1750–1914* (New York: Cambridge University Press, 2007).

11. See Falcone, "The Rocket's Red Glare," chap. 1. See also David Zimmerman, *Top Secret Exchange: The Tizard Mission and the Scientific War* (Montreal: McGill University Press, 1996).

12. Kevin Brown, *Penicillin Man: Alexander Fleming and the Antibiotic Revolution* (Stroud, UK: Sutton, 2004); Alexander Fleming Laboratory Museum, "The Discovery and Development of Penicillin"; Peter Neushul, "Science, Government, and the Mass Production of Penicillin," *Journal of the History of Medicine and Allied Sciences* 48, no. 4 (1993): 371–395.

13. Sir Henry Tizard, "Note on the Manufacture of British Secret Equipment in the United States," November 7, 1940, AVIA [Aviation] 10/2, the National Archives, Kew (hereafter TNA). For more on industry's tensions with the New Deal, as well as industrial self-presentation and discourses of patriotism, see Kim Phillips-Fein, *Invisible Hands: The Businessmen's Crusade against the New Deal* (New York: W. W. Norton, 2009). See Mark Wilson, *Destructive Creation: American Business and the Winning of World War II* (Philadelphia: University of Pennsylvania Press, 2016); James Sparrow, *Warfare State: World War II Americans and the Age of Big Government* (New York: Oxford University Press, 2011); Jackson Lears, *Fables of Abundance: A Cultural History of Advertising in America* (New York: Basic Books, 1994); Roland Marchand, *Advertising the American Dream: Making Way for Modernity, 1920–1940* (Berkeley: University of California Press, 1985).

14. Marc Landas, *Cold War Resistance: The International Struggle over Antibiotics* (Lincoln: University of Nebraska Press, 2020), 43; John A. Heitmann, *Scaling Up: Science, Engineering and the American Chemical Industry* (Philadelphia: Chemical Heritage Foundation, 1985), 17.

15. Falcone, "The Rocket's Red Glare," chap. 2.

16. Alexander Fleming Laboratory Museum, "The Discovery and Development of Penicillin."

17. Neushul, "Science, Government, and the Mass Production of Penicillin," 382; Alexander Fleming Laboratory Museum, "The Discovery and Development of Penicillin."

18. Penicillin was hardly the only program for which industrialists refused funding in order to avoid strings being attached to their activities; R&D on adrenal steroids represents another noteworthy case. See Rasmussen, "Of 'Small Men,' Big Science and Bigger Business."

19. Neushul, "Science, Government, and the Mass Production of Penicillin," 385; Robert D. Coghill, "Penicillin: A Wartime Accomplishment," *Chemical and Engineering News* 23 (1946): 2310–2316.

20. Neushul, "Science, Government, and the Mass Production of Penicillin"; Alexander Fleming Laboratory Museum, "The Discovery and Development of Penicillin."

21. John Patrick Swann. "The Search for Synthetic Penicillin during World War II," *British Journal for the History of Science* 16, no. 2 (1983): 154–190.

22. Plant estimates came from expected demand, and industry's tendency was to inflate these. Neushul, "Science, Government, and the Mass Production of Penicillin"; W. H. Helfand, H. B. Woodruff, K. M. H. Coleman, and D. L. Cowen, "Wartime

Industrial Development of Penicillin in the United States," in *The History of Antibiotics: A Symposium*, ed. John Parascandola (Madison, WI: American Institute of the History of Pharmacy, 1980), 31–56.

23. Rasmussen, "Of 'Small Men,' Big Science and Bigger Business." For more on Bush, the OSRD, and the corporate-friendly associationalism of the wartime state R&D apparatus, see Larry Owens, "The Counterproductive Management of Science in the Second World War: Vannevar Bush and the Office of Scientific Research and Development," *Business History Review* 68, no. 4 (1994): 515–576; Nathan Reingold, "Vannevar Bush's New Deal for Research: Or, The Triumph of the Old Order," *Historical Studies in the Physical and Biological Sciences* 17, no. 2 (1987): 299–344.

24. William Rosen, *Miracle Cure: The Creation of Antibiotics and the Birth of Modern Medicine* (New York: Viking, 2017), chap. 5.

25. The standard contract for industry also allowed companies to submit progress reports to trusted intermediaries on their own terms rather than directly to officials. It is worth noting that the army, navy, and other federal agencies allowed academics to retain their own intellectual property. Rasmussen, "Of 'Small Men,' Big Science and Bigger Business," 8; see also Owens, "The Counterproductive Management."

26. Note also that in 1945 Andrew Moyer—the researcher at Peoria who had worked with Florey and Heatley—filed his own patents in both the United States and Britain. Moyer did so, notably, while never mentioning the role of either the government or the Oxford group. Much has been written about the duplicity (or not) of Moyer in the intervening decades. Ronald Bentley, "Different Roads to Discovery: Prontosil (hence Sulfa Drugs) and Penicillin (hence β-Lactams)," *Journal of Industrial Microbiology and Biotechnology* 36, no. 6 (2009): 775–786.

27. Adams, "The Penicillin Mystique and the Popular Press."

28. Alexander Fleming Laboratory Museum, "The Discovery and Development of Penicillin."

29. Kevin Brown, *Fighting Fit: Health, Medicine, and War in the Twentieth Century* (Stroud, UK: History Press, 2008), 135. For a case study of similar transnational and corporate dynamics involving US firms and the German chemical industry, see Kathryn Steen, "German Chemicals and American Politics, 1919–1921," in *The German Chemical Industry in the Twentieth Century*, ed. John E. Lesch (Dordrecht: Kluwer Academic Publishers, 2000); Steen, "Confiscated Commerce: American Importers of German Synthetic Organic Chemicals, 1914–1929," *History and Technology* 12, no. 3 (1995): 261–284.

30. See Neushul, "Science, Government, and the Mass Production of Penicillin"; and Albert Elder, ed., *The History of Penicillin Production* (New York: American Institute of Chemical Engineers, 1970). Rasmussen supports Neushul's and Elder's contention, with some qualifications.

31. Bentley, "Different Roads to Discovery."

32. Robert Bud, "Penicillin and the New Elizabethans," *British Journal for the History of Science* 31, no. 3 (1998): 305–333.

33. Neushul, "Science, Government, and the Mass Production of Penicillin"; Bentley, "Different Roads to Discovery"; and Rosen, *Miracle Cure*.

34. United States Department of Justice, *Investigation of Government Patent Practices and Policies: Report and Recommendations of the Attorney General to the President* (Washington, DC: Government Printing Office, 1947). Numbers from Neushul, "Science, Government, and the Mass Production of Penicillin."

35. Neushul, "Science, Government, and the Mass Production of Penicillin."

36. Neushul, "Science, Government, and the Mass Production of Penicillin." Elder charged that Richards overemphasized synthesis and that it was only the prudence of the army in seeking the safety of maintaining parallel programs that ensured the success of fermentation. Note that Swann disputes Neushul, arguing that despite the lack of investment, it was nevertheless Richards who had convinced firms to join the fermentation program in the early days; Swann, "The Search for Synthetic Penicillin," 395.

37. Neushul, "Science, Government, and the Mass Production of Penicillin," 395.

38. R. B. Woodward, "Recent Advances in the Chemistry of Natural Products," Nobel Lecture, December 11, 1965, https://www.nobelprize.org/uploads/2018/06/woodward-lecture.pdf; Swann, "The Search for Synthetic Penicillin."

39. Neushul, "Science, Government, and the Mass Production of Penicillin."

40. Alexander Fleming Laboratory Museum, "The Discovery and Development of Penicillin."

41. One billion units = 1,000 serious cases, meaning that supplies on D-Day were enough to treat 100,000 infected soldiers and growing; Coghill, "Penicillin." Totals from A. N. Richards, "Production of Penicillin in the United States (1941–1946)," *Nature* 201, no. 441–445 (1964), https://www.nature.com/articles/201441a0.pdf; and Alexander Fleming Laboratory Museum, "The Discovery and Development of Penicillin."

42. Aide memoire, British Embassy, Washington, DC, May 19, 1943, AVIA 10/129, TNA.

43. See, for example, Letter, Secretary Frank Knox to British Purchasing Commission, October 30, 1940, AVIA 10/129, TNA; and Letter, Secretary Frank Knox to British Purchasing Commission, January 9, 1941, AVIA 10/129, TNA.

44. Memorandum, by J. Foster to Sir Henry Tizard, September 26, 1940, AVIA 10/2, TNA; and Memorandum, K. E. Shelley, K.C., July 19, 1941, AVIA 10/129, TNA.

45. Letter, Frank Knox to British Purchasing Commission, Washington, DC, January 9, 1941, AVIA 10/129, TNA; Memorandum, Ministry of Aircraft Production to British Embassy, Washington, DC, November 16, 1940, AVIA 10/129, TNA; Memorandum, Ministry of Aircraft Production to British Air Commission, Washington, DC, January 10, 1941, AVIA 10/129, TNA; Memorandum, British Air Commission, Washington, DC, to Ministry of Aircraft Production, June 3, 1941, AVIA 10/129, TNA.

46. Letter, Karl T. Compton to Carroll L. Wilson, August 14, 1941, OSRD, Records of the Liaison Office, RG227, Entry 168, Box 22, US National Archives II, College Park, Maryland.

47. Not until the Kennedy administration were many of these issues finally resolved, by then mostly through legislation, case law, and executive policy, though significant issues were left unresolved; Sean M. O'Connor, "Taking, Tort, or Crown Right? The Confused Early History of Government Patent Policy," *John Marshall Review of Intellectual Property Law* 12 (2012): 146–204.

48. O'Connor, "Taking, Tort, or Crown Right?"

49. Alex Wellerstein, "Patenting the Bomb: Nuclear Weapons, Intellectual Property, and Technological Control," *Isis* 99, no. 1 (2008): 57–87.

50. Lend-Lease Act, March 11, 1941, Pub. L. No. 11, 55 Stat. 31 (1941), https://tile.loc.gov/storage-services/service/ll/llsl//llsl-c77s1/llsl-c77s1.pdf.

51. "Memorandum by Mr. Elbridge Durbrow of the Division of European Affairs," December 23, 1943, in United States Department of State, *Foreign Relations of the United States: Diplomatic Papers, 1943*, vol. 3, *The British Commonwealth, Eastern Europe, The Far East*, ed. William M. Franklin and E. R. Perkins (Washington, DC: Government Printing Office, 1963), https://history.state.gov/historicaldocuments/frus1943v03/d648.

52. Lend-Lease Act.

53. "Preliminary Agreement between the United States and the United Kingdom, February 23, 1942," Avalon Project, Yale Law School, 2008, https://avalon.law.yale.edu/20th_century/decade04.asp.

54. P. H. Goffey to L. J. Douglas-Mann, June 4, 1942, AVIA 10/129, TNA.

55. Ralph L. Chappell and W. Houston Kenyon Jr., "Patent Costs of Military Procurement in Wartime," *Law and Contemporary Problems* 12 (1947): 704; O'Connor, "Taking, Tort, or Crown Right?"; text of the actual act: Royalty Adjustment Act of 1942 ("Use of Inventions for Benefit of U.S."), October 31, 1942, Pub. L. No. 768, 56 Stat. 1013 (1942), https://tile.loc.gov/storage-services/service/ll/llsl//llsl-c76s2-s3/llsl-c76s2-s3.pdf.

56. US Department of Justice, *Investigation of Government Patent Practices and Policies*, 336.

57. "Proposed Joint Determination of the Governments of the United States and the United Kingdom Pursuant to a Certain Agreement Dated January 25, 1946," FD 1/5341, TNA.

58. "Statement of Case Submitted by the Therapeutic Research Corporation of Great Britain Ltd to the Medical Research Council," March 6, 1953, FD 1/5341, TNA.

59. "Supplement to the Statement of Case Submitted by the Therapeutic Research Corporation of Great Britain Ltd to the Medical Research Council on 6th March 1953," January 7, 1957, FD 1/5341, TNA.

60. Text of the amendment: August 21, 1941, Pub. L. No. 239, 55 Stat. 657 (1941), https://tile.loc.gov/storage-services/service/ll/llsl//llsl-c77s1/llsl-c77s1.pdf; Daniel

P. Gross, "The Consequences of Invention Secrecy: Evidence from the USPTO Patent Secrecy Program in World War II," Working Paper 19-090, Harvard Business School and National Bureau of Economic Research, May 12, 2019.

61. Gross, "The Consequences of Invention Secrecy."

62. Wellerstein, "Patenting the Bomb," 73.

63. Assistant Controller (R&D), "Exchange of Technical Information with U.S.A., Proposed Statement of Position and Admiralty Policy," September 20, 1944, BT 305/3, TNA; Draft Interim Report, Interdepartmental Committee on Safeguarding Post-war Rights in Inventions, 1944, BT 305/3, TNA; and Memorandum, L. J. Douglas-Mann for Director General, B.A.C., "In regard to negotiations between the U.S. and U.S. governments on the subject of the free use . . . ," April 26, 1942, AVIA 10/129, TNA; Aide memoire, British Embassy, Washington, DC, May 19, 1943; and Letter, Henry Stimson to Honorable Sir Clive Baillieu, September 18, 1941, AVIA 10/129, TNA.

64. Robert S. Pasley and John TeSelle, "Patent Rights and Technical Information in the Military Assistance Program," *Law and Contemporary Problems* 29 (1964): 566–590.

65. Pasley and TeSelle, "Patent Rights and Technical Information," 570.

66. Publication of Inventions Act, July 1, 1940, Pub. L. No. 700, 54 Stat. 710, https://tile.loc.gov/storage-services/service/ll/llsl//llsl-c76s2-s3/llsl-c76s2-s3.pdf. For more on expanding secrecy restrictions through time, see Peter Galison, "Secrecy in Three Acts," *Social Research* 77 (2010): 941–974.

67. This law predated the August 1941 law (see note 60), which imposed a blanket restriction on foreign filing unless so authorized.

68. Gross, "The Consequences of Invention Secrecy," 2.

69. "Secrecy Order for Serial No. 70,412 filed Mar. 23, 1936, Applicant William F. Friedman and Frank B. Rowlett," Department of Commerce, May 19, 1947 (digitized by NSA), https://www.nsa.gov/Portals/70/documents/news-features/declassified-documents/friedman-documents/patent-equipment/FOLDER_088/41702069074139.pdf.

70. Wellerstein, "Patenting the Bomb."

71. US Department of Justice, *Investigation of Government Patent Practices and Policies.*

72. Bentley, "Different Roads to Discovery."

73. Coghill, "Penicillin."

74. "Agreement between the Governments of the United States and the United Kingdom of Great Britain and Northern Ireland Respecting the Exchange of Information on Penicillin," January 25, 1946, 60 Stat. 1485 (1946), 1488, https://tile.loc.gov/storage-services/service/ll/llsl//llsl-c79s2/llsl-c79s2.pdf.

75. Gross, "The Consequences of Invention Secrecy," 11; Sabing H. Lee, "Protecting the Private Inventor under the Peacetime Provisions of the Invention Secrecy Act," *Berkeley Technology Law Journal* 12, no. 2 (1997): 345–411.

76. Adams, "The Penicillin Mystique and the Popular Press."

77. Coghill, "Penicillin."

78. "Penicillin for Export to Foreign Countries," *Foreign Commerce Weekly*, July 15, 1944, 16–17.

79. Letter, Sir Nigel Campbell to Secretary, Ministry of Production, June 12, 1944, CAB [Cabinet Papers] 110/73, TNA.

80. Mark Merrell, "We Send Penicillin to Foreign Lands," *Journal of International Economy*, April 6, 1946, 3–4, 45–47.

81. USLON Washington to Air Ministry for Whitehall J[oint] A[merican] S[ecretariat], September 9, 1944, CAB 110/73, TNA; Alan P. Dobson, "The Export White Paper, 10 September, 1941," *Economic History Review* 34, no. 1 (1986): 59–76.

82. Office of War Information, *A Handbook of the United States: Pertinent Information about the United States and the War for Use Overseas* (Washington: OWI Overseas Branch, January 1944), 123, https://collections.nlm.nih.gov/ext/dw/01210260R/PDF/01210260R.pdf.

83. Testimony of Milo Perkins to Woodrum Committee, June 23, 1943, in *Hearings before a Subcommittee of the Committee on Appropriations, United States Senate, Seventy-Eighth Congress, First Session, on H. R. 2968* (Washington, DC: Government Printing Office, 1943), 3, https://hdl.handle.net/2027/uiug.30112120068355.

84. Nechama Janet Cohen Cox, "The Ministry of Economic Warfare and Britain's Conduct of Economic Warfare, 1939–45" (PhD diss., King's College London, 2001), 216.

85. Coghill, "Penicillin."

86. "Penicillin for Export to Foreign Countries."

87. Merrell, "We Send Penicillin to Foreign Lands"; "Penicillin for Export to Foreign Countries."

88. Copy of Mr. T.W. Child's Memo to Mr. R.H. Brand and Dr. J.R. Mote, June 17, 1944, CAB 110/73, TNA.

89. British Colonies Supply Mission, Washington to S[ecretary] of S[tate] Colonies, "Penicillin," September 5, 1944, CAB 110/73, TNA.

90. JAS to RAFDEL Washington, June 27, 1944, CAB 110/73, TNA; telegram, USLON 306, British Colonies Supply Mission, Washington, D.C. to Air Ministry, Whitehall, September 9, 1944, CAB 110/73, TNA; and telegram, British Colonies Supply Mission, Washington D.C., to S. of S., Colonies, September 6, 1944, CAB 110/73, TNA.

91. Telegram, USLON 306, British Colonies Supply Mission, Washington, D.C. to Air Ministry, Whitehall, September 9, 1944, CAB 110/73, TNA; and Telegram, British Colonies Supply Mission, Washington D.C., to S. of S., Colonies, September 6, 1944, CAB 110/73, TNA.

92. BSM Washington to M of Supply, "Medical Supplies—Penicillin," June 26, 1944, CAB 110/73, TNA.

93. BSM Washington to M of Supply, "Medical Supplies—Penicillin," June 26, 1944, CAB 110/73, TNA.

94. Ministry of Production, "Penicillin—Secret," June 1–June 12, 1944, CAB 110/73, TNA.

95. Coghill, "Penicillin."

96. Telegram, Ministry of Supply to B.S.M. Washington, March 9, 1944, CAB 110/73, TNA.

97. British Colonies Supply Mission, Washington, to Secretary of State Colonies, September 5, 1944, CAB 110/73, TNA; USLON 306, from Washington to Air Ministry, Whitehall for JAS, September 9, 1944, CAB 110/73, TNA.

98. British Colonies Supply Mission, Washington, to Secretary of State Colonies, "Reply Urgently Required," September 5, 1944, CAB 110/73, TNA.

99. W. K. Hancock and M. M. Gowing, *British War Economy* (London: HMSO and Longmans, Green, 1949), 516–533, https://www.ibiblio.org/hyperwar/UN/UK/UK-Civil-WarEcon/UK-Civil-WarEcon-18.html.

100. Moore to Secretary, June 15, 1944; Campbell proposal on Letter, Sir Nigel Campbell to Secretary, Ministry of Production, June 12, 1944, CAB 110/73, TNA.

101. The Ambassador in the United Kingdom (Winant) to the Secretary of State, September 7, 1941, in United States Department of State, *Foreign Relations of the United States: Diplomatic Papers, 1941*, vol. 3, *The British Commonwealth; The Near East and Africa*, ed. N. O. Sappington, Francis C. Prescott, and Kieran J. Carroll (Washington, DC: Government Printing Office, 1959), https://history.state.gov/historicaldocuments/frus1941v03/d23; for more about the Export White Paper, and its place in the overall government effort to avoid a postwar economic catastrophe given the liquidation of British assets and decimation of exports, see Dobson, "The Export White Paper."

102. Dobson, "The Export White Paper."

103. H. Y. Hodson to Secretary, June 16, 1944, CAB 110–73, TNA.

104. Child to Brand and Mote, June 17, 1944, CAB 110–73, TNA.

105. Hancock and Gowing, *British War Economy*, 516–533.

106. Alexander Fleming Laboratory Museum, "The Discovery and Development of Penicillin."

107. Joint American Secretariat to RAFDEL Washington, September 12, 1944, CAB 110/73, TNA.

108. Joint American Secretariat to RAFDEL Washington, June 27 1944, CAB 110/73, TNA.

109. RAFDEL, Washington, to Air Ministry, Whitehall for JAS, July 1, 1944, CAB 110/73, TNA.

110. Warburton to Weir, "Medical Supplies—Penicillin," June 27, 1944, CAB 110/73, TNA; John T. Connor to Vannevar Bush, October 28, 1952, Vannevar Bush Papers, box 27, LoC.

111. Robert Bud, "Upheaval in the Moral Economy of Science? Patenting, Teamwork and the World War II Experience of Penicillin," *History and Technology* 24, no. 2 (2008): 173–190; "Supplement to the Statement of Case Submitted by the

Therapeutic Research Corporation of Great Britain Ltd to the Medical Research Council on 6th March 1953," January 7, 1957, FD 1/5341, TNA.

112. Therapeutic Research Corporation of Great Britain Ltd to Medical Research Council, "Statement of Case," March 6, 1953, FD 1/5341, TNA.

113. *Hansard Parliamentary Debates*, 404 Parl. Deb. H.C. (5th ser.), November 2, 1944, cols. 1093–1100, "Penicillin (Supplies)"; *Hansard Parliamentary Debates*, 398 Parl. Deb. H.C. (5th ser.), April 5, 1944, cols. 2002–2003, "Oral Answers to Questions."

114. Bud, "Penicillin and the New Elizabethans."

115. Jonathan Liebenau. "The British Success with Penicillin." *Social Studies of Science* 17, no. 1 (1987): 69–86.

116. *Hansard Parliamentary Debates*, 601 Parl. Deb. H.C. (5th ser.), March 12, 1959, cols. 1518–1519.

117. Judy Slinn, "Patents and the UK Pharmaceutical Industry between 1945 and the 1970s," *History and Technology* 24, no. 2 (2008): 191–205.

118. Letter, Vannevar Bush to James B. Conant, October 29, 1952; and Letter, Vannevar Bush to John T. Connor, October 29, 1952, box 27, Vannevar Bush Papers, LoC.

119. E. M. Hugh-Jones to James B. Conant, July 4, 1954, box 27, Vannevar Bush Papers, LoC, reproduced in John T. Connor to Vannevar Bush, October 28, 1952, box 27, Vannevar Bush Papers, LoC. For "scholars since," see especially Bud, "Penicillin and the New Elizabethans." The justifiability of the payment has been a subject of debate. Robert Bud writes that the notion that America "stole" penicillin was an exaggerated discourse, but notes also that Glaxo did pay Merck half a million pounds for penicillin and streptomycin know-how through 1956, when its annual net profits were £1.5 million; Bud, *Penicillin: Triumph and Tragedy* (Oxford: Oxford University Press, 2007), 72. Daniel Podolsky quotes the John T. Connor counterargument, but notes that Florey's greatest regret in later years was that he had failed his colleagues and his laboratory by not seeking a patent on the penicillin extraction process; Podolsky, *Cures out of Chaos: How Unexpected Discoveries Led to Breakthroughs in Medicine and Health* (Amsterdam: Harwood Academic Publishers, 1997), 206. Bentley writes that British companies paid many millions of pounds in licenses over the years, but that these scaled down over time as penicillin prices came down and did not limit British sales rights outside the United States and Canada; Bentley, "Different Roads to Discovery." See also Trevor Illtyd Williams, *Howard Florey: Penicillin and After* (New York: Oxford University Press, 1984); David Wilson, *In Search of Penicillin* (New York: Knopf, 1976).

120. Bentley, "Different Roads to Discovery"; Bud, "Innovators, Deep Fermentation and Antibiotics"; Therapeutic Research Corporation of Great Britain Ltd to Medical Research Council, "Statement of Case"; and Letter, A. Lansborough Thomson to Chairman, Research and Development Board, Washington, D.C., March 29, 1950, FD 1/5341, TNA.

121. Bud, "Penicillin and the New Elizabethans"; see also, for example, Edgerton, *Warfare State*; David Edgerton, *Science, Technology, and the British Industrial*

"Decline," 1870–1970 (Cambridge: Cambridge University Press, 1996); Jim Tomlinson, "Thrice Denied: 'Declinism' as a Recurrent Theme in British History in the Long Twentieth Century," *Twentieth Century British History* 20, no. 2 (2009): 227–251; Richard English and Michael Kenny, "British Decline or the Politics of Declinism?," *British Journal of Politics and International Relations* 1, no. 2 (1999): 252–266.

122. Shama, "The Role of the Media in Influencing Public Attitudes."

123. Cheryl R. Jorgensen-Earp, and Darwin D. Jorgensen. "'Miracle from Mouldy Cheese': Chronological versus Thematic Self-Narratives in the Discovery of Penicillin," *Quarterly Journal of Speech* 88, no. 1 (2002): 69–90, at 82. The authors are here citing a formulation from Malcolm Smith, *Britain and 1940: History, Myth, and Popular Memory* (New York: Routledge, 2000).

124. J. C. Cain, "£1,000,000 for University Inventions," *Inventions for Industry* (National Research and Development Corporation) 42 (1975): 4–5. Reprinted in *Hearings before the Subcommittee on Monopoly and Anticompetitive Activities*, May 22–23, June 20–21, 26, 1978, 95th Cong., Part 2: *Appendix* (Washington, DC: Government Printing Office, 1978), 1359, https://congressional.proquest.com/congressional/docview/t29.d30.hrg-1978-sbs-0026?accountid=15172.

125. Patents Act, 1949, 12–13 and 14 Geo. 6, c. 87, https://www.legislation.gov.uk/ukpga/Geo6/12-13-14/87/contents; and Bud, "Penicillin and the New Elizabethans."

126. Committee on Science and Astronautics, U.S. House of Representatives, "Appendix A: National Research Development Corporation," 107th Cong., May 11–12, 1961 (Washington, DC: Government Printing Office, 1961).

127. Bentley, "Different Roads to Discovery."

128. As Florey observed in 1949, "too high a tribute cannot be paid to the enterprise and energy with which the American manufacturing firms tackled the large-scale production of the drug. Had it not been for their efforts there would certainly not have been sufficient penicillin by D-Day in Normandy in 1944 to treat all severe casualties, both British and American"; Alexander Fleming Laboratory Museum, "The Discovery and Development of Penicillin." While Bush and Connor at the time, and some observers since, have credited the work done in industrial laboratories as the most important advance once penicillin had been brought to the United States, Neushul ("Science, Government, and the Mass Production of Penicillin") argues that it was the work done at the NRRL in Peoria under Coghill that made by far the most significant advances toward mass production, despite limited resources and ongoing opposition from the synthesis proponents. With the WPB and military bringing its research to fruition, the NRRL's collaboration with the Oxford group was perhaps the most important American contribution to the drug's development.

129. Letter, Vannevar Bush to James B. Conant, October 29, 1952; and Letter, Vannevar Bush to John T. Connor, October 29, 1952, box 27, Vannevar Bush Papers, LoC.

130. Nelson Kardos and Arnold L. Demain, "Penicillin: The Medicine with the Greatest Impact on Therapeutic Outcomes," *Applied Microbiology and Biotechnology* 92, no. 4 (2011): 677–687; Basil Achilladelis, "Innovation in the Pharmaceutical Industry," in *Pharmaceutical Innovation: Revolutionizing Human Health*, ed. Ralph Landau, Basil Achilladelis, and Alexander Scriabine (Philadelphia: Chemical Heritage Press, 1999), 61.

131. Merrell, "We Send Penicillin to Foreign Lands."

132. Daniele Cozzoli, "Penicillin and the European Response to Post-war American Hegemony: The Case of Leo-penicillin," *History and Technology*, 30, no. 1–2 (2014): 83–103.

133. Cozzoli, "Penicillin and the European Response."

134. Merrell, "We Send Penicillin to Foreign Lands."

135. Lewis Paul Todd, *The Marshall Plan: A Program of International Cooperation* (Washington, DC: Economic Cooperation Administration, 1950), 27; Cozzolli, "Penicillin and the European Response"; Bud, *Penicillin*, 85–88; Carl Bruch, Ross Wolfarth, and Vladislav Michalcik, "Natural Resources, Post-Conflict Reconstruction, and Regional Integration: Lessons from the Marshall Plan and Other Reconstruction Efforts," in *Assessing and Restoring Natural Resources in Post-Conflict Peacebuilding*, ed. David Jensen and Stephen Lonergan (New York: Routledge, 2012), 357.

136. Merrell, "We Send Penicillin to Foreign Lands."

137. Coghill, "Penicillin."

138. Joint American Secretariat to RAFDEL, Washington, June 27 1944, CAB 110/73, TNA.

139. See for example Bud, "Penicillin and the New Elizabethans"; Shama, "The Role of the Media in Influencing Public Attitudes"; Shama, "Auntibiotics: The BBC, Penicillin, and the Second World War," *British Medical Journal* 337 (December 2008): a2746.

140. These two subfields framed the story as an antagonistic set of relations between government and industry, which subsequent scholars have nuanced by instead showing that there is a great degree of interdependence between the two, characterized by variation in terms of the extent to which government or corporations set the agenda and reaped the benefit—that is, which was the tail and which was the dog being wagged. Scholars doing important work at the intersection of capitalism and state-building histories include Mark Wilson, Jennifer Ott, Gary Gerstle, Nancy Fraser, Bethany Moreton, and Louis Hyman. The subject is receiving increasing attention among historians of US foreign relations, as well: for example, Megan Black, David Painter, and Robert Vitalis.

Chapter Four

Dangerous Calculations

The Origins of the US High-Performance Computer Export Safeguards Regime, 1968–1974

MARIO DANIELS

In the 1970s, the exports of high-performance computers (HPCs) became more strictly controlled than the sales of nuclear reactors. The United States and its allies in the Coordinating Committee for Multilateral Export Controls (CoCom)—the Western multilateral institution for the control of exports, especially to the Eastern Bloc countries—established a system of elaborate "safeguards" that aimed at the postshipment monitoring of the use of large computers *within* the Soviet Union and the Iron Curtain countries. These safeguards included an array of technical surveillance measures to forestall the "diversion" for military purposes of computers that had been exported for civilian applications. HPCs were, along with nuclear technologies and inertial navigation equipment, the only technologies protected by special safeguards.[1] Thus, in the computer field, the export control system, well established since the 1940s, witnessed a remarkable bureaucratic innovation that considerably extended the extraterritorial reach of US export controls well beyond the Iron Curtain. Most importantly, the United States was able to install a system of on-site inspections that pre-dated nuclear facility inspections in the Soviet Union by one and a half decades. During most of the Cold War, the Soviet Union staunchly opposed the presence of foreign inspectors because it saw them as potential spies. But in order to receive large computer systems, the Soviet Union as well as its Eastern European allies accepted inspections apparently without much resistance. Hence, HPC safeguards are an extraordinary example of Cold War diplomacy, the far and deep reach of US export controls and the innovative ways American policymakers conceived of the regulation of technology sharing in the international realm.

The invention of safeguards around 1970 was part of the larger debates about the international institutionalization of nonproliferation policies in the postwar era. Clearly, HPC safeguards were part of the debates about the role and rights of the International Atomic Energy Agency (IAEA), as well as about the stipulations of international agreements like the nonproliferation treaty (1968–1970).[2] HPCs were understood as a key technology for the analysis of nuclear test data and nuclear weapons design. At the same time, HPCs were a dual-use technology whose significance went far beyond nuclear weapons. HPCs had many military applications. For example, they were used for data processing and data storage in complex communication and intelligence systems (not least in the space program), weapons system design and production, military procurement, and the administration of the vast US military apparatus. But even though the military origins of modern computer technology were still much more obvious than they are today, the 1960s and 1970s also were the period when the commercialization and the use of computers in the civilian sphere made great strides. While the everyday use of computers steadily expanded so that they became a ubiquitous consumer good, the computer industry also became much more international. US companies like IBM or Control Data Corporation (CDC) dominated international markets. Yet they faced competitors especially from Western Europe who vied for market shares in the West—and increasingly also in the Eastern Bloc. The thaw in East-West relations in the second half of the 1960s opened up promising market opportunities. The export of computers, however, was still tightly controlled by CoCom as well as by the United States, which was the main source for European producers of technologies and components. In the spirit of political détente and bent on conquering new markets, America's Western partners began to press for a liberalization of computer export controls. As this chapter shows, HPC safeguards were one way the United States had of managing these political and commercial pressures and pushing against a complete erosion of technology controls. I argue that even though safeguards were a result of export control relaxation, they considerably expanded the extraterritorial reach of US technological regulations and carried them far into Soviet territory.

This expansion became possible through a creative combination of export control regulations and contracts between companies and governments and, indirectly, between governments themselves. The civil law instrument of contracts enhanced and boosted the statutory export control laws and administrative regulations. Indeed, the US government has developed a complex system of contracts that complement other legal instruments of knowledge and information regulation. Classification rules, for example, are translated into contractual relations between the federal government and its employees. National-security reviews of foreign direct investment is another case in point. The Committee on Foreign Investment in the United States (CFIUS) can protect American know-how by dint of so-called mitigation measures. These are contracts by which the foreign company that acquires a US firm can be forced to divest certain parts of its acquisition; to assure the appointment to the board of directors only a limited number of non-US citizens; or to concede to the federal government intrusive monitoring rights over the company's daily activities. An analysis of safeguards promises insights about the role of contractual instruments in the United States' control over the sharing of technological knowledge.[3]

This chapter is the very first study that analyzes the complex origins of HPC safeguards and their implementation in the late 1960s and 1970s. Secondary literature virtually does not exist. It will especially focus on and contextualize the intense controversies about this new way of controlling the international sharing of cutting-edge technologies. Indeed, as with détente in general, the policies and measures of computer safeguards were highly controversial and led to great tension between the US government and the business community, but particularly within the US political institutions. Indeed, controversies about the right assessments of the technological, political, and national-security risks of sharing HPCs with Cold War adversaries are found at the very beginning of the invention of HPC safeguards. Increasingly, the focus of these debates shifted from proliferation concerns to the complex question of the dangers of technology transfers in the dual-use realm. The pressing questions were what and how much technological know-how the Soviets would acquire if they imported HPCs, and to what extent their climbing the learning curve would mean enhancing their military posture and their political stability.

British Computers for Serpukhov

The export control case that led to the establishment of HPC safeguards began in late 1969 when the British company International Computers Ltd. (ICL) negotiated with the Soviet Union about the sale two HPCs of the type 1906-A.[4] The computers were meant to be used at the Institute of High Energy Physics at Serpukhov, sixty miles south of Moscow and since 1967 the site of the world's largest proton synchrotron. Like the European Organization for Nuclear Research (Conseil Européen pour la Recherche Nucléaire,

CERN) in Switzerland, the Serpukhov particle accelerator was the site of international scientific cooperation. The ICL computer was supposed to support the institute's large research community that encompassed about three hundred foreign scientists, including Americans.[5]

ICL had been established in 1968 through a merger of three British companies and was the biggest computer company in Europe and outside the United States. The British government had pushed for the merger, seeking to position the new company as the United Kingdom's "national champion" in the international markets dominated by American competitors.[6] The Serpukhov deal was worth $11 million and was thus highly attractive for ICL, whose annual sales target was $240 million.[7] But the Soviet market promised to be much more than the one-off sale of a large scientific computer. After 1945, computers were at first not high on the Soviet priority list of technological development. From the late 1950s, however, the Soviets slowly reassessed their understanding of the technological, economic, and military significance of computers and invested increasing resources in attempts to build up a national computer industry. These efforts put much emphasis on the copying of Western (especially US) technology in order to leapfrog to the technological state of the art.[8] The main challenge was to translate domestic and acquired knowledge into mass production, particularly in the field of transistors and integrated circuits. Accordingly, the Soviet Union embarked, as the CIA put it, on a "campaign to acquire from the Free World the modern electronics production technology which it desperately needs to support its computer industry of integrated circuits and high-precision printed circuit boards."[9]

The "computer gap" between the United States and the Soviet Union was staggering. Whereas US companies built 16,000 computers in 1970, the Soviet Union produced only 800.[10] And while in the Soviet Union there were an estimated 5,000–6,000 units in use, around 63,000 computers were set up in the United States.[11] There was also a huge quality gap. The Soviet computer design was "at least a generation behind" the United States. In short, the computer industry was a "troubled and lagging sector of the Soviet economy."[12]

ICL wanted to take advantage of the Soviet technological lag, and its Serpukhov deal coincided with growing Soviet efforts to catch up. Since the early 1960s, the Soviets had especially zeroed in on computerized data processing and record keeping. Accordingly, legal computer imports from the West, which had commenced in 1959, began to "increase sharply from 1964 onward," and most active were British companies.[13] At the same time, a Soviet initiative to build a family of upward-compatible computers similar to the IBM 360 faltered, paving the way to the "Ryad" project. The Soviet Union pursued it from 1967–1968 in cooperation with its allies of the Council for Mutual Economic Assistance (Comecon) and derived its technological setup directly from IBM designs.[14]

ICL hoped to benefit from these developments as well as from the sale of large scientific computers like the 1906-A. In April 1968, the company sent a team to Moscow to talk about cooperating in the development of the Ryad series. The Soviet planning agency Gosplan painted a rosy picture and spoke about spending £15–20 billion in 1969 alone and the need for 20,000–30,000 computers in the next decade.[15] ICL apparently planned to cooperate with the Soviets in the development of software for the Ryad series, "an area in which the USSR was especially weak." Clearly, they had a foot in the door. It was the only Western computer company that had been permitted to open a sales office in Moscow.[16]

US-UK Disputes over High-Performance Computer Export Controls

But ICL had a daunting problem: US export controls.[17] Before ICL could go ahead and sell computers to the Soviets it needed the permission of CoCom as well as from the US government. Despite pressures toward liberalization and the fact that computer sales were picking up steam, CoCom still regulated larger computers quite stringently. According to CoCom's principles and rules, the export of a controlled item was possible only after the unanimous approval of all member states (i.e., all NATO states, minus Iceland, plus Japan) of a formal request by the exporting country for an exception. The main actor here was the United States, which was in the late 1960s, before embarking on détente itself, still the most vociferous proponent of a broad and stringent technology embargo. Moreover, in the computer field the United States was so dominant that virtually all European computers were based on technologies or included components of US origin. That meant they were subject to US reexport regulations and could be sold to Eastern Bloc countries only after receiving a license from the US government (i.e., usually from the Office of Export Control within the Department of Commerce). The $11 million ICL 1906-A, for example, incorporated $3 million worth of American components.[18]

Thus, in August 1970 the British government submitted on behalf of its national champion a memorandum outlining the planned sale and asking for US agreement before the submission of the case to CoCom. The US answer, sent in October 1970, was negative. The main problem was that the ICL 1906-A's computing power far exceeded that of the biggest Soviet-grown system and also was far beyond the threshold where the CoCom embargo kicked in. CoCom's cut-off point was at a "processing data rate" (PDR) of eight million bits per second (mbs) whereas each of the two ICL

1906-A operated at a PDR of 44 mbs. Depending on the method of measuring computing capabilities, the largest Soviet model, the BESM-6, was two to ten times slower than the British computer. Even though the US cabinet-level Export Administration Review Board (EARB)—in charge of formulating export control policy—had in early 1970s somewhat relaxed its standards, the ICL1906-A was still too powerful to be shared. The EARB had determined that systems operating at 15–30 mbs could be sold to the Soviet Union—but only if the exports were limited to *one* computer per year. For the export control community, "the risk would be 'significant' if more than one of that size were exported. A computer having a PDR of more than 30 mbs would require the endorsement of the EARB as a 'blue ribbon' case." The decisive US argument, however, was that there was a high risk the Serpukhov computer would not be used for civilian scientific research but could easily be "diverted" for military purposes like nuclear tests—a dangerous abuse that could not be forestalled by any regulatory means.[19]

This US opposition to the British computer export was directly based on deliberations within the US executive about an earlier proposed computer sale. Indeed, the import of an ICL 1906-A had not been the Soviet's first choice for Serpukhov. They would very much have preferred to buy the much more powerful CDC 6600, the flagship of the US firm CDC and today often referred to as the "first supercomputer." It was only after the export of the CDC computer ran into difficulties in Washington that the Soviet Union began to negotiate with ICL.[20] CDC had developed its machine with support from the Atomic Energy Commission (AEC), whose labs in Livermore and Los Alamos were among the first users. Soon, CDC also sold the computer internationally, especially to scientific institutions in France, West Germany, Canada, Sweden, Italy, the United Kingdom, and Australia, in many cases for atomic energy research. One of the customers was CERN, in Geneva. Here, Soviet researchers cooperated with Western partners and arguably had access to the CDC 6600.[21]

The Soviets attempted to trade access to its Serpukhov synchrotron and the basic scientific research produced there for Western technology. When in February and March 1969 five US high-energy physicists traveled to Serpukhov to assess the chances for American-Soviet cooperation there, they learned that the Soviet government made access for US scientists contingent on the sale of a CDC 6600 or a computer of a comparable size. Similarly, French access was made dependent on the provision of a bubble chamber, a device for the detection of subatomic particles. Remarkably, the Soviets communicated their willingness to "accept some U.S. controls on the use" of the computer—and thus kicked off the discussion about HPC safeguards.[22] The Soviets said, as Admiral Hyman G. Rickover put it, “We just want to use it. You can put a padlock on it, and we won’t steal the technology or anything like that.”[23]

Despite the strict US and CoCom export controls over large computers and despite the fact that the CDC was then the largest computer in the world—roughly five times faster than the ICL 1906-A and up to forty-six times faster than biggest Soviet computer, the BESM-6—the US government did not roundly dismiss the Soviet offer.[24] In late summer 1969 the AEC began a study “of the pros and cons of placing a CDC computer 6600 computer at Serpukhov.”[25] One result of these investigations was the very first catalog of concrete safeguards, submitted in a report to the president’s Office of Science and Technology (OST) in November 1969. These safeguards were a mixture of on-site monitoring and the continuous recording and expert analysis of the actual calculations done with the CDC 6600. The agreement stipulated

1. Closed shop, with systems programming done by US personnel only, adequate surveillance and assurance of non-use during any off hours.
2. Batch computing only (no terminals).
3. Fortran[26] programs only, with adequate documentation required to ensure efficient operation and for verification purposes.
4. Complete recording of input and sampling of output.
5. Sampling of internal computer executions.
6. Creation in the United States of a part-time group of high-energy physicists, computer center managers, and weapons designers to analyze a small sample of recorded data.[27]

From a technical standpoint these measures appeared to promise quite effective guarantees against the “diversion” of the Serpukhov computers for nuclear test calculations. That was at least the position of a panel of computer experts, who concluded in January 1970 that the installation of the CDC 6600 could “proceed without threat to national security provided the conditions” listed “prevailed.”[28]

Not everyone agreed, and a formidable set of political opponents gathered their forces. One of the most prominent critics was Admiral Hyman G. Rickover, best known for his advocacy of a nuclear-powered navy. In a memorandum to the AEC, he stated that those who supported the HPC export “underestimated the risks.” He stated apodictically that it was “not in our national interest to permit Soviet purchase of our high-speed, high-performance” and added a jab against safeguards: “with or without controls.” One of Rickover’s concerns was the transfer of technological knowledge to the Soviets. He conceded that the “fact you get a computer doesn’t mean you can manufacture one yourself. It isn’t quite that easy.”

But he claimed nevertheless that an export would speed up Soviet technological progress. Emphasizing "that the fundamental element in modern technology is computers," he warned, "If you give your best computer away, you are giving the competition the opportunity to develop his technology just as rapidly as you do."[29]

These statements were part of a long testimony Rickover gave to the Joint Committee on Atomic Energy of the US Congress in a hearing that dealt mainly with nuclear propulsion technology. In this larger context, the admiral time and again attacked the Department of State (DoS) for efforts to share technology with Western European states, or as he put it, "to give our nuclear propulsion technology away." By selling submarines, the United States would transmit knowledge: "Once that is done, away it goes." And with the knowledge, the United States would lose the technological lead its naval power rested on. Finally, there was the danger that US technology, once in foreign hands, would end up in the Soviet Union. But even among allies the United States "should not go out of our way to give information away or to use this information to buy friendship. History should have taught us by now that you do not 'buy' friendship from your allies especially when the price is your own security. . . . It is time that the Secretary of State stopped the 'do-gooders' in his Department from being more concerned with their client countries than they are with the security of the United States." What was true for allies and submarines certainly applied clearly also to enemies and computers. But Rickover feared that by buying a HPC, the Soviets would not only gain technology. He argued that even if the Serpukhov computer was used just for civilian purposes, it would "probably free up other computers of Soviet design for weapons use."[30]

The CDC 6600 deal finally hit the wall because the Joint Committee on Atomic Energy made an effort to stop it.[31] In summer 1970 the export license for the CDC 6600 was officially denied. At the same time the sale of a smaller CDC 6400 for use at the synchrotron in Yerevan was halted.[32] The CDC cases and debates thus had a significant overlap with the ICL case and shaped the way the British export was conceived. In early 1970, the debate about safeguards and the exports of HPCs seemed to have reached a clear-cut conclusion, summed up by the EARB. Safeguards were impossible to implement and had serious technical flaws. It "would be extremely difficult with current technology to detect diversion without full time, expert, on-site monitoring and that proper monitoring of the work of the computer would require constant sampling of the sub-routines for evaluation by an expert." And even if a close monitoring process could be established, the situation was "further complicated by the fact that high energy physics programs are similar to those for weapons development and very few experts are capable of distinguishing between them. Bubble chamber experiments could resemble experiments on nuclear weapons effects, for example."[33]

Department of State vs. Department of Defense: Do Safeguards Work? Or Are They a Bad and Dangerous Idea?

The British government, however, was not willing to accept the US government's decision against the ICL deal, communicated in October 1970, only a few months after the denial of the CDC license. The timing for losing an $11 million deal was decidedly bad. In 1970–1971 the global computer industry slipped into its first recession, mainly caused by the slowing of the US economy since 1969. The computer sales in the United States dropped by 20 percent and the markets in Western Europe and the United Kingdom showed a similar negative trend.[34] After the United States had in early November 1970 also objected at CoCom against the ICL 1906-A sale, the United Kingdom protested in an aide memoire to the United States that "it could not accept US statements particularly with respect to the danger of diversion." During a meeting with President Richard Nixon in December, the British prime minister Edward Heath stressed once again the British "interest in getting US approval of the transaction."[35]

This was the beginning of a tenacious British diplomatic campaign to salvage the ICL deal, which the United Kingdom reportedly complemented by putting pressure on the other CoCom member states to relax the restrictions on HPCs.[36] Henry Kissinger, the president's assistant for national security affairs, opened a back door to revise the export license denial. Following up the Heath visit, Kissinger commissioned from the Under Secretaries Committee yet another study of the United Kingdom's computer case and stipulated that it should be based on a technical assessment conducted by the OST.[37]

An intense controversy ensued, revolving around three questions. Could safeguards actually be implemented? If so, would they be "effective in reducing the risks of misuse of the computers?"[38] And what was the political effect on CoCom export control policy if the export license was granted? At its core, the debate was about differing conceptions of the relationship of economic interests and national security, that is, the benefits and potential repercussions of selling cutting-edge technology. As often happens in the export control field, it became apparent how exceedingly difficult it was to assess the dangers of technology sharing in a complex inter- and transnational environment.

The results of the review process, submitted in March 1971, exposed a deep rift within the US government.[39] Two groups of departments and agencies were pitted against each other with the DoS in favor of a safeguarded sale, supported by the OST, the Treasury Department, the Council of Economic Advisors, and the United States Information Agency. The Department of Defense (DoD) was the most adamant detractor of the HPC export, backed up by two other central institutions of the national-security state, the AEC and the Joint Chiefs of Staff, but also by the special trade representative. The Commerce Department positioned itself somewhere in between, but thought the deal could be allowed though it emphasized the necessity of "very explicitly articulated safeguards."[40] Again, the immense tension between the DoS and the DoD that had Admiral Rickover alluded to before Congress shaped the bureaucratic tug of war over export control policy.

The interdepartmental debate revolved around a new British proposal. Since the failure of their sales initiative in CoCom, the British government had, in cooperation with ICL, refined its outline for a safeguard regime. It suggested five verification mechanisms:

1. the planning of as full a research schedule as possible for the machines
2. the necessity of written supporting documentation for individual programs runs
3. contractual rights of free access by specialists of free world parties, which have cooperation agreements with Serpukhov (presently the United States, the United Kingdom, and France)
4. ten-year control over spares and on-site maintenance
5. ICL willingness to obtain Soviet agreement before export of the computers to permit UK personnel to empty memory cores on demand of their stored informational contents and to transmit them for UK (and US) governmental analysis.[41]

These stipulations put more emphasis than did the first proposals on a safeguards regime of on-site inspections, which would give Western governments unprecedented rights of access on Soviet state territory. The surveillance through the physical presence of specialists with credentials from Western governments was to be complemented by close technical monitoring of the potentially dangerous calculations the computers would execute on site.

Both the free access rights and the fifth item on the list were exceedingly intrusive. They would give the US and the UK governments direct and routine access to data produced at a Soviet research institution. The OST panel that prepared the technical assessment of the National Security Council (NSC) report saw this method, referred to as data core "dumps," as the best precaution against the "clandestine diversion" of calculation power. Even though the OST admitted that a "complete elimination of risk" was impossible, "dumps" would "expose the Soviets to a finite danger of discovery." Moreover, the OST's idea of acceptable risks was not just a function of the case *that* diversion of calculations could happen. It had a distinct idea of *how much* redirected data was dangerous. It stated that "the diversion of less than of the order of 25 percent of capacity for two or more years would not be worth the effort required of the Soviets to effect clandestine diversion. The panel believes that this much diversion of the computers' time from legitimate work needs would likely be sensed by foreign specialists working at the Institute."[42]

The DoS's position was based on this risk assessment, adding that "Soviets would be much more likely to use an additional Soviet-built BESM-6, which has approximately 25 percent of the power of the total British system, for weapons calculations than to try to divert time from an installation not completely under their control." Closely watched, it was also unlikely that the Soviets would use the ICL 1906-A for classified calculations. Altogether, the DoS's position was that the proposed safeguards package would "bring the risk of misuse to an acceptably low level."[43]

The DoD vehemently disagreed. It believed the proposed safeguards were "ineffective since they offer no high probability" that diversion would be detected. The DoD discarded outright the first three safeguards on the list, claiming that they could not even detect "a diversion of approximately half of the computational capacity." As for US scientists' monitoring the computers on site, the DoD apparently did not really trust their willingness to do that, not least because that would mean "extended tours of duty" as these controls were effective only with "U.S. presence . . . on all shifts." Withholding spares—number 4 on the list—was not really a safeguard, but a sanction after a diversion had already happened. Only the "dumping" of computer cores for government analysis had "any teeth" at all. But the proposed dumps left considerable loopholes. If only the internal memory of the central processing units was scrutinized, not all processed information was disclosed. The DoD subscribed to a worst-case scenario befitting the image of the Soviets as a nefarious enemy relentlessly plotting against the United States: "It pointed out that schemes are possible to 'capture' the [computers'] executive system and replace it with one which 'looks' the same externally . . . but which allows 'hidden' programs to be run." Suspicious in this context was also that the units the Soviets wanted had "unusually large external memories" that did not correspond to the stated uses at Serpukhov. The DoS, by contrast, had explained that the memory capacities that would be provided were similar to those at comparable Western scientific projects. But most importantly, the DoD was convinced that there was in fact simply "no presently developed methodology for analyzing the contents" of core dumps.[44]

The DoS, the DoD, and their respective supporters in the US executive not only disagreed about the technical aspects, they also clashed over differing political interpretations of the ICL case. The central question was if granting ICL an export license would establish a precedent for US and CoCom decision making on future HPC sales, and what repercussions or positive effects setting such a precedent could have.

The DoS, again in lockstep with OST, claimed that the ICL sale constituted "a unique situation which justifies a single exception." Serpukhov had an outstanding international stature and scientific significance. Moreover, ICL was the only non-US company striving for business beyond the technological threshold defined by CoCom regulations. The criteria applied in this case were not directly comparable to the American CDC case and would therefore not affect future CDC sales. Their HPCs were much more sensitive because of their importance for US weapons design. Thus, neither for CoCom nor for US export controls would the case establish a precedent. But more importantly, selling the computer would have a decidedly positive impact on US diplomacy. It would strengthen US-UK cooperation, whereas a continued American opposition would likely alienate the British and give them reason to press even harder within CoCom for a relaxation of the multilateral control rules. Finally, against the backdrop of the nonproliferation treaty negotiations, the DoS touted a Soviet agreement to US on-site inspections and safeguards as setting a powerful political precedent for future arms control negotiations.[45]

For the DoD, and also for the AEC, issuing a license would be tantamount to a breach in the national-security dike. ICL's export would encourage US and Western firms to push for more HPC deals with the enemy. "Each request will claim the end-user to be deserving; each will assert the improbability of diversion; and each will propose to include safeguards similar to those accepted for Serpukhov." Since reliable safeguards did in fact not exist, the "net result" of such a pressure on the export control standards "will be the rapid destruction of existing U.S. and COCOM controls"—and potentially not only in the realm of HPCs but in regard to other strategic technologies as well.[46]

Thus, in March 1971, the ICL case had, once again, reached an impasse. By now, only President Richard Nixon could resolve the interdepartmental gridlock. Based on the NSC report's elaborate presentation of the pros and cons, Nixon opted for a position that was closer to the DoS than to the DoD but that nevertheless tried to rein in British and the computer industry's expectations regarding future HPC business on the other side of the Iron Curtain. To calm the DoD worries, Nixon's decision also tried to forestall repercussions of the ICL case for CoCom. In May, the Serpukhov license was granted, "on the conditions that (1) the U.K. agree to effective implementation of the proposed safeguards, particularly joint procedures for the inspection and random core dump provisions including the transmission of the core printouts to the U.S., and (2) the U.K. agree to support continuation of tight controls on computers and technology at the next Coordinating Committee List Review."[47] In summer 1971, after roughly twenty months of negotiations, the export control case of two the British ICL 1906-A HPCs could be closed.[48]

The green light for the ICL computers certainly did not mean that all problems of export control policy for HPCs were resolved. The long-term effects of the ICL case were decidedly ambiguous—because it, indeed, established a precedent, both in sense of what the DoS had hoped and what the DoD had feared. In the context of US-Soviet relations, the Serpukhov computer export established for the first time in theory and practice the principle of on-site inspection on Soviet soil—well before the Soviet Union was willing to accept nuclear and arms control verification inspections.

Since the onset of the Cold War, the Soviet Union had perceived foreign inspections on its own soil with utter distrust as they touched upon matters of national security and sovereignty. The harshest critics denounced on-site inspections as a form of "legalized espionage."[49] Even though the debates about nuclear safeguards had begun right after World War II and the United States and its allies had been pushing for inspections, the Soviets still adamantly opposed them well into the 1970s. In fact, the lowest common denominator was that the IAEA safeguards regime excluded all nuclear weapons countries from inspections, except on a voluntary basis. It was applied only to IAEA members who were not in the nuclear club.[50]

In 1976, however, for the very first time the Soviet Union accepted as part of the Peaceful Nuclear Explosions Treaty on-site inspections in a bilateral treaty with the United States. In practice, these inspections would have been quite limited. The treaty stipulated the presence of inspectors for underground nuclear explosions of more than 150 kilotons and provided that observers may be allowed for explosions above 100 kilotons. Since neither the Soviets nor the United States had tested atomic devices larger than 150 kilotons for years, it was unlikely that inspections would actually take place. But the Soviets accepted them in principle, and this was indeed a considerable breakthrough.[51] It would not last long. Embroiled in intense debates about the Carter administration's failed attempts to negotiate a comprehensive test-ban treaty with the Soviets, the Senate did not ratify the Peaceful Nuclear Explosions Treaty (in fact this happened only in 1990).[52] Even though the Soviets sent signals after this setback in 1982 that they would be willing to negotiate about IAEA-style inspections, real success was forthcoming only after Gorbachev had commenced his reform policies. In 1985, the Soviet government signed an agreement with the IAEA that accepted safeguards, including inspections, for certain nuclear facilities. Finally, in July 1986 an international team that included Western experts conducted for the first time an inspection on Soviet territory.[53]

Against this backdrop of the complicated history of nuclear safeguards, it was a stunning development in Cold War relations that, as early as 1971, the Soviets accepted a whole catalog of intrusive safeguards for HPCs—reportedly even "without demur."[54] Moreover, the 1971 agreement between the United Kingdom, the United States, and the Soviet Union was not a one-off event but set a precedent for HPC exports, though not in the way the DoS had envisioned. It did not have an immediate impact on international arms control practice. In the field of the trade with HPCs across the Iron Curtain, however, Serpukhov became the model for the use of safeguards. Until East-West relations degenerated again into a period of high tension at the end of the 1970s—the so-called Second Cold War—safeguards became a standard feature of sales contracts with Eastern Bloc states and of US and CoCom export controls. Did this imply an erosion of the traditionally strict US export denial policy for HPCs, as the DoD feared?

Détente and Computer Export Control Liberalization

As it happens, in the early 1970s it was not only the HPC controls that seemingly eroded. The entire export control system had entered a period of profound change. It had shrunk considerably. Hence, the discussions about the exports of the CDC 6600 and the ICL 1906-A were part and parcel of a much broader trend in the United States toward a liberalization of export controls. Pressures on the system had markedly increased in the second half of the 1960s and then gained additional momentum with the onset of US-Soviet détente during the Nixon administration. The HPC cases exemplifies the enormous challenges this shift posed for the US export control system and CoCom.

The central milestone of this liberalization trend was the Export Administration Act (EAA) of 1969, signed into law in December 1969 at the same time that the HPC issues occupied the US and the British governments. For the first time since 1949, this latest rendition of US export control legislation relativized the regulation of technology trade with an explicit commitment to the facilitation of trade relations. The EAA "encourage[d] trade with all countries with which we have diplomatic or trading relations, except those countries with which such trade has been determined by the President to be against the national interest."[55] This did not sound like a dramatic and radical change, but the EAA was the key prerequisite for cutting back controls and opening up to the Soviet Union. The new act replaced the much stricter and broader Export Control Act of 1962.[56] The change of statutory language also signified the changed outlook of Congress. The emphasis shifted from "control" of exports to their "administration."

This reorientation was the result of several closely linked economic, political, strategic, and conceptual changes. First of all, the European push toward détente set in much earlier than in the United States. The CoCom partners increasingly perceived economic relations as the primary vehicle of East-West rapprochement. Establishing closer trade ties promised to enhance political stability on the continent while generating revenues for companies and governments. The British were not alone with their advocacy for export control relaxation. Other states chimed in, not least West Germany whose "Ostpolitik" had strong economic and trade components. While Western European companies reaped more and more benefits, the tight US export control policy relegated American companies to the sidelines. This appeared less and less acceptable, not least because the United States was confronted with a deteriorating balance of payments and an economic downswing. Questions of technology trade and transfer played a key role in these political developments.[57]

The pushes toward détente facilitated a reconsideration of the goals and strategies of export controls. The perception of the threat that emanated from the Eastern Bloc lost some of its urgency and weight as the East and the West negotiated a wide range of agreements on trade, scientific exchange, and arms control. After twenty years, influential voices within the US government increasingly questioned the effectiveness of the Cold War embargo policy against the Soviet bloc. The DoS's Policy Planning Council under Walt Whitman Rostow, for example, produced as early as 1963 a report that paved the way for a movement away from ideas of economic warfare to concepts of "tactical linkage" that emphasized the use of economic relations and especially technological dependencies as political levers of influence on the Soviet alliance. This trend was certainly not uncontroversial but it slowly opened up the issue of East-West trade for Nixon and Kissinger's détente policy in the 1970s.[58]

Finally, US considerations on trade liberalizations were also pushed by technological change. The United States incrementally, if slowly, lost the overpowering technological dominance it had enjoyed in the world after 1945.[59] No doubt, the Western Europeans were catching up technologically—even though they chafed in the late 1960s at the "technological gap" between them and the United States. While Jean-Jacques Servan-Schreiber's international bestseller *The American Challenge* warned of the dangers of being dependent on American computer technology and becoming a kind of high-tech colony of the United States, European high-tech firms emerged as competitors US companies worried about.[60] As traditionally US exports controls were stricter than those of other CoCom states, the EAA of 1969 relaxed licensing requirements for technologies, which were also available from non-US sources in order to strengthen US industry on the international markets.[61]

Even though the US move toward détente and export control liberalization was in the making for several years and translated into statutory law by the end of the 1960s, there was a distinct delay until 1972 in the implementation of the new understanding of the role and scope of export controls.[62] The CDC and the ICL cases marked exactly this period of transition and show how much pain and controversy it caused within the US government. But while Nixon as the Republican candidate for the presidency had opposed the EAA of 1969, in 1972 his administration fully embraced closer economic and technological ties to the Soviet Union.[63] In October 1972, the United States and the Soviet Union signed a comprehensive trade agreement, which was followed by a massive reduction of the export control list of the Department of Commerce from 550 unilaterally controlled categories to only seventy-three just a half year later.[64] The agreement was accompanied by no fewer than eight agreements on scientific-technological cooperation, signed between June 1972 and July 1973.[65]

The Codification of High-Performance Computer Safeguards

At the same time as the scientific-technological cooperation agreements were signed, the Soviets finally accepted, in March 1973, the official contractual agreement on the safeguards on the two ICL computers at Serpukhov. After they were installed, the British Atomic Energy Commission took charge of the monitoring procedures. They "proved to be expensive, costing £80,000 per annum." Also, it was difficult to find "people to operate the monitoring system." But the US government was satisfied with the implementation of the safeguard regime.[66] One year later, in March 1974, the Nixon administration took the next step und codified the safeguard regime in the National Security Decision Memorandum (NSDM) 247.

At first sight, this memorandum was an extension of the broader export control liberalization trend to the HPC field. The threshold for computers that could be exported to communist countries was raised from a PDR of 8 mbs to 32 mbs. Only the sales of computers beyond this level would still be "reviewed on a case-by-case basis and strictly limited to demonstrably peaceful applications." There was also a limit restriction for the "export of completed hardware for use as part of computer systems produced by Communist countries . . . in number, performance, and presale conditions." But all electromechanical peripheral equipment, including spares and technological data, was "decontrolled," that is, eliminated from the control list.[67]

Notwithstanding these relaxations, the HPC regulations remained strict overall. Explicitly excluded was the "export or transfer of computer technology, production facilities, specialized programs and comprehensive programming service. These restraints include those placed on the means to design, develop, and produce computers, peripheral storage devices and storage media, displays, high speed memories, and electronic components." In short, the liberalization pertained only to hardware within a limit that was in fact well below the level of the ICL 1906-A (which had a PDR of roughly 44 mbs) or the much more powerful CDC 6600.[68] More importantly, the decontrols included only hardware. They still prohibited the sharing of "know-how," that is, the knowledge and skills needed to design and produce high-technology goods. Indeed, the regulation of knowledge in the form of data and services allowing for interpersonal teaching and learning were at the very center of the HPC export controls—and would soon be defined as the core principle of the entire US export control system, as laid out in the so-called Bucy Report of 1976.[69] Accordingly, the president commissioned in NSDM 247 a study on the "technical and administrative issues relating, in particular, to safeguard procedures and the means to control the export of technical information, training services, and software."[70]

Yet even computers below the 32 mbs threshold were still subject to safeguards with a strong emphasis on on-site inspections. Licenses would be granted only if the exporting country assured CoCom (1) that the computer would be used for peaceful purposes; "(2) that the seller would visit (and report on the continuing end-use) the computer facility monthly (quarterly for lower performance computers . . .)"; and (3) that the exporting country submitted "a signed statement from the importing agency or end-user providing assurance of peaceful end-use, the right of access to the computer facility, and the assumption of the responsibility to report any significant change to the fact presented."[71] In practice, the threshold for quarterly visits was a computing power of 13 mbs, for monthly visits at 20 mbs. For large computers below 13 mbs the "right to access" was part of contractual agreements.[72]

Even though the CDC 6600 and the ICL 1906-A cases appeared to be the beginning of a larger liberalization trend of HPC export controls, the fears in the DoD of an erosion of the entire control system were, at least in this technological field, unwarranted. There was no great demand for Western computers, and the overall development of the Eastern Bloc market turned to be rather disappointing for Western companies. Between 1970 and 1975 only about "200 Western computers were sold," among them merely about fifty "relatively large machines." The other 150 computers were small machines and microcomputers. As the leading trade journal of the data-processing industry, *Computerworld*, summed it up, the total value of all these computers was "in the order of $150 million realized over five years by at least 20 manufacturers from the United Kingdom, France, Germany, Sweden, Japan and the United States. This just about averages at $1.5 million per year per manufacturer, which is less than what some salesmen are selling within a single block of Manhattan."[73]

Conclusion

The debate on the exports of Western HPCs to the Soviet Union in the late 1960s and early 1970s changed the US and CoCom export control policy in several, seemingly contradictory ways.

At first sight, the sale of two British ICL 1906-A computers to the Soviet Institute of High Energy Physics in Serpukhov was part of a larger trend toward general export control policy liberalization under the auspices of détente. It also appeared to be a case that paved the way for a reform of HPC export controls in the interest of more East-West trade in this technological field. When President Nixon decided to override the concerns of the Department of Defense in May 1971, national security appeared to have been put on the back burner. The safeguard regime developed for this case was from the perspective of the DoD only a questionable trick to circumvent and soften existing limitations to technology transfer. It would not work and it would establish—despite of claims that it would be a unique, isolated case—a dangerous precedent.

The DoD was partly right at least about the second point. Safeguards became the standard operating procedures for the West-East sale of large computer systems and were codified in March 1974 through NSDM 247. But they were not the breach in the national-security dam the DoD had warned of. They had a rather paradoxical effect. While they looked like a liberalization measure, they in fact enormously expanded the extraterritorial reach of US and CoCom export controls into the geographical territories of the Soviet Union and its satellites. On-site inspections and technical surveillance missions were put in place that differed little from those that the Soviet's had fought against in regard to the IAEA's nuclear nonproliferation inspection regime. Because the Soviet Union wanted to have access to a technology that only the United States possessed, it was willing to accept far-reaching intrusions into its national sovereignty—not just once, but up to fifty times between 1970 and 1975 alone. In contracts with Western companies—and by extension with the US government and CoCom—the Soviets accepted quarterly, often even monthly, inspections by Western experts, who checked on the use of computers at places like Serpukhov and removed "core data dumps" for investigation by the US intelligence community. In any other policy field in East-West relations, such practices were unheard of. And this was not just a random policy field: HPCs were deemed a modern key technology and highly sensitive because of their broad spectrum of national-security applications, including the design of nuclear weapons.

NSDM 247 not only codified safeguards. It also put strong emphasis on export controls over know-how and paved the way for the groundbreaking reform of the control system after the publication of the Bucy Report. Certainly, it was not by accident that Fred Bucy, the main advocate of know-how regulation, was as chief operating officer of Texas Instruments a prominent figurehead of the US computer industry. He would soon become a vocal critique of safeguards because he thought they did not forestall the transmission of technical know-how to the enemy.

The DoD was wrong on another account, too. As controversial as they were, safeguards worked surprisingly well in practice. In 1977, the CEO of CDC, Robert D. Schmidt, reported to Congress that there had been "not a single instance of failure on the part of the Soviets to comply with the required procedures." He added, "I was recently asked how the Soviet customers react to" safeguards. "After all, it was pointed out, they bought and paid for the computer but we are telling them what they can and cannot do with it. My answer was that the customer doesn't like it at all, no more than you or I would if someone were continually looking over our shoulder with a constant reminder of their distrust of our integrity. Nevertheless, they acknowledge those agreed procedures as a condition of having the computer and they submit to them, although not cheerfully."[74]

Notes

1. Office of the Director of Defense Research and Engineering, *An Analysis of Export Control of U.S. Technology—A DoD Perspective: A Report of the Defense Science Board Task Force on Export of U.S. Technology* (Washington, DC: US Department of Commerce National Technical Information Service, 1976) (hereafter Bucy Report), 24.

2. For the larger context, see Elisabeth Roehrlich, "Negotiating Verification: International Diplomacy and the Evolution of Nuclear Safeguards, 1945–1972," *Diplomacy and Statecraft* 29, no. 1 (2018): 29–50.

3. Brandt J. C. Pasco, "*United States National Security Reviews of Foreign Direct Investment*: From Classified Programs to Critical Infrastructure, This Is What the Committee on Foreign Investment in the United States Cares About," *ICSID Review* 29, no. 2 (2014), 350–371.

4. Anthony Astrachan, "U.S. Veto Stops Sale of Computers to Soviet Union," *Washington Post*, May 24, 1971, A10; CIA, *ICL Computers for the USSR*, February 22, 1971, 1, https://www.cia.gov/readingroom/docs/DOC_0000969851.pdf. The CIA

report mentions a different sales package but all other public sources consistently mention two 1906-A computers as the main issue.

5. "Pooling Brains to Study the Atom," *Business Week*, August 22, 1970.

6. Alvin Shuster, "British Will Create a Computer Giant to Assist Exports," *New York Times*, March 22, 1968, 69, 79; James W. Cortada, "Public Policies and the Development of National Computer Industries in Britain, France, and the Soviet Union, 1940–1980," *Journal of Contemporary History* 44, no. 3 (2009): 493–512, at 502.

7. "NSC Under Secretaries Committee Report," March 16, 1971, in *Foreign Relations of the United States* (hereafter *FRUS*), *1969–1976*, vol. 4, *Foreign Assistance, International Development, Trade Policies, 1969–1972*, ed. Bruce F. Duncombe (Washington, DC: Government Printing Office, 2002), 933–941, at 933 (doc. 372, enclosure); Shuster, "British Will Create a Computer Giant," 79.

8. See Seymour E. Goodman, "Soviet Computing and Technology Transfer: An Overview," *World Politics* 31, no. 4 (1979): 539–570.

9. CIA, Directorate of Intelligence, *Intelligence Memorandum: Production of Computers in the USSR*, July 1971, 2, https://www.cia.gov/readingroom/docs/CIA-RDP85T00875R001700010088-3.pdf.

10. CIA, Directorate of Intelligence, *Intelligence Memorandum: Production of Computers in the USSR*, 1.

11. [CIA], *Foreign Computer Capabilities*, September 25, 1969, 1, https://www.cia.gov/readingroom/docs/DOC_0005577292.pdf; Ivan Berenyi, "Computers in Eastern Europe," *Scientific American* 223, no. 4 (1970): 102–109, at 102.

12. CIA, Directorate of Intelligence, *Intelligence Memorandum: Production of Computers in the USSR*, 3.

13. Berenyi, "Computers in Eastern Europe," 108; Goodman, "Soviet Computing and Technology Transfer," 546, 548.

14. Goodman, "Soviet Computing and Technology Transfer," 548, 551–552, 554.

15. Frank Cain, "Computers and the Cold War: United States Restrictions on the Export of Computers to the Soviet Union and Communist China," *Journal of Contemporary History* 40, no. 1 (2005): 131–147, at 141.

16. CIA, *ICL Computers for the USSR*, 4.

17. For the best introduction to export controls in the 1960s, and in general, see Harold J. Berman and John R. Garson, "United States Export Controls—Past, Present, and Future," *Columbia Law Review* 67, no. 5 (1967): 791–890.

18. "NSC Under Secretaries Committee Report," 933 (doc. 372, enclosure).

19. CIA, *ICL Computers for the USSR*, 1–3, and Attachment A, "Operating Characteristics of Selected Computers."

20. CIA, *ICL Computers for the USSR*, 4; Wikipedia, s.b. "History of Supercomputing," last modified July 19, 2021, 19:34, https://en.wikipedia.org/wiki/History_of_supercomputing. For a history of the technical development of the CDC 6600, see James E. Thornton, "The CDC 6000 Project," *Annals of the History of Computing* 2, no. 4 (1980): 338–348.

21. Vice Admiral H. G. Rickover, Testimony, March 19 and 20, 1970, in *Naval Nuclear Propulsion Program—1970: Hearings before the Joint Committee on Atomic Energy, Congress of the United States, Ninety-First Congress, Second Session on Naval Nuclear Propulsion Program* (Washington, DC: Government Printing Office, 1970), 14–16. See also Donald MacKenzie, "The Influence of Los Alamos and Livermore National Laboratories on the Development of Supercomputing," *Annals of the History of Computing* 13, no. 2 (1991): 179–201.

22. Robert S. Allen and John A. Goldsmith, "Speedy Computer Wanted by Soviets," newspaper article of unknown origin (syndicated column "Inside Washington"), September 9, 1970, https://www.cia.gov/readingroom/docs/CIA-RDP7200337R000200190021-4.pdf.

23. Rickover, Testimony, 14.

24. For the comparison of computing power see CIA, *ICL Computers for the USSR*, Attachment A.

25. Rickover, Testimony, 14.

26. Fortran is a programming language developed by IBM.

27. CIA, *ICL Computers for the USSR*, Attachment C, "Proposed Safeguards for High Performance Computers."

28. CIA, *ICL Computers for the USSR*, Attachment C.

29. Rickover, Testimony, 14–15.

30. Rickover, Testimony, 8–10, 17.

31. Allen and Goldsmith, "Speedy Computer Wanted by Soviets."

32. CIA, *ICL Computers for the USSR*, Attachment D, "Recent Soviet Efforts to Obtain Large Western Computers."

33. CIA, *ICL Computers for the USSR*, 2–3.

34. Cortada, "Public Policies and the Development of National Computer Industries," 502.

35. CIA, *ICL Computers for the USSR*, 3.

36. Robert S. Allen and John A. Goldsmith, "Computers for Russia: Tense U.S., Britain," *Northern Virginia Sun*, November 3, 1970, https://www.cia.gov/readingroom/docs/CIA-RDP89B01354R000100060019-5.pdf.

37. "Memorandum from the President's Assistant for National Security Affairs (Kissinger) to the Chairman of the National Security Council Under Secretaries Committee (Irwin)," January 25, 1971, in *FRUS, 1969–1976*, 4:926 (doc. 369).

38. "NSC Under Secretaries Committee Report," 933 (doc. 372, enclosure).

39. Aspects of this debate are also presented by Cain, "Computers and the Cold War," 144–146, but without discussion of the fact that safeguards were new and a major political and regulatory innovation.

40. "NSC Under Secretaries Committee Report," 933 (doc. 372, enclosure).

41. "NSC Under Secretaries Committee Report," 933 (doc. 372, enclosure).

42. "NSC Under Secretaries Committee Report," 934 (doc. 372, enclosure).

43. "NSC Under Secretaries Committee Report," 933–934 (doc. 372, enclosure).

44. "NSC Under Secretaries Committee Report," 934–936 (doc. 372, enclosure).

45. "NSC Under Secretaries Committee Report," 936–939 (doc. 372, enclosure).

46. "NSC Under Secretaries Committee Report," 937–939 (doc. 372, enclosure).

47. "Memorandum from the Staff Director of the National Security Council Under Secretaries Committee (Hartman) to the Members of the Under Secretaries Committee," May 13, 1971, in *FRUS, 1969–1976*, 4:942 (doc. 374). See also "Letter from President Nixon to Prime Minister Heath," May 12, 1971, in *FRUS, 1969–1976*, 4:941 (doc. 373).

48. Robert G. Kaiser, "Soviets Are Buying 2 British Computers," *Washington Post*, July 3, 1971, A17.

49. Warren Heckrotte, "A Soviet View of Verification," *Bulletin of the Atomic Scientists* 42, no. 8 (1986): 12–15, at 13.

50. "Prepared Statement of Harry A. Finley, Associate Director, International Division, U.S. General Accounting Office," in *IAEA Programs of Safeguards: Hearing before the Senate on Foreign Relations, Ninety-Seventh Congress, First Session, December 2, 1981* (Washington, DC: Government Printing Office, 1982), 30–35, at 31; Stephen Gorove, "Maintaining Order through On-site Inspections: Focus on the IAEA," *Western Reserve Law Review* 18, no. 5 (1967): 1525–1547, at 1541; Mikhail Kokeyev and Andrei Androsov, *Verification: The Soviet Stance; Its Past, Present and Future* (New York: United Nations, 1990), 5; Nuclear Energy Policy Study Group, *Nuclear Power Issues and Choices* (Cambridge, MA: Ballinger, 1977), 292; International Atomic Energy Agency, "List of Member States," https://www.iaea.org/about/governance/list-of-member-states.

51. David Binder, "U.S. and Soviet Sign a Pact that Limits Atomic Tests," *New York Times*, May 29, 1976, 1, 5.

52. Thomas Graham and Jamien J. LaVera, *Arms Control Treaties in the Nuclear Era* (Seattle: University of Washington Press, 2002), 435–436; Joel S. Wit, "Who's Afraid of On-Site inspection?," *Christian Science Monitor*, July 13, 1982.

53. IAEA, *The Evolution of IAEA Safeguards* (Vienna: IAEA, 1998), 68–69; "Agreement between the Union of Soviet Socialist Republics and the International Atomic Energy Agency for the Application of Safeguards in the Union of Soviet Socialist Republics," February 21, 1985, INFCIR/327, https://www.iaea.org/publications/documents/infcircs/text-agreement-21-february-1985-between-union-soviet-socialist-republics-and-agency-application-safeguards-union-soviet-socialist-republics; William J. Broad, "Westerners Reach Soviet to Check Atom Site," *New York Times*, July 6, 1986, 1, 12.

54. Nicholas Wade, "Computer Sales to U.S.S.R.: Critics Look for Quid Pro Quos," *Science*, 183, no. 4124 (1974): 499–501, at 500.

55. Export Administration Act [EAA] of 1969, Pub. L. No. 91-184, 83 Stat. 841 (December 30, 1969), sec. 3(1), https://www.govinfo.gov/content/pkg/STATUTE-83/pdf/STATUTE-83-Pg841.pdf, in *International Legal Materials* 9, no. 1 (1970): 192–196, at 192.

56. Export Control Act of 1949, amendment, Pub. L. No. 87-515, 76 Stat. 127 (July 1962), https://www.govinfo.gov/content/pkg/STATUTE-76/pdf/STATUTE-76-Pg127.pdf.

57. Alan P. Dobson, *US Economic Statecraft of Survival, 1933–1991: Of Sanctions, Embargoes and Economic Warfare* (London: Routledge, 2002), 171–173; Jeffrey W. Golan, "U.S. Technology Transfers to the Soviet Union and the Protection of National Security," *Law and Policy in International Relations* 11, no. 3 (1979): 1037–1107, at 1047; Angela Stent, *From Embargo to Ostpolitik: The Political Economy of West German–Soviet Relations, 1955–1981* (Cambridge: Cambridge University Press, 1981); Randall Newnham, "Economic Linkage and Willy Brandt's Ostpolitik: The Case of the Warsaw Treaty," *German Politics* 16, no. 2 (2007): 247–263.

58. Michael Mastanduno, *Economic Containment: CoCom and the Politics of East-West Trade* (Ithaca, NY: Cornell University Press, 1992), 131–134, 143–144.

59. Richard R. Nelson and Gavin Wright, "The Rise and Fall of American Technological Leadership: The Postwar Era in Historical Perspective," *Journal of Economic Literature* 30, no. 4 (1992): 1931–1964.

60. On the debate about the "American challenge" and the "technological gap," see Jean-Jacques Servan-Schreiber, *The American Challenge* (New York: Atheneum 1968); Bernard Nossiter, "Europe's Technology Gap," Parts 1–5, *Washington Post*, February 12–16, 1967; Robert Gilpin, "European Disunion and the Technology Gap," *The Public Interest* 10 (1968): 43–54.

61. Mastanduno, *Economic Containment*, 141. The EAA's text is not overly explicit about the problem of what the export control community refers to as "foreign availability," but in two sections it is flagged as an important concern; EAA of 1969, sec. 2(4) and 4(2)(b).

62. Mastanduo, *Economic Containment*, 144.

63. Golan, "U.S. Technology Transfers," 1047n60.

64. Mastanduno, *Economic Containment*, 146–147. On the trade agreement, see also Kazimierz Grzybowski, "United States–Soviet Union Trade Agreement of 1972," *Law and Contemporary Problems* 37, no. 3 (1972): 395–428; Philip J. Fungiello, *American-Soviet Trade in the Cold War* (Chapel Hill: University of North Carolina Press, 1988), 176–184.

65. Loren R. Graham, "Aspects of Sharing Science and Technology," *Annals of the American Academy of Political and Social Science* 414, no. 1 (1974): 85–95.

66. Cain, "Computers and the Cold War," 146.

67. National Security Decision Memorandum (NSDM) 247: U.S. Policy on the Export of Computers to Communist Countries, March 14, 1974, 1, https://fas.org/irp/offdocs/nsdm-nixon/nsdm_247.pdf (hereafter NSDM 247).

68. CIA, *ICL Computers for the USSR*, Attachment A.

69. Discussed in detail in Mario Daniels and John Krige, *Knowledge Regulation and National Security in Postwar America* (Chicago: University of Chicago Press, 2022), chap. 4.

70. NSDM 247, 2; Bucy Report.

71. NSDM 247, 2.

72. See the graphic "Controls on Computer System Exports to the Communist Countries," in *Computer Exports to the Soviet Union: Hearing before the Subcommittee on International Economic Policy and Trade of the Committee on International Relations, House of Representatives, Ninety-Fifth Congress, First Session, June 27, 1977* (Washington, DC: Government Printing Office, 1978), 40.

73. Bohdan Szuprowicz, "Chances Slim for Penetration of Soviet Market," *Computerworld* 11, no. 37 (September 12, 1977), 94. For assessing the computing power of sold models compare the computers listed in this article with those in CIA, *ICL Computers for the USSR*, Attachment A.

74. "Statement of Robert D. Schmidt, Executive Vice President, Control Data Corp.," in *Computer Exports to the Soviet Union*, 25.

Chapter Five

Regulating the Transnational Flow of Intangible Knowledge of Space Launchers between the United States and China in the Clinton Era

JOHN KRIGE

Accident Inquiries as Sites of Sensitive Knowledge Transfer

In summer 1992, China's Long March 2-E rocket successfully launched the first of Hughes Space and Communications' (hereafter Hughes) Optus series of satellites for use in Australia and New Zealand. However, the launcher carrying the second satellite, Optus B2, which lifted off in December 1992, exploded during ascent from the base at Xichang in China, destroying both the rocket and the satellite. History repeated itself on January 26, 1995, when an Apstar-2 communications satellite manufactured by Hughes for APT Satellite Holdings, which served parts of the Asian market, was also lost soon after launch. In both cases the American engineers attributed the accident to the failure of the rivets joining the two hemispheres of the rocket fairing (a cone surrounding the satellite to protect it during launch) under adverse weather conditions.[1] On both occasions, the Chinese investigators disagreed, insisting that the fairing was not at issue. Rather, high-altitude shear winds had damaged the connection between the satellite and the rocket, an interface that was Hughes's responsibility.

On February 15, 1996, about a year after the Apstar launch failure, a Long March LM-3B rocket carrying an Intelsat 708 satellite built by Space Systems/Loral (hereafter Loral) on its maiden flight veered off course and flew more or less horizontally off its launch pad for twenty-two seconds before smashing into a village on a hillside near Xichang. As many as a hundred people were killed: the numbers remain in dispute. Insurance companies and underwriters, concerned about the risks if the launcher were used for an upcoming launch of another Hughes satellite on the Long March 3 series, insisted that an independent review was needed (as was usual whenever a commercial launch failed). By the late 1990s they had paid out $3.8 billion of launch insurance—almost as much as the $4.2 billion paid in insurance premiums. Intelsat 708 was insured for $204.7 million.[2] Insurers were emphatic that they would not cover the next launch until the cause of the Intelsat 708 launch failure had been established by an international committee, and accepted and acted on by the Chinese launch operator. Faced with this pressure, Loral set up the International Review Committee (IRC) with experts from its own ranks, along with engineers from Hughes, Daimler-Benz Aerospace, General Dynamics, Intelsat, and British Aerospace. Western and Chinese experts again disagreed about the causes of the accident. Both traced its origins to the inertial guidance platform that steered the rocket's trajectory. They reached different conclusions as to the precise fault depending on the amount of telemetry data they used to analyze its malfunction. After extensive discussion a compromise was reached and the IRC's arguments prevailed.[3]

The IRC came to its conclusions after two major face-to-face meetings with Chinese engineers and project managers. The first was at Loral's office in Palo Alto, California, from April 22 to 24, 1996, and the second a week later, from April 30 to May 1, this time in Beijing. There were four Chinese representatives present at the first, almost two dozen at the second. The Western experts raised a number of technical questions about the voluminous material offered to them by their hosts, and visited assembly and test facilities for guidance and control equipment. They spent some time discussing their findings in their hotel rooms, which were probably bugged.[4] It was more than likely that they had divulged sensitive information to Chinese experts who worked with both civilian and military space technologies.

This chapter describes the intense and at times virulent debates over the threat to national security posed by the American satellite manufacturers and their Western colleagues who exchanged information and ideas with their Chinese counterparts in these accident investigations.[5] The furor was triggered when an official in the Defense Technology Security Administration—an arm of the Department of Defense (DoD) specifically concerned with the national security risks associated with critical technology—read of the exchanges in a technical publication. No one accused the companies of exporting critical hardware to China. At stake rather was the risk that Western experts had transferred sensitive, unclassified dual-use knowledge and know-how to their Chinese colleagues—

knowledge and know-how that could improve the performance and reliability of civilian space launchers *and* of military ballistic missiles.

The face-to-face meetings held between Western and Chinese experts in the United States and in Beijing are at the focal point of this chapter. For the American authorities, "exports" of sensitive intangible knowledge occur whenever and wherever experts from the United States and foreign nationals meet together, be it in a corporate boardroom in Palo Alto, in a hotel room in Beijing, or on a rocket launch pad in Xichang. The border across which the knowledge flowed is, as I stress in the introduction to this volume, a legal boundary constructed by institutions (here, the Departments of Commerce, of Defense, and of State) to regulate the sharing of knowledge with representatives of foreign entities from a country of concern, here the China Great Wall Industry Corporation (CGWIC, which managed launch services) and the China Academy of Launch Vehicle Technology (CALT, which operated them). The charge against the companies was that they had shared sensitive unclassified knowledge that posed a risk to national security without first seeking an export license to do so.

The analysis in this chapter lies at the intersection of four distinct historical narratives. The first is the history of US export controls, with an emphasis on their extension beyond simple trade in commodities to embrace the sharing of unclassified sensitive knowledge and know-how with foreign entities. The second is the history of American foreign policy and trade relations with China, notably in the space sector, that began in the Reagan years and that were liberalized during President Clinton's two terms in office, exposing him to charges that he put business interests ahead of national security. The third is the history of the domestic investigations by committees of Congress and the Senate into the behavior of Hughes and Loral in their dealings with China and with the American administration. Here I am interested in how the investigation established that intangible knowledge had moved *across* national borders by opening the black box of transnational transactions *at* borders, and in the policies put in place by Congress to control the dissemination of something as elusive as intangible know-how and tacit knowledge. Fourth, I engage with the historical roots of the Chinese leaders' ambivalent attitude to international collaboration in science and technology, and its impact on the official interpretation of the findings of one of the American investigative committees.

Many transnational studies are limited by their sources and are not able to analyze the refashioning of knowledge that has moved across borders at the "receiving" end. However, in this case we do have responses by the Chinese authorities to charges that they had "stolen" highly sensitive knowledge from the United States. Granted the intensely polemical nature of these exchanges—the Chinese authorities denied that they had learned anything new from the Western-led accident investigations[6]—and the complexity of the knowledge transfer process, it is difficult to draw hard and fast conclusions from this official document. All the same it is of methodological interest. By describing the structure and domestic context of the arguments used by the Chinese State Council I hope to throw some light on the diverse political and cultural considerations that can shape knowledge exchanges between very different cultures. The field of science, technology, and society (STS) has taught us that knowledge claims cannot be detached from the individuals who make them. They are also the bearers of the power relations between the parties concerned in face-to-face interactions, along with the historical legacies of oppression that those relations sometimes embody.

Export Controls as Instruments to Regulate Transnational Flows of Knowledge and Know-How: A Very Short History

In the late 1970s the United States government made a major, far-reaching change to the scope of export controls.[7] Merriam-Webster's dictionary defines the verb "export" as to "carry or send (something, such as a commodity) to some other place (such as another country)." The legal instruments regulating such exports, as defined in the 1950s, always went beyond commodities as such to include the technical data needed to use them properly, typically in a manufacturing process. By the mid-1970s, however, the increasing pressure of economic competition between the advanced industrial countries, particularly in the semiconductor and related industries, expanded the reach of export controls to explicitly embrace know-how—tacit or intangible knowledge embodied in heads and hands.

Industrial know-how had by no means been ignored when the export regime was first established. But its perceived importance grew sharply in the wake of the liberalization of trade that accompanied the policy of détente with the Soviet Union beginning in the late 1960s. In 1974 the DoD and conservative members of Congress insisted that the Soviets were taking advantage of the relaxation in tension "to acquire U.S. technological know-how [that] had important military applications under what [were] supposed to be commercial agreements."[8] Malcolm J. Currie, an applied physicist and the director of defense research and engineering, asked Fred J. Bucy, the executive vice president of Texas Instruments Inc., to define appropriate policies to meet the threat. The report of his task force, comprising representative from four key high-tech sectors (airframes, aircraft jet engines, instrumentation, and solid-state devices) "proved to be among the most influential documents produced on U.S. export control policy," providing the impetus for "a conceptual shift in the overall approach of American officials."[9]

Bucy had always opposed the relaxation of US-Soviet trade ties by the Nixon administration. He shared Currie's view that existing export control legislation was inadequate to deal with the acquisition of dual-use manufacturing "technology" by the communist bloc. His own rather unique definition of technology reflected his corporate (i.e., not academic) background. Technology, Bucy writes, "is the specific know-how required to define a product to fulfill a need, to design it, and to manufacture it."[10] What bothered Bucy most about "technology transfer" was that "once released, technology can neither be taken back nor controlled. Its release is an irreversible decision."[11] His vision extended beyond an emphasis on the loss of know-how as such. He realized that, for the first time since World War II, the United States was faced with severe international competition from Japan and Western Europe, who were eager to trade with the Soviet bloc in key high-tech sectors. The "technological gap" that had separated the US economy from the rest of the industrialized world was closing rapidly in what historians now see as the onset of a new phase of globalization.[12] Export controls could no longer be conceptualized solely as instruments to isolate the communist economy. Their goal had to be redefined. They were necessary to maintain US *strategic lead-time* by restricting technology transfer with allies and enemies alike. Controlling the transnational flows of technology—defined pithily by Bucy as the "detail of how to do things"—was just as important as regulating the export of goods to maintain the United States' comparative technological advantage.[13]

It proved extremely difficult to implement the recommendations of the Bucy Report. In practice it inspired the construction of a Militarily Critical Technologies List that had grown to 800 pages by the mid-1980s. The conception that had inspired the list remained intact, however, that is, that the United States should restrict the transfer of dual-use knowledge and know-how of emerging technologies that secured its strategic lead-time in global markets. Its importance was amplified in the 1980s when Japanese domination of the semiconductor industry alerted the intellectual and political elite that American hegemony could not be secured by military power alone: it needed to be underpinned by a competitive, export driven high-tech manufacturing industry.[14] The concept of national security was expanded to embrace economic security. It provided a guiding principle for the Clinton administration's enthusiastic embrace of trade liberalization with China in the 1990s, and it remained a pillar of national security strategy for decades. The Trump administration's key policy paper of December 2017 stated unambiguously that "economic security is national security."[15] Clinton saw trade with China as a means to enhance the competitiveness of

US high-tech industries in civilian domains in the name of economic security. His critics saw his liberal export policies for those self-same dual-use technologies, technical data, knowledge, and know-how as posing a dire threat to national security.

US-China Space Collaboration from Reagan to Clinton

Soon after entering the White House in 1981, President Ronald Reagan was persuaded that a strategic alliance with Beijing could exploit the Sino-Soviet split to the United States' advantage, even at the cost of closer US relations with Taiwan.[16] The People's Republic of China (PRC) was classified as a "friendly non-allied country" and trade relations were rapidly expanded. In particular it was agreed that, subject to various guarantees and safeguards, American satellite manufacturers could use China's Long March series of rockets to increase their launch options after the *Challenger* accident in January 1986 grounded the space shuttle fleet. These agreements remained in force through the Bush and Clinton eras. Their implementation was restricted by sanctions imposed on China after the violent clampdown on prodemocracy protestors in Tiananmen Square in June 1989 and by charges that China was proliferating missile technology to unstable regions of the globe. However, neither president had much trouble persuading Congress to waive these restrictions on a case-by-case basis.

The collapse of the Soviet Union, the rival who had provided the single most important reason for restricting trade in the name of national security, led the American government to reduce the scope of export controls. The number of dual-use licenses issued by the Department of Commerce dropped precipitously from nearly 100,000 in 1989 to just 8,705 in 1996.[17] This was not because they were difficult to get, but because licenses were not required any longer. Clinton's campaign slogan "It's the economy, stupid" encouraged the implementation of a neoliberal market-driven economy in which the pursuit of economic opportunity was seen not to threaten national security but actually to enhance it. This was particularly true as regards emerging technologies in the defense sector. The revolution in military affairs that was field-tested in Operation Desert Storm against Saddam Hussein's invasion of Kuwait in 1991 was crucial in this regard. It transformed the future "substructure of war, that [would] become information domination, and its primary building blocks [were] computers, communications systems, satellites and sensors."[18] It also impacted the federal military procurement system. A so-called segregated, ghettoized defense industrial base was merged with the civil sector whenever possible. In 1994 Anita Jones, the DoD's director of research and engineering in Clinton's administration, told Congress that defense contractors would have to learn to serve "multiple customers, not just one, to produce market products rather than to respond to specifications, and to regard cost as important as performance."[19] In the field of satellite technology, which concerns us here, William Reinsch, the under secretary for export administration in the Department of Commerce, explained to Congress in 1998 that "as the lines between military and civilian technology become increasingly blurred, a second-class commercial satellite industry means a second-class military satellite industry as well. The same companies make both products. And they depend on exports for their health and for the revenues of the next generation of products."[20] Using Chinese launchers to put US-built commercial satellites into orbit would secure a significant market share for American manufacturers, increase profits, enhance economic security, and by extension strengthen national security.

During the Reagan years all satellites had been treated as munitions, and export licenses were granted by the State Department that put a priority on national security. To facilitate exports to China, first President Bush, and then President Clinton, gradually reduced the range of performance-enhancing technical features that impeded the unrestricted export of American communications satellites for civilian use. They also transferred authority over their licensing to the Department of Commerce, which privileged business interests. Intense lobbying by the satellite (and high-performance computer) industries eager to get a major foothold in the Chinese market had paid off, but at a cost. Trade liberalization in these dual-use technologies produced blowback from conservatives in Congress who felt that Beijing should be penalized for its violations of human rights, for its treatment of Tibet, for its bellicose behavior in the Taiwan Strait and for its proliferation policies. It was in this fraught domestic political context that American manufacturers launched their satellites in China in the 1990s, and helped identify the causes of launch failures when the rockets exploded in flight.

US Inquiries into the Knowledge Sharing with China: Moving Intangible Knowledge across Borders

In September 1997 the Department of Justice began criminal investigations into allegations by the State Department that Hughes and Loral, in collaborating actively in three accident investigations of the Long March rocket, had provided technical assistance to China without an export license.[21] Six months later, in mid-February 1998, and while the investigation was still under way, President Clinton waived post-Tiananmen sanctions imposed on China and authorized the Commerce Department to grant a license to Loral to export its China-8 satellite for a Long March launch. He did so over the objections of the Justice Department, which warned him that his action could bias the jury in its case against the company. A highly critical article in the *New York Times* in April 1998 suggested that Clinton had behaved recklessly because Loral's chairman, Bernie Schwartz, was also the largest personal donor to the Democratic Party during his 1996 reelection campaign.[22] The president faced a powerful assault by a group of Washington "insiders" deeply hostile to closer engagement with the PRC who accused him of being indifferent to China's aggressive behavior and of putting business interests and political favors ahead of national security. At the same time, he was also being investigated for irregularities in campaign financing and was embarrassed by revelations of his sexual relations with White House intern Monica Lewinsky. In a toxic political environment that eventually led to the president's impeachment in 1998, an inquiry was launched by a Republican-dominated Congress into the behavior of the American satellite firms in China. Speaker Newt Gingrich (R-GA) set up a congressional select committee cochaired by Christopher Cox (R-CA) and Norm Dirks (D-WA) to investigate the matter. The Senate Select Committee on Intelligence also debated the issue. Its report was published in May 1999.[23] An unclassified version of the 930-page "Cox Committee Report" was released that same month.[24]

The Cox Report dealt with the theft of US technology in a number of sectors (high-performance computers, advanced machine tools, jet engines, and space). At the last minute the committee added a chapter accusing China of stealing information on America's most advanced nuclear warheads that could be launched with a new generation of ballistic missiles. The public response to the report was dominated by the claims made in this section. Nuclear fears fed apocalyptic speculation: Dana Rohrabacher (R-CA) said that the Cox Committee had established that "Some of the industry people . . . had provided upgrades to Communist Chinese rockets that threaten to incinerate millions of Americans."[25]

The Cox Committee Report is mostly remembered for its polemical tone and many factual errors.[26] However, it also provides valuable information extracted from corporate minutes of meetings and acquired in interviews with key stakeholders in the space sector that helps throw light on the claims that satellite companies shared sensitive knowledge and know-how with their Chinese counterparts. In what follows I treat this document, and the contemporaneous Senate Intelligence Committee report, as primary sources that offer a rare insight into the dynamics of knowledge sharing at the interface between Western and Chinese aerospace scientists and engineers.

The CIA described "space launch vehicles as ballistic missiles in disguise."[27] With this in mind, the main preoccupation of the members of the Cox Committee and of the Senate Select Committee on Intelligence was whether American satellite manufacturers and Western experts had shared knowledge with Chinese launch providers that could improve the reliability and performance of ballistic missiles. If they had, Hughes and Loral ought to have applied to the State Department for export licenses before they helped Chinese engineers figure out the reasons for the Long March launch failures carrying their satellites. The analyses made by the lawyers and engineers who explored these issues on behalf of the government give us an idea of the fuzziness of the concept of intangible knowledge, provide a fascinating insight into some of the paths along which it moves in face-to-face encounters, and alert us to the difficulty of assessing how effectively it is taken up at the "receiving end."

Long before the Cox Committee began to look into the risks posed by "technology transfer" to China in the late 1990s, Henry Sokolski had pointed out the dangers to national security of using the Long March launcher. Sokolski served in the Bush administration as the deputy for nonproliferation in the DoD. As early as 1994 he emphasized that if Chinese launchers were used, "none of the hardware is, in fact, as likely to fall into Chinese hands or to be as militarily significant as is the intangible know-how many US firms want to transfer, in order to ensure the successful launch of the satellites."[28] In practice, the issue was even more complicated. As Sokolski explained in written testimony to a congressional committee in 1998,

> "Know-how" conveys a given technical procedure, such as satellite integration for a particular satellite and rocket launcher. "Know why" goes further to explain in engineering and scientific terms why a given procedure is arranged the way it is (i.e. why certain steps must be followed in a given order by others and what fundamental problems or risks these procedures are designed to mitigate or resolve). Such "know why" would enable the Chinese on their own to engineer around such problems for other rocket or satellite systems. In short know how is relevant only to a particular system, know why empowers the student to engineer around similar problems for a variety of systems.[29]

Sokolski argued that theoretical techniques like coupling-load analysis exemplified the potential to generalize from a specific case. Coupling-load analysis was critical to ensure that a multistage launcher passed through the sequence of ignition, stage separation, motor cut-off, and so on, without unduly shaking or shattering its sensitive payload. Sharing the techniques and procedures of coupling-load analysis with the Chinese could help them redesign not only their rather rigid satellite launchers to ensure that they safely orbited their most fragile and sensitive US satellites—but also their missiles topped with nuclear warheads. Even more than "know-how," this portability of "know-why" across the civil-military divide was at the heart of accusations against Hughes and Loral of failing to protect national security.

The investigators into the accident enquiries took note that highly sensitive hardware had been lost on the crash site in Xichang, even if none had officially changed hands between the American companies and the Chinese launch providers.[30] What particularly bothered them, however, was the sharing of technical data and, above all, intangible knowledge. Remarking on the investigations conducted by Hughes, the State Department observed that "transfer of technical data is only one sensitive issue. Just as important is US (HAC) [Hughes Aircraft Company] procedural 'know-how' and systems testing/launch 'philosophy' learned over decades of trial and error."[31] In a similar vein, the Cox Committee observed that "there is as much experienced-based art as science in the successful application of the well-established numerical analysis and design methods available," and that were widely known.[32] It added that "the successful application of these theories and methods in design often require know-how and engineering judgment derived from experience."[33] What bothered them was "the benefit of this experience and know-how that Hughes engineers could have made available to their PRC counterparts" after the Apstar accident, a benefit that would "stand them in good stead in developing fairings (or shrouds) for ballistic missiles."[34]

The dynamics of knowledge transfer was also of some concern. It would not have mattered if the Western aerospace engineers had simply pointed out certain facts that had a bearing on identifying the cause of an explosion. They had done far more, however: they had explained how to fix the problem. Thus the Cox Report complained that the IRC set up at the behest of the insurance companies had gone beyond "an independent assessment of the most probable cause or causes of failure" by reviewing and "mak[ing] assessments and recommendations concerning the corrective measures *to remove the causes of failure*."[35] Sometimes these improvements amounted to little more than practical recommendations, like "Review designs and avoid single point failures—increase redundancy" or "Improve quality control in manufacturing."[36] On other occasions there was a more sophisticated learning process involved. For example, "A person with technical expertise or experience may guide or shape a discussion, leading it in some way by using the public domain information that is being provided."[37] The Cox Report expanded rather eloquently on how knowledge moves in such discussions:

> The search for the true failure mode in an accident investigation is not a simple, straightforward procedure. In some respects, it is like finding the way through a maze. It is all too easy to start down the wrong path, and stay on it for too long.

> Insights, hunches and clues based on technical judgments and experience in prior failure mode analyses, simulations, and accident investigations can be helpful [particularly if they come from individuals or groups outside the organization].[38]

Systems/launch "philosophy" learned over decades of trial and error, the experienced-based art required to apply coupling-load analysis, integrated design know-how and know-why, insights, hunches, clues—all were intangible forms of knowledge, all had been internalized over long years of professional activity in the field, and all had been shared with the Chinese aerospace community to help them make sense of the raw data they had at hand.

What persuasive strategies had been used to ensure that the Chinese engineers and managers actually accepted the arguments made by the Western experts? The Cox Report remarked that the IRC had not simply asked the PRC engineers to reconsider their original diagnosis of the accident. They had steered them away from their "protracted narrow focus on the wrong failure mode." They had insisted that the Chinese approach was incomplete and unconvincing. Their "continuing skepticism" with the Chinese interpretation, and their "persistent calling attention" to the details of the telemetry data of the Intelsat 708 launch had eventually enabled their Chinese colleagues to think out of the box.[39] The discussions around the tables in Palo Alto and Beijing involved far more than a collective inquiry and exchange of information into the cause of the accident. They were more like a seminar in which the Western experts used the Socratic method to transfer intangible knowledge that enabled the Chinese to see matters differently and to accept the analysis made by the Western experts.

Did this pose a threat to national security? Sokolski was emphatic that the sharing of know-why with Chinese engineers had markedly improved the performance and reliability of their satellite launchers. He produced a chart at a congressional hearing showing that, while 78 percent of Chinese launches between 1970 and 1996 had not met their performance objectives, every single one of the next ten launches had been a success. These even involved more complex launch systems and satellite payloads than before, indicating that the engineers and managers had not only broadened but also deepened their understanding of what makes for a successful launch campaign.[40]

Had this better understanding actually moved across the civil-military divide to missiles tipped with nuclear warheads, however? Though everyone agreed that such transfer was possible, many were reluctant to assert that US national security had actually been jeopardized. A group at Stanford's Center for International Security and Cooperation hedged its bets. It concluded, "It is clear that, not mainly the information transmitted, but the example of rigorous, objective fault analysis, management attention, and quality control given by Western engineers *may be* of use to the Chinese in designing future launch vehicles and missiles."[41] The Senate select committee was also careful. It remarked that even though there was significant technology transfer during the launch failure analyses, "the integration of U.S. technology and know-how into the PRC's ICBM force may not be apparent for several years if at all."[42] Even then, it went on, "indigenous improvements and improvements derived from non-U.S. foreign sources will make it difficult to detect and measure to what extent technology transfers from American sources may have helped the PRC." In fact, it boldly asserted that the extensive assistance the Chinese had received from non-US sources "probably is more important for the PRC ballistic missile development program than the technical knowledge gained during the American satellite launch campaigns."[43] Even an interagency review committee asked by the members of the Cox Committee to look into the risks to national security arising from the help given by Western engineers to the Chinese launch provider was prudent. It highlighted the potential transferability of "Western diagnostic processes" by Hughes and Loral but went on cautiously to say that they "*could* improve PRC pre-flight and post flight failure analysis for their ballistic missile programs" and that that, in turn "*could* increase future ballistic missile reliability" (my emphasis).[44]

The focus of these inquiries was on what the Chinese launch providers could have *learned* from Western experts in face-to-face interactions and whether or not they actually used that new know-how and know-why to *improve* the performance of their rocket and ballistic missile fleets. The technical experts advising the Cox Committee and the Senate Intelligence Committee echoed Bucy's concerns expressed two decades before, namely, that teaching people "the detail of how to do things" enabled them to apply what they had learned in new ways and in related technical domains. They explained the multiple paths along which know-how could travel in face-to-face exchanges, guided by intangible hunches, experience, and professional judgment. And they stressed how difficult it was to assess the extent to which Western experts had in fact upgraded the knowledge base of their Chinese counterparts. This difficulty of objectively assessing how much new knowledge Chinese engineers acquired derived from the ephemeral, intangible nature of know-how itself and the complex gestures associated with its flow across "borders." Only the Chinese knew, with any certainty, what gaps and deficiencies in their understanding had been filled.

The Department of Justice did not file criminal charges against either Hughes or Loral.[45] The State Department pursued its charges that export regulations had been violated, however. In January 2002 Loral paid $20 million in fines, $6 million of which was retained to improve its in-house export control compliance system. In March 2003, Boeing, which had acquired Hughes Space and Communications Company in 2000, along with Hughes, agreed that participation in the Apstar-2 and Intelsat 708 accident investigations violated export regulations. They were fined $32 million, of which $20 million was paid to the government, $4 million was "forgiven," and $8 million was devoted to improving export compliance measures in the company.

The Cultural Politics of Knowledge Sharing in the Contact Zone: China's Response

When knowledge and know-how cross national borders in face-to-face interactions, they sometimes engage actors, not only from different countries but also with different political, ideological, and cultural worldviews. These different attitudes are deeply embedded in the identities of the individuals who negotiate transnational transactions and can profoundly affect their representations of the knowledge transfer process. In this case the American and Chinese versions of what knowledge moved between experts in the two countries were diametrically opposed to each other. Driven by political and ideological agendas on both sides, and constructed partly for domestic consumption, their statements obviously have to be treated with immense care. That granted, what interests me here is the rhetorical strategies used by the Chinese authorities to contest the Cox Committee Report, strategies that throw light on some of the deep-seated "cultural" variables that can impact the process of knowledge circulation and production at the transnational border.

The Cox Committee Report criticized both China and the US satellite manufacturers Hughes and Loral for behaving illegally. It accused the Chinese partners of "stealing" sensitive knowledge, and it chided the US corporations for sharing sensitive knowledge without first getting an export license from their government to do so. The Chinese Information Office of the State Council published a long rebuttal within two months of the report's release.[46] It ignored the criticisms of the American satellite manufacturers and focused exclusively on the accusations made by the Cox Committee against Chinese engineers and managers.

The State Council dismissed the Cox Committee Report out of hand as a pack of lies whose main aim was "to fan anti-China feeling and undermine Sino-US relations."[47] In particular it was uncompromisingly hostile to accusations that China had "stolen" secret information on nuclear warheads from the United States or that it had acquired technological know-how from Western engineers that helped it improve its missile and space programs. It did not try to refute such charges by proving them false. Instead, its counterattack rested on two related claims: first, that the Chinese had succeeded technologically on their own, without foreign help and, second, that claims to the contrary were expressions of racial prejudice. Thus, beginning in the 1950s, China had "successfully overcome a series of difficult technological problems and mastered nuclear technology within a reasonably short time, by relying on its own forces, on its large number of talented scientists full of creative spirit, and on the energetic support of the people throughout the country."[48] Similarly, the State Council insisted that "with more than 40 years of independent development of rockets and missiles, China has possessed a complete set of reliability design methods and failure diagnosis treatment regulations, as well as a strict quality control system" built up over a history of more than a hundred flights.[49] They were furious that the Cox Committee Report, instead of recognizing these indigenous achievements, had accused all actors engaged in Sino-US collaboration—the government, research institutes, business agencies, official and nonofficial Chinese representatives in the United States, and American Chinese and Chinese students in America—of espionage activities. This "underestimated the creativity of the Chinese people and Chinese scientists," and was "typical racial discrimination, and a deadly insult to the Chinese nation."[50] It marked a resurgence of the McCarthyism of the 1950s and it exposed the "aberrant personality of some American politicians hostile to China's development and becoming powerful."[51] If the State Council is to be believed, then, no sensitive knowledge and know-how was shared with Chinese engineers and project managers because they had nothing to learn from the Western experts.

Chinese accusations of American racial discrimination are difficult to evaluate, though the famous case of Wen Ho Lee, who was mistakenly accused of stealing nuclear secrets from Los Alamos in 1999, suggests that they may not have lacked foundation.[52] By contrast, there is no doubt that the Chinese report exaggerated the autonomy of China's scientific and technological development. The Mao regime received a considerable amount of technical support and training from the Soviet Union until August 1960, when Premier Khrushchev unilaterally withdrew all assistance from the PRC. When the brilliant Chinese rocket engineer Qian Xuesen eventually returned to China from Caltech's Jet Propulsion Laboratory in the mid-1950s, his initial enthusiasm at being back home changed to disappointment when he realized the abysmal lack of resources needed to develop missiles. He admitted that "I was worried to death about it . . . I didn't know how to struggle in a difficult environment . . . how to start from scratch."[53] The army quickly gave him more than 3,000 trained technical professionals and cadres, industry supplied him with 300 engineers, and the government sent increasing numbers of students to Soviet universities to learn aeronautical engineering.[54] The State Council ignored this foreign assistance in the response to the Cox Report, just as it was ignored in a museum exhibit celebrating Qian Xuesen's extraordinary achievements in missile, rocket, and satellite development at Jiao Tong University in Shanghai.[55]

This emphasis on indigenous technological development is indicative of the singular importance the Chinese authorities attached to technology as a motor of economic and military achievement. It is an expression of their "technonationalism," that is, the conviction that technological self-reliance is central to the country's national security, economic prosperity and national prestige.[56] Michael Adas has stressed the tight coupling, since at least the end of the nineteenth century, of technological "progress," national pride, and nation building. After the disastrous setbacks of the Great Leap Forward and of the Cultural Revolution during almost three decades of Mao's rule, his successors announced the four pillars of modernization in 1976. They were agriculture, industry, science and technology, and defense. The emphasis on science and technology to overcome China's "backwardness" with respect to advanced Western, capitalist powers was central to the construction of a powerful modern state and was at the core of successive national economic plans.

The ideology of technonationalism was co-constructed with the quest for self-sufficiency, that is, with the determination to depend as little as possible on foreign powers to achieve national scientific and technological goals. Contemporary Chinese uneasiness regarding foreign influence goes back to foreign incursions in the mid-nineteenth century, when it was obliged to accept the presence of imperial powers on its territory and to suffer the humiliation and social disruption of the Opium Wars. It was amplified by the sudden withdrawal of Soviet technical support in 1960. As one Western observer writes, the Chinese "are aware of the potential adverse consequences of extensive borrowing of technology and foreign involvement in their domestic affairs: excessive penetration of their economy by external forces; potential destruction of domestic industry; and corruption and crime."[57] It was with some hesitation, then, that Chinese premier Deng Xiaoping opened out to the West in the late 1970s, reassuring his audience in a speech in March 1978 that "independence does not mean shutting the door on the world nor does self-reliance mean blind opposition to everything foreign. Science and technology are a kind of wealth created in common by all mankind. Any nation or country must learn from the strong points of other nations and countries, from their advanced science and technology."[58] A 1987 report by the US Office of Technology Assessment noted that, in implementing this policy, the Chinese authorities deliberately sought intangible knowledge, including advanced management methods, new design principles, and know-how to overcome their "backwardness" vis-à-vis leading Western capitalist powers. As the report put it,

> Chinese policy has discouraged the acquisition of complete plants and equipment and has stressed the acquisition of know-how, "acquiring the hen and not just the egg," as the Chinese put it. Thus, modes of technology transfer that offer more intimate interactions with foreign technical personnel have come to be preferred. . . . As a result of this change, a much greater proportion of the technology imported since the end of the 1970s has been "unembodied" technology, or pure know-how.[59]

We have reason to believe, then, that a considerable amount of know-how was acquired by Chinese aerospace engineers from their Western counterparts, but that the cultural barriers of technonationalism and self-sufficiency made it difficult for them or their authorities to admit it openly. American charges that the Chinese had stolen technology only exacerbated the tension between the protagonists and triggered accusations of racial discrimination.

These barriers were also at play in exchanges between Hughes engineers and engineers from the CGWIC and the CALT when they discussed the causes of the failed launch of the Long March LM-2E rocket in December 1992. As we saw, the American engineers were emphatic that transonic buffeting by winds at high altitudes had torn apart the two hemispheres of the rocket fairing. The PRC representatives would not acknowledge any fault. In its response to the Cox Committee Report, the State Council insisted that "China has mature experience in fairing design. . . . The fairing of the LM-2E China uses in commercial launches was designed and produced on the basis of 10 successful flights of that used for the rocket series." China, the response went on, "has relied on its own strength to accomplish improvements and developments."[60]

Technonationalism and self-sufficiency also impacted the terms of the official accident report adopted by the two parties at a meeting in Beijing on May 11 and 12, 1993. In this case we have direct access to some of the minutes themselves and to verbatim responses by American engineers who were interviewed by members of the Cox Committee. We learn that both sides used "an analysis of the Launch Vehicle telemetry, inspection of the Launch Vehicle fairing debris, and special tests" to come to quite different conclusions. The CGWIC/CALT claimed that "there was no design or manufacturing or integration flaw in the *Launch Vehicle or the fairing* which caused the failure." Hughes accepted this conclusion, which exonerated the launch provider from any blame. For its part,

Hughes's team "determined that no design or manufacturing flaw can be found in the *spacecraft* which caused the failure." The Chinese parties accepted this conclusion, which exonerated the satellite company of any blame.[61] In other words, Hughes accepted publicly that the Chinese launcher was not at fault, while the Chinese accepted that the American company was not at fault. Hughes's director of launch service acquisition told the Cox Committee inquiry that "politically we could not write down on paper that the fairing had failed and that they were at fault. It was a nonstarter in China." Behind the scenes the agreement was "Now, go fix the fairing."[62]

Very different considerations led the Chinese and the Hughes engineers and managers to accept this public statement. Long-term financial considerations were important to the American side. As one Hughes official pointed out, the company wanted to keep using the Long March because it was cheaper than the European alternative, Ariane. If it insisted that the fairing was at fault it would be far more difficult to insure the rocket with foreign underwriters for future campaigns that were in the pipeline. This would lead to conflict with both the CGWIC and "China in general" that would hurt the company's overall business in the country for years to come. Hughes's representative in Beijing concluded that "if we swallow this one and let our Chinese friends off the hook, it will actually do more good for Hughes."[63]

There is no evidence to suggest that financial concerns also lay behind the position taken by the Chinese negotiators. From the outset they had been authorized by US negotiators to sell their launches well below market prices (much to the frustration of American launch providers), and they could look forward to increased demand for their Long March rocket. A quite different reason for their refusal to accept direct criticism of their space program can be found in the domestic political context in the late 1990s. There was, at that time, a revival of the "Two Bombs, One Satellite" ideology, a phrase that referred to China's outstanding achievements in the nuclear and space sectors in the 1960s and 1970s. The Chinese authorities were actively promoting it as "a role model and source of inspiration to help guide the embarkation of the new great leap forward in science and technology."[64] The principles that defined this "spirit" were spelled out in his twenty-four-character statement by Chinese premier Jiang Zemin in 1999. They insisted on national renewal through self-reliance and indigenization, and strongly advocated coordination and cooperation across the economy, facilitated by knowledge diffusion. They also provided the guiding principle for China's medium- and long-term plan for science and technology in the early 2000s. The inquiries into the launch failures occurred in a domestic context in which the tension between foreign "dependence" and self-reliance again loomed over the planning process. Technological achievements in the nuclear and space sectors were said to confirm that Chinese scientists and engineers could achieve long-term goals without foreign assistance. To accept publicly that their rocket program was bedeviled by flaws pointed out by Western experts would dilute its persuasiveness as a "role model" for the next great leap forward.[65]

In my introduction to this book I discuss the concept of contact zones introduced by Mary Louise Pratt and adapted by Kapil Raj. I describe them as local sites of spatial and temporal copresence and of knowledge co-construction in which tacit skills and intangible knowledge are transferred between different cultures in an asymmetric field of power. The brief analysis here exemplifies the dynamics of such intercultural "clashes" in an asymmetric field comprising experts from a global technological power and their homologues from an "emerging" country sensitive to foreign oppression leading back to at least the Opium Wars of the mid-nineteenth century. The managers at Hughes crafted a report that did not offend their partners so as to ensure further access to the lucrative Chinese market. Their Chinese counterparts crafted a report that inspired confidence in the guiding principles of their "Two Bombs, One Satellite" ideology and associated planning policies.

Policing the Border

When the Reagan administration agreed that American satellite manufacturers could use Chinese rockets to launch their payloads, it was obviously aware of the risks of "technology transfer" beyond essential information of "form, fit, and function" that was needed to mate the satellite to the launcher. To that end, the two governments signed a Memorandum of Agreement on Satellite Technology Safeguards (ratified on March 16, 1989). It was intended to curb the proliferation of know-how from the civilian to the military sector. The application and enforcement of nonproliferation safeguards was, of course, a common practice in the nuclear domain. It was invoked in other situations when the presence of a human monitor was deemed essential to ensure that sensitive knowledge and know-how was not shared with the United States' adversaries.

In his contribution to this volume, Mario Daniels (chapter 4) describes the introduction of on-site inspections by Western experts to "safeguard" against the "diversion" of high-performance computers sold to the Soviet Union in the early 1970s from civilian to military programs. The same approach was embraced here. The Sino-American agreement on satellite technology safeguards made allowance for "(a) round-the-clock surveillance by US security personnel from the satellite's arrival in China to its launch into space; (b) the presence of US government officials at all satellite/launch vehicle technical coordination meetings; and (c) prior approval by US government of all data and information provided to the Chinese side by US satellite manufacturers."[66] This agreement was signed when satellites were treated as munitions for export-control purposes. The gradual lifting of restrictions on their export during the 1990s, and the shift of authority over their export licensing from the State Department to the Department of Commerce, drew the sting from the monitoring program, which ended up being honored more in the breach than in the observance. The members of the Cox Committee and of the Senate Select Committee on Intelligence deplored this relaxation of on-site controls. In fact, even before their investigations were completed, the Strom Thurmond National Defense Authorization Act for Fiscal Year 1999, voted in September 1998, returned jurisdiction over satellite exports to the State Department and imposed strict constraints on knowledge sharing for launch campaigns in China. It also made monitoring of launches from all foreign countries (excepting members of NATO and major non-NATO American allies) obligatory.

The scope of the now-mandatory monitoring program by the DoD was spelled out in section 1514 of the Act.[67] It covered, but was not limited to, "technical discussion and activities including the design, development, operation, maintenance, modification and repair" of the technical infrastructure of a launch campaign. Monitors had to be present when the satellite was transported to a launch base, integrated with the rocket, and throughout the launch phase itself, from testing and checkout prior to launch, through the launch itself, to the return of the equipment to the United States. Any disruption of the launch campaign, including a launch failure and its investigation, had to be closely monitored as well. The cost of the monitoring services had to be reimbursed to the government by the entity that received such services.

The implementation of these provisions was explained to a congressional oversight committee on satellite export controls in June 2000. James Bodner, the DoD's principal deputy under secretary of defense for policy, described the significant impact the new legislation was having. Previously, missile engineers were sent off on temporary duty assignment to monitor meetings and launch sites. This gave the impression that the government did not take monitoring seriously. Now, by contrast, the DoD had "a dedicated team of people. I think we have 33 today. We will be up above 40 next year." These monitors already had considerable experience in US military and commercial satellite launch campaigns. "When it comes to China and Russia," Bodner went on, "they attend every technical meeting where there might be transfer of tech data. That is certainly one of the most vulnerable points. It is not just at the launch site or in the case of failure. It is in the design of the system in the first place because some of the most critical losses are the tech data that might be lost [*sic*]."[68] James Lewis of the Center for Strategic and International Studies elaborated on the intrusive new powers now available to the government. Access by DoD monitors to the launch campaign of a US-built satellite in some foreign countries was authorized "during the construction of the satellite, including the participation of monitors in telephone conferences, prior review of data to be exchanged and access to the manufacturers' databases."[69] Bodner put the point succinctly: his DoD teams "monitor these things from the cradle to the grave."[70]

Concluding Remarks

The extraterritorial reach of American export controls on sensitive data and know-how into the core of a launch campaign with Chinese rockets entirely decouples the national "border" across which knowledge moves from the territorial limits of the United States. This border, constructed by legislation, implemented by various departments of the administration, and policed by monitors from the DoD, is temporarily erected whenever any American expert engages in technical discussions with a foreign national at home or abroad. It is imposed by self-censorship and by monitors that stop the flow of technical data and "the detail of how to do things" in face-to-face exchanges with adversaries from "countries of concern." The acceptance of such intrusive government surveillance into everyday corporate practice, as well as at highly sensitive launch sites in China, attests to the power that the national-security state could wield to protect American interests, and the willingness of corporate entities and foreign governments to accept it, if they wanted to do business together.

This case study extends the transnational analysis of knowledge flows to a microstudy of the processes of intangible knowledge sharing/denial that occurs in contact zones. The immense difficulty that the US government had in trying to establish if dual-use intangible knowledge was indeed acquired by Chinese aerospace engineers shows that it is one thing to follow the movement of knowledge across national borders, it is another to analyze whether it was successfully appropriated and selectively adapted to a new context at the receiving end. That analysis is all the more difficult when the transfer of intangible know-how and tacit skills is at issue. This is one reason why American investigators were often very careful in their discussion of this question, all the more so as it was at the heart of the accusations that Hughes and Loral had breached national security.

The insistence by the State Council that Chinese aerospace engineers had learned nothing new from the Western experts only muddied the waters even more.

In a RAND report in 1981, Thane Gustafson remarks that "in the end the transfer of technology depends less on the fact that knowledge and skills have been divulged than on the fact that the receiver knew how to make creative use of them."[71] The Senate Intelligence Committee realized this, if only implicitly. Hence it went to considerable effort to describe the broader institutional, industrial, and political context in which the Long March engineers practiced their craft. A cluster of interrelated factors convinced them that national security had indeed been jeopardized: "The substantial similarities between space launch vehicles and ballistic missile technology, . . . the integration of the PRC space launch and ballistic missile industries, the PRC's intention to modernize and upgrade its ballistic missile force, evidence that U.S know-how was incorporated into the PRC's space launch program," and the committee's assumption that "any improvements in the PRC's space launch vehicle will be incorporated whenever practicable in the PRC's military ballistic missile program."[72] Taken together, these considerations led them to believe that Hughes and Loral had helped China improve the reliability and performance of its ballistic missile fleet. They also led the drafters of the Strom Thurmond Act to insist on the need for DoD monitors to police the boundary between knowledge sharing/denying from "the cradle to the grave" in any future launch campaign of an American-built satellite in China.

Notes

1. The Hughes engineers claimed that, under certain weather conditions, the fairing was unable to withstand "aerodynamic forces, buffeting, and aeroelastic . . . effects that are encountered as the rocket enters the transonic phase of flight." These effects were accentuated by high winds aloft on the day of the launch. Select Committee on National Security and Military/Commercial Concerns with the People's Republic of China, H.R. Rep. No. 105-851 (January 2, 1999) (hereafter Cox Committee Report), 2:71, https://www.govinfo.gov/content/pkg/GPO-CRPT-105hrpt851/pdf/GPO-CRPT-105hrpt851.pdf.

2. Cox Committee Report, 2:270, 300.

3. The members of the IRC disagreed with the PRC engineers' assessment that the accident was caused by a broken wire (or flawed solder joint) in the inner frame of the inertial platform. This finding, they insisted, was based on an analysis of only the first seven seconds of the telemetry data. It was not compatible with the telemetry data covering the full twenty-two seconds of flight, which pointed to an open circuit

in the follower frame as being at fault. After some resistance, the PRC experts agreed that the "absence of current output from the servo-loop of the follow-up frame of the inertial guidance platform" was indeed to blame; Cox Committee Report, 2:157.

4. Cox Committee Report, 2:109.

5. A far briefer account of this issue, focused on export control policies, is in John Krige, ed., *How Knowledge Moves: Writing the Transnational History of Science and Technology* (Chicago: University of Chicago Press, 2019), chap. 2. For a more extensive study, see Mario Daniels and John Krige, *Knowledge Regulation and National Security in Postwar America* (Chicago: University of Chicago Press, 2022), chap. 9. This version is deliberately framed in terms of the methodology developed in the introduction to this volume, and includes the Chinese perspective.

6. Government of the People's Republic of China (PRC), Information Office of the State Council, *Facts Speak Louder than Words and Lies Will Collapse by Themselves: Further Refutation of the Cox Report*, July 15, 1999, Federation of American Scientists, https://fas.org/sgp/news/1999/07/chinacox/index.html.

7. The history of US export controls is treated in detail in Daniels and Krige, *Knowledge Regulation and National Security.*

8. "Détente: A Trade Giveaway?," *Business Week*, January 12, 1974, 64–66.

9. Office of the Director of Defense Research and Engineering, *An Analysis of Export Control of U.S. Technology—A DOD Perspective: A Report of the Defense Science Board Task Force on Export of U.S. Technology* (Washington, DC: US Department of Commerce National Technical Information Service, 1976) (hereafter Bucy Report); see also J. Fred Bucy, "On Strategic Technology Transfer to the Soviet Union," *International Security* 1, no. 4 (1977): 25–43. For the citation about the importance of the report, see Michael Mastanduno, *Economic Containment: CoCom and the Politics of East-West Trade* (Ithaca, NY: Cornell University Press, 1992), 187.

10. Bucy, "On Strategic Technology Transfer," 28.

11. Bucy, "On Strategic Technology Transfer," 28.

12. For one now standard work, see Niall Ferguson, Charles S. Maier, Erez Manela, and Daniel J. Sargent, eds., *The Shock of the Global: The 1970s in Perspective* (Cambridge, MA: Belknap Press of Harvard University Press, 2010).

13. Bucy Report, v.

14. Mario Daniels, "Japanese Industrial Espionage, Foreign Direct Investment and the Decline of the U.S. Industrial Base in the 1980s," *Bulletin of the German Historical Institute* 63 (2018): 45–66, provides a valuable overview. See also Daniels and Krige, *Knowledge Regulation and National Security.*

15. White House, *National Security Strategy of the United States of America* (Washington, DC: White House, December 2017), 17, attributed to President Donald Trump, November 2017.

16. For a comprehensive analysis see Hugo Meijer, *Trading with the Enemy: The Making of US Export Control Policy toward the People's Republic of China* (Oxford: Oxford University Press, 2016).

17. Statement of Gary Milhollin, executive director, Wisconsin Project for Nuclear Arms Control, in *U.S. Export Control and Nonproliferation Policy and the Role and Responsibility of the Department of Defense: Hearing Before the Committee on Armed Services, United States Senate, One Hundred Fifth Congress, Second Session, July 9, 1998* (Washington, DC: Government Printing Office, 1998), 29.

18. Michael J. Mazarr, *The Revolution in Military Affairs: A Framework for Defense Planning* (Carlisle, PA: Strategic Studies Institute, US Army War College, June 10, 1994), sec. VII.

19. Statement of Anita Jones, director, Defense Research and Engineering, Department of Defense, in *America's Dual-Use Technology Future: Are We Prepared? Hearing before the Subcommittee on Technology, Environment, and Aviation of the Congressional Committee on Science, Space and Technology, One Hundred Third Congress, Second Session, May 17, 1994* (Washington, DC: Government Printing Office, 1994), 41.

20. William Reinsch, under secretary for export administration, Department of Commerce, *Hearing Before the Committee on Commerce, Science and Transportation*, 105 Cong. 22 (September 17, 1998).

21. For the dramatic events summarized here see Robert D. Lamb, *Satellites, Security and Scandal: Understanding the Politics of Export Control* (College Park, MD: Center for International and Security Studies, 2005).

22. Jeff Gerth and Raymond Bonner, "Companies Are Investigated for Aid to China on Rockets," *New York Times*, April 4, 1998, A1, A3.

23. Select Committee on Intelligence, United States Senate, *Report on Impacts to U.S. National Security of Advanced Satellite Technology Exports to the People's Republic of China (PRC), and Report on the PRC's Efforts to Influence U.S. Policy* (Washington, DC: Government Printing Office, 1999) (hereafter Senate Intelligence Committee).

24. See note 1 above.

25. Quoted by Lamb, *Satellites, Security and Scandal*, 53.

26. Joseph Cirincione, "Cox Report and the Threat from China," presentation to the CATO Institute, June 7, 1999, https://carnegieendowment.org/1999/06/07/cox-report-and-threat-from-china-pub-131; M. M. May, ed., *The Cox Committee Report: An Assessment* (Stanford, CA: Center for International Security and Cooperation, December 1999); John M. Spratt Jr,, "Keep the Facts of the Cox Report in Perspective," *Arms Control Today* 29, no. 4 (1999): 24–25, 34; Nicholas Rostow, "The 'Panofsky' Critique and the Cox Committee Report: 50 Factual Errors in the Four Essays," unpublished manuscript, ca. 2000.

27. Senate Intelligence Committee, 6.

28. Henry Sokolski, "U.S. Satellites to China: Unseen Proliferation Concerns," *International Defense Review* 4 (1994): 23–26.

29. *United States Policy Regarding the Export of Satellites to China: Joint Hearings before the Committee on National Security Meeting with the Committee on International*

Relations, 105th Cong. 396 (June 1, 18, and 23, 1998) (written response by Henry Sokolski, executive director, Nonproliferation Policy Education Center).

30. They were distressed to learn that the Chinese authorities had denied official US observers access for five hours to the crash site after the Long March rocket veered off course and smashed into the hillside village. They also established that radiation-hardened encryption chips had disappeared from a command box found among the debris. This made them only more determined to improve the monitor program (see below).

31. Senate Intelligence Committee, 12.

32. Cox Committee Report, 2:85.

33. Cox Committee Report, 2:93.

34. Cox Committee Report, 2:85, 86.

35. Cox Committee Report, 2:206; emphasis added.

36. Cox Committee Report, 2:169–170.

37. Cox Committee Report, 2:164.

38. Cox Committee Report, 2:213.

39. Cox Committee Report, 2:212–213.

40. Statement of Henry Sokolski, executive director, Nonproliferation Policy Education Center, in *United States Policy Regarding the Export of Satellites to China: Joint Hearings before the Committee on National Security Meeting Jointly with Committee on International Relations, House of Representatives, One Hundred Fifth Congress, Second Session, June 17, 18 and 23, 1998* (Washington, DC: Government Printing Office, 1999), 10–14.

41. May, *The Cox Committee Report*, 18; emphasis added.

42. Senate Intelligence Committee, 11.

43. Senate intelligence Committee, 12.

44. Cox Committee Report, 2:98, 160.

45. See Lamb, *Satellites, Security and Scandal*, 51 for this paragraph.

46. Government of the PRC, *Facts Speak Louder than Words*.

47. Government of the PRC, *Facts Speak Louder than Words*, preamble.

48. Government of the PRC, *Facts Speak Louder than Words*, I.

49. Government of the PRC, *Facts Speak Louder than Words*, III.

50. Government of the PRC, *Facts Speak Louder than Words*, I.

51. Government of the PRC, *Facts Speak Louder than Words*, V.

52. I refer to the case of Wen Ho Lee, unjustly accused in 1999 of giving China the "crown jewels" of the US nuclear weapons program; see Dan Stober and Ian Hoffman, *A Convenient Spy: Wen Ho Lee and the Politics of Nuclear Espionage* (New York: Simon and Schuster, 2001). See also the assault by the FBI on many Chinese academic researchers as described in Daniels and Krige, *Knowledge Regulation and National Security*, epilogue.

53. Iris Chang, *Thread of the Silkworm* (New York: Basic Books, 1995), 208.

54. Chang, *Thread of the Silkworm*, 214–215.

55. John Krige, "Representing the Life of an Outstanding Chinese Aeronautical Engineer: A Transnational Perspective," *Technology's Stories* 6, no. 2 (2018), https://www.technologystories.org/chinese-engineer/.

56. Tai Ming Cheung, *Fortifying China: The Struggle to Build a Modern Defense Economy* (Ithaca, NY: Cornell University Press, 2009), 237. Robert Manning defines technonationalism as "a set of industrial policies aimed at self-sufficiency, cultivating 'national champions' in tech sectors while curbing foreign competition just as a new era of advanced technology is unfolding," in Evan A. Feigenbaum, "Soldiers, Weapons, and Chinese Development Strategy: The Mao Era Military in China's Economic and Institutional Debate," *China Quarterly* 185 (May 1999): 285–313.

57. Denis Fred Simon, "China's Drive to Close the Technological Gap: S&T Reform and the Imperatives to Catch Up," *China Quarterly* 119 (September 1989): 589–630, at 619. See also Julian Gewirtz, *The Remaking of China: Myth, Modernization, and the Tumult of the 1980s* (Cambridge, MA: Belknap Press of Harvard University Press, 2022).

58. Cited by Mary Brown Bullock, "The Effects of Tiananmen on China's International Scientific and Educational Cooperation," in *China's Economic Dilemmas in the 1990s: The Problems of Reforms, Modernization, and Interdependence*, Study Papers Submitted to the Joint Economic Committee Congress of the United States (Washington, DC: Government Printing Office, 1991), 2:611–628, at 612.

59. US Congress, Office of Technology Assessment, *Technology Transfer to China* (Washington, DC: Government Printing Office, 1987), 41.

60. Government of the PRC, *Facts Speak Louder than Words*, II.

61. "Minutes of Meeting Held in Beijing on 11 to 12 May 1993 between Hughes and CGWIC Regarding the Conclusion of the Optus B2 Failure Investigations," Cox Committee Report, 2:22; emphases added.

62. Cox Committee Report, 2:23.

63. Cox Committee Report, 2:33.

64. Cheung, *Fortifying China*, 239.

65. It could be argued that Hughes simply respected the need for their Chinese partners to "save face." This cultural characteristic is widely used in the West to "explain" Chinese behavior. Doubtless this deeply entrenched cultural value was at play here, but I want to move beyond cultural generalizations to identify the specific domestic concerns that animated the Chinese engineers and authorities in this case.

66. Meijer, *Trading with the Enemy*, 95.

67. Strom Thurmond National Defense Authorization Act for Fiscal Year 1999, Pub. L. No. 105-261, 112 Stat. 2175 (1998), Title XV, Subtitle B, National Security Controls on Satellite Export Licensing, sec. 1514(a)2(B), (i)–(iv).

68. Statement by James Bodner in *Oversight of Satellite Export Controls: Hearing before the Subcommittee on International Economic Policy, Export and Trade*

Promotion of the Committee on Foreign Relations, United States Senate, One Hundredth Sixth Congress, Second Session, June 7, 2000 (Washington, DC: Government Printing Office, 2000), 30–31.

69. James Andrew Lewis, *Preserving America's Strength in Satellite Technology: A Report of the CSIS Satellite Commission* (Washington, DC: Center for Strategic International Studies, 2003), 22.

70. Bodner, statement, 31.

71. Thane Gustafon, *Selling the Russians the Rope? Soviet Technology Policy and U.S. Export Controls* (Santa Monica, CA: RAND Report R-2649-ARPA, April 1981), vii.

72. Senate Intelligence Committee, 6.

Part II

Facilitating Transnational Knowledge Flows

Chapter Six

Beyond Borlaug's Shadow

Mexican Seeds and the Narratives of the Green Revolution

GABRIELA SOTO LAVEAGA

> Basically, today's varieties all have blood from Mexican wheat in them, in one way or another . . .[1]
>
> R. S. PARODA, former director general of the Indian Council of Agricultural Research

In late summer 1965, Raúl Valdés, a diplomat with the Mexican embassy in New Delhi, stood in a farmer's field, far removed from his typical social and political duties. Standing before Indian farmers, he encouraged them to look carefully at what he held in his cupped hands. In one hand he held local wheat seeds, and in the other Mexican high-yielding wheat seeds. With the use of a translator, he tried unsuccessfully to explain why Mexican seed varieties were better than local ones, which produced tall, handsome stalks of wheat that nonetheless were prone to lodging (falling over). In an interview decades later, Valdés confessed that introducing new seed varieties "was not easy."[2] Tasked by the ambassador to supervise a team of Mexican embassy employees who, like him, had never before set foot in farm fields, he nonetheless believed they were an essential but unacknowledged cog in the initial transfer of agricultural technology from Mexico to South Asia. He also vividly recalled how the growing famine injected urgency to the task of convincing Indian farmers to switch to high-yielding seed varieties (HYV).

Yet in dozens of publications about disseminating high-yielding seed knowledge among Indian farmers, Dr. M.S. Swaminathan, scientist and the architect of the Green Revolution's infrastructure in India, never mentioned the role of Mexicans, either agronomists or diplomats, in educating the nation's farmers. In fact, when recently interviewed, the then ninety-four-year-old scientist seemed surprised and adamantly insisted that Mexico had never helped "beyond providing the first seeds." It had all been, he explained, Indian or American know-how.[3]

Nevertheless, Mexican diplomatic memorandums from the 1960s reveal the presence of Mexican agronomists on the ground in India, evidence that complicates what we understand as mid-twentieth-century technical assistance, which nations get credit for transferring knowledge, and how we tell histories of the Green Revolution. In the historical narrative these seeds, universally known as "Mexican seeds," perform a seemingly impossible act: they explicitly reference Mexico yet at the same time shed themselves of any affiliation with Mexican expertise and domestic science. Ironically, when Mexican seeds arrived in India, they were divested of association with national agrarian reform programs and instead became a stand-in for capitalist development and the transformative American promise of technology. Indeed, the story of the arrival of high-yielding Mexican seeds to India has become a tale not of a nation's agrotechnological prowess but rather of a singular man, Norman Borlaug, and his outsized role in mid-twentieth-century agricultural practices.[4]

In fact, these seeds were also referred to as "Borlaug's seeds" by Indian and other researchers. Indeed, the charisma and indisputable work ethic of Borlaug, a beloved mentor and scientist, tower over narratives of the Green Revolution, making it at times difficult to retrieve other stories from his benevolent shadow. Yet we now understand that "magic" seeds alone could never have succeeded in foreign soils. More than germplasm, it was the inputs—i.e., government subsidies and investment in fertilizer, pesticides, irrigation, and, most important, seed research—that assured successful harvests. Less attention has been paid to the crafting of initial (and persuasive) narratives of scientific success that omitted or glossed over doubts about planting seeds tailored for Mexican soils in foreign lands.[5] In her study on the circulation of seeds, historian Courtney Fullilove reminds us, "cultivated seeds are not products of nature but deep-time technologies . . . and [they] simultaneously contain and obscure the complex social relations required for their production."[6]

When speaking about transnational flows of knowledge we often think of ruptures or obstacles to that flow. Yet how do we examine histories of scientific agriculture when the knowledge celebrated is *itself the impediment* to lasting progress? Put differently, knowledge of, and easy access to, wheat seeds coupled with an unprecedented investment in high-yield seed research became the impediment to develop strong, locally driven, eco-friendlier solutions to farming in India. And, in the process of acclaiming the role of Norman Borlaug and the Rockefeller Foundation (RF), the vital role of Mexico and the participation of other nations was obscured. Faulty, incomplete narratives such as these impede a broader understanding of knowledge that does not pass professed traditional centers of knowledge production.

Much has been written about the many socioeconomic and ecological failures of the so-called Green Revolution, from the subsequent degradation of local ecologies and pollution of groundwater to its role in the declining standards of living of small-plot farmers across the globe. Many of these studies focus on the successful spread of high-yielding hybrid seeds in South Asia and the role of American technocrats in implementing capital-driven agricultural practices in much of the world. The persistent focus on American development aid in the guise of agricultural technology, however, yields incomplete histories that impede a broader understanding of historical participants who do not fit tidily into an international development narrative. In this chapter I argue that expanding the story of the participation of knowledge transfer to include relegated figures, such as 1960s diplomats and Mexican agronomists in South Asia, reveals a more nuanced understanding of "development" aid and questions the power and problematic nature of early hunger narratives that came to define *how* we tell histories of the Green Revolution today.

Explicitly, this chapter focuses on two seemingly disparate sources: first, 1960s Mexican diplomatic memos about wheat seeds in India written by one of the twentieth century's leading essayists, the ambassador to India when the seeds arrived and future Nobel laureate Octavio Paz; and second, a 1990 transcript of Indian scientists discussing their firsthand role in the introduction of new seeds in the 1960s. In these sources there are echoes of or direct mention of the RF, the Ford Foundation, and Norman Borlaug, but none of them drives the narrative of technological change. To put it differently, these are not stories of development aid, they are histories of national innovation. This chapter is divided into three sections: the first briefly situates the emergence of HYV wheat seeds within a Sonoran context; the second and third sections detail the arrival in India of HYV seeds from two points of view, as reported in Mexican diplomatic reports and in the memories of Indian scientists. To get to those histories we must first give a quick overview of why Mexican seeds were in Indian soil in the first place.

The Yaqui Valley

To expand the historical narrative of the Green Revolution we have to start from the premise that the place where high-yielding seeds were developed in Mexico, the Yaqui Valley, was not an accidental location. Southern Sonora was home to some of the nation's wealthiest and politically best-connected farmers, including two former Mexican presidents,

Álvaro Obregón (1920–1924) and Plutarco Elías Calles (1924–1928), whose vast agrobusinesses were binational and bilingual in nature.[7] In fact, it was not uncommon for children of wealthy northern Mexican farm families to attend college in the Unites States and elsewhere, such as Mexican president (1911–1913) and prosperous landowner from the neighboring state of Coahuila, Francisco I. Madero, who briefly studied agricultural technology at the University of California, Berkeley and, like his brothers, business administration in Paris.

In addition to the native Yaqui who still inhabited it, though their fertile lands had been taken from them decades earlier, the Yaqui Valley was also populated by foreign small-plot and commercial farmers and entrepreneurs, including many Americans, whose land would be expropriated during the reforms of the late 1930s.[8] The proximity to the United States—both territorially and in business—meant that the Yaqui Valley was also a busy corridor for new technologies and ideas from its northern neighbor.[9] By most accounts, though geographically removed from Mexico City, this progressive farming region was well connected not only to the rest of the nation but also to the states across the border and beyond.

The reality is that when agronomist Norman Borlaug arrived in the Yaqui Valley in the early 1940s, he was not arriving in a technological backwater. The farmers he befriended were well organized, well funded, and business savvy, and, most important, they relied on and trusted agricultural experimentation.[10] A decade earlier, for example, state governor Rodolfo Elías Calles, son of the former president of Mexico, founded an agricultural bank (Banco Agrícola de Sonora) and an agricultural experimental station, basing both in Ciudad Obregón in the Yaqui Valley. More importantly, he selected Edmundo Taboada Ramírez, one of Mexico's leading plant geneticists, to run the station. Since the early 1900s, farmers had organized in associations for better profits but also, as in the case of president Álvaro Obregón, to produce better crops. While the majority of the valley's farmers were neither affluent nor influential, Borlaug swiftly grasped the importance and political power of Sonora's agrarian elite. Many of these families would come to consider Borlaug a close personal friend.[11]

The origins of the Green Revolution are thus rooted in and shaped by the farming practices of Mexico's wealthy northern farmers. Yet most histories of the Green Revolution begin not in Ciudad Obregón, where the main experimental station for HYV wheat seeds and Borlaug's home base were located. Instead, these narratives usually originate with the road trip of Henry Wallace to tour the poverty of central and southern Mexican farmers, who neither contributed nor benefited from seed experimentation that would take place in northern Mexico.[12] The need to develop a backward, agriculturally impoverished region by importing science makes for a better beginning to a story of global germplasm exchange and experimentation. Yet perspective matters for a more accurate story.

The (Summarized) Official Narrative

In the early1940s an internal assessment of the global public health programs of the RF, in place since 1913, revealed that despite some successful vertical health campaigns (i.e., vaccination, education, interventions at the government level, grants, etc.) the overall health of the world's poorest citizens who had come into contact with RF services had not improved dramatically. The report determined that it was not a question of health but, rather, an equally fundamental issue: hunger and its associate, malnutrition.[13] The report's findings led to a shift in the RF's focus from health to agriculture, specifically research to produce better crops that could, in turn, yield more food. Influencing this shift in research was World War II and the realization that the United States needed to strengthen its food trade with countries of the Western hemisphere. Conveniently, Mexico had undergone extensive land expropriations in favor of small farmers in the late 1930s, yet the country had shifted politically under new president's Manuel Ávila Camacho counterreform, which sought to focus on industrializing and mechanizing larger commercial farmers. When Vice President–elect Henry A. Wallace toured rural Mexico in 1940, the country was at a domestic crossroads: would its agricultural output be based on government investment in small, subsistence farmers or in commercially driven farms? Wallace returned home to present a general assessment of the economic and technological potential of Mexican countryside to board members of the RF and shortly thereafter American scientists set out to also tour Mexico's agricultural lands. Out of these conversations emerged a partnership with Mexico's powerful Ministry of Agriculture and Livestock (SAG) and the RF, which jointly created the Office of Special Studies for agricultural research in Mexico.

When the RF arrived in 1943, Mexico's National School of Agriculture (Escuela Nacional de Agricultura) had worked on developing hybrid crops, in particular corn, the staple crop of Mexico, but plant breeding often lacked funding and personnel, and though experimental stations existed throughout the nation, they had struggled to remain fully equipped and staffed.[14] An insufficient supply of trained agronomists was also a problem. For example, in the 1920s, the National School of Agriculture had a new plant-breeding degree but was slow in graduating many students.[15] Then head of the General Office of Agriculture (Dirección General de Agricultura), Edmundo Taboada Ramírez, underscored that "in that era experimental work had not quite reached scientific levels."[16]

Taboada himself, as already mentioned, would be key in the institutionalization of agricultural science in Mexico. He had studied at Cornell University and in 1932 was named agricultural attaché of the Mexican embassy in Washington, and in that capacity toured both the United States and Canada. It was in Canada that he allowed himself to envision the "impossible dream" of creating a chain of experimental stations, such as the ones he had seen in the nations to the north of Mexico.[17] Taboada had a chance to implement what he had learned at Cornell and at the Central Experimental Station in Parm, Ottawa, when in 1934 he became director of the experimental station in the Yaqui Valley. In addition, from 1936 to 1939 he functioned as the general director of agricultural experimental stations for the National Agricultural Credit Bank (Banco Nacional de Crédito Agrícola) and designed core classes in plant genetics and agricultural research and development for the National School of Agriculture. Norman Borlaug's famed shuttle breeding—a methodology which relied on planting seeds in locations that differed in latitude, altitude, and rainfall to produce photoperiod-insensitive and "widely adapted germplasm"[18]—depended on this web of existing Mexican experimental stations.

Foreigners, especially those with plans to make crops yield more in this arid land, were not new to Sonorans. Historical evidence contradicts the oft-repeated account that Yaqui Valley farmers "had little faith in agricultural science and scientists."[19] In fact, southern Sonora, in particular the Yaqui Valley, was planned to attract nineteenth-century foreign investment. Hence the government incentivized agricultural projects, especially from those willing to invest in irrigation canals and crop research. The Yaqui Valley's first crop experimental station was built in 1909 by American-owned irrigation and development company Compania Richardson, yet the station was one of several found scattered throughout Mexico's microclimates.[20] Moreover, it was not unusual for *haciendas*, Mexico's vast rural estates, to experiment with their commercial crops, though most did not have on-site experimental stations, as advocated in 1919 by a California resident and Mexican crop enthusiast who wrote to a Sonoran farmer about planting eucalyptus groves in the state.[21]

Cotton and rice were the mainstay crops of the region, though graphs of total wheat production in the country as early as 1934 reveal that wheat yields picked up as corn production declined in Mexico. As disclosed in *Agricultura*, a magazine published by the Ministry of Agriculture and Development, in the early 1930s Sonora did not make it into the top ten wheat-producing states in the country.[22] The emphasis a decade later by scientists of the Office of Special Studies on wheat-focused research paired nicely, however, with changing trends in Mexican crop production and food consumption. By 1956, Mexico was wheat self-sufficient, due in large part to the strains created by the researchers at the Office of Special Studies. By 1964, 94 percent of "wheat growing land in Mexico" was planted with improved varieties developed in the Yaqui Valley.[23] When these dwarf wheat seeds were developed in Mexico in the 1940s and 1950s there were 118 agricultural scientists working with the Rockefeller Foundation–Mexico Office of Special Studies, and one hundred of them were Mexican.[24]

As wheat research and production boomed in 1960s Mexico, monsoon rains failed in India.

Mexican Wheat Seeds Arrive in India

In 1962 the Indian Agricultural Research Institute (IARI) reached out, via the Ministry of Food and Agriculture, to the RF, seeking some dwarf seed varieties. In March 1963, RF agronomist Norman Borlaug visited the leading wheat breeding centers in India. In November of that same year, the first shipment of Mexican seeds arrived for trials in five Indian experimental fields: Delhi, Ludhiana, Pusa, Kanpur, and Pantnagar. In short, the sample varieties "showed that the dwarf material, like Lerma Rojo, had distinct advantages over the local material."[25]

In particular, two Mexican semidwarfs (Sonora 64 and Lerma Rojo 64) "outyielded all Indian cultivars by at least 30%."[26] These new semidwarf wheats served as the foundation for India's new national wheat production program. The program, run by the IARI, consisted of the following steps: test new wheats at both research stations and under agronomy trials in various locations; ensure multiplication of the improved seed; hold farmer demonstrations in India's wheat-producing regions; and continue "importing massive amounts of wheat seed from Mexico."[27]

Farmer demonstrations were crucial because Mexican seeds in Indian soils required a wholly new approach to planting. Seeds needed a new planting depth (no more than 5 cm, as opposed to the usual 10–15 cm), in addition to up to a month delay in planting (from regular late October to even mid-December) and irrigation needed to start sooner (at twenty-one days instead of the traditional thirty to thirty-five days).[28] Crucially, farmers were encouraged "to apply 120 to 140kg/ha of nitrogen, instead of the usual 40kg/ha."

This new style of farming required, as in Mexico, the foundation of a series of agencies and a new approach to plant research in the country. Agencies were needed to multiply the seed, fertilizer companies to provide double—in some instances triple—the previously needed fertilizer, and, in addition, new networks were needed with recently created agricultural universities. In other words, hybrid dwarf seeds could not *and did not* travel alone. They could not divest themselves of the complicated networks of expert knowledge exchange that existed in Mexico. Those networks needed to be replicated for the seeds to succeed on foreign ground. Hence agricultural universities (such as Punjab Agricultural University in Ludhiana and Uttar Pradesh Agricultural University in Pantnagar) began to multiply the new wheat seeds. Since production in research centers could not meet demand, India continued to import seed from Mexico. To address this shortage, by 1964 the Indian government created the National Seed Corporation, and the following year large-scale demonstrations were taking place throughout Indian fields.

Crucial to the spread of Mexican varieties was to cement the idea that locality was not essential for generating strong genetic traits in new lines. Indian scientists, such as Dilbagh Singh Athwal, often called "the father of Wheat Revolution," quickly modified Mexican seeds to create, for example, in 1966 PV 18, which used Mexican and Indian germplasm to produce varieties that better thrived in foreign soils. Using the first field tests, Dr. Athwal, then head of the Department of Plant Breeding of Punjab Agricultural University, continued selecting as reported, "from among the thousands of Mexican breeding lines and came up with another dwarf that averaged 30 percent higher yields than the first Mexican import."[29] For the 1966 harvest, the eighty Punjab farmers who tested PV 18 in their fields "produced 2.5 to 3 tons an acre," compared to 1,100 pound average in the Punjab.[30] Athwal later developed Kalyansona (also referred to as Kalyan 227), India's most popular HYV wheat seed, and named after the village, Kalyanpur, where Athwal was born in 1928. In both Punjabi and Hindi the name means "salvation." In the summer of 1966, as famine spread in India, Mexico sent 200 tons of Sonoran farmers' seed to India or, as a newspaper of the time described HYV seeds, "five thousand acres' worth of 'salvation' goes out to the Punjab farmers. . . . At the same time, 100,000 acres' worth of PV 18" was also distributed in the Punjab.[31] An increase in black market prices for wheat or "Athwal seed grains" in the Punjab revealed Indian farmers effusive adoption of the new seeds.[32]

Sacks of HYV wheat seed in transit from Mexico to India made for captivating photographs. Images of seed sacks on trucks, in warehouses, in the air being hoisted onto a waiting freight ship, in the ship's hold, and so on, are found by the dozen in the Rockefeller Archive Center's holdings. Hybrid dwarf wheat, stout, close to the ground, and resistant to rust, so different from its lithe and fragile antecessor, also made for compelling visuals. These striking pictures made demonstrations imperative. Farmers needed to touch the seed, walk through furrows of swaying wheat, and, most important, see the scientists, strikingly different in appearance from the farmers, standing authoritatively next to experimental seed and stalks.

As with most narratives of crops, farmer field days became essential to make the connection between seed and farmer.

In one of the first communications with Indian scientists, Borlaug set the measure of success when noting a less-than-stellar outcome in the first trials. "I feel, therefore," he stated, "that this variety suffered severely from lack of adequate irrigation."[33] This letter from the leading authority on high-yielding dwarf varieties would have a tremendous impact on how seeds were planted in India. It *was not* (faulty) seeds or the (faulty) science of wide adaptability but rather the receiving nation's infrastructure (i.e., irrigation), inability to embrace technology (i.e., traditional farming ways), improper agricultural practice or translation of standard norms (i.e., inadequate use of fertilizer). The narrative that the science of HYV seeds was infallible and would fail only if "reluctant," "traditional," or "stubborn" locals did not follow procedures remained uncontested and became part of the foundational narrative. In this account, scientifically driven yield outputs would be the only measure of success.

In his 1964 response to Borlaug's assessment, Swaminathan repeats that the "yields of Lerma Rojo 64, Sonora 64, Sonora 63, Variety V 17 . . . should be at least 6 tons per hectare." And he reinforced this as the gold standard by adding, "Until this is attained it should be recognised that there are some other factors which are limiting their yield potential and studies should be undertaken. . . . Unless we realise the full potentialities of a variety in yield trials it will not be possible to know which factor is limiting."[34] Curiously, a mere paragraph after this assessment, Swaminathan reveals the following: "One striking factor which has come to light from the maximum yield potential trial carried out at Delhi is that the varieties which have done well at 120 lbs of nitrogen have also been the top yielders at the lower doses of fertilizer application." While this did not happen with varieties that were designed for lower doses of fertilizer, it reveals that in Indian soils the correlation between seed and fertilizer did not always correspond to the predicted yields.

Some seed varieties (such as V 17) could be introduced directly while others, the majority, like the Sonora variety, which became the favored seed of Mexican scientists and Indian scientists, needed to be tweaked genetically or coaxed with larger amounts of fertilizer to produce as effectively as it had in its place of origin, Sonora.

The Indian Memos of Octavio Paz

Octavio Paz arrived in India in 1962, having previously been stationed in New York, Paris, and Tokyo. In his collection of musings and memories of his time in the country, *In Light of India*, he reveals what many Mexicans before and after him also discovered: an unexpected familiarity and affinity with Indian society, history, and culture. Despite different colonial masters, the two countries shared, in addition to a rich and complicated preimperial history, a fervent belief in the power of science and technology to socially transform them once they were newly independent nations. They also shared religious and cultural attitudes that included, according to Paz, food. Entire passages of his book are devoted to the resemblance between, say, *mole* and curry. While his personal writings tend to gravitate toward a celebration of cultural closeness, his embassy memos focus on the dramatic political time in India punctuated by increasing social upheaval due to growing hunger. Paz was already stationed in Delhi when Mexican dwarf seeds first made their appearance in the country. His memos, not surprisingly, lend a different nuance to the arrival of seeds from Mexico.

Paz's keen attention to detail and a needle-fine analysis of sociopolitical and cultural practices later earned him the 1990 Nobel Prize in Literature. His near-daily diplomatic reports are, not surprisingly, a remarkable sampling of how to reflect in writing on a society in transition. A single-spaced, three-page report on the splintering of India's communist party, for example, is a mini–master class on "the politics which communists must follow in developing countries," and it illustrates Paz's erudite nature and his strong anchor in the region's history.[35] Paz's memos on Mexican seeds in India stretch from the early 1960s to late 1968, when he resigned his diplomatic post in protest at the government-orchestrated killing of university students during a Mexico City demonstration. His resignation as ambassador effectively ended his diplomatic career.

From the outset, the Mexican embassy in India had an outsized mandate. Established in 1951, the embassy represented for Mexico more than diplomatic and economic ties with South Asia. The Mexican nation felt great affinity with India (and expressed it often) and believed that Mexico together with India could forge an alliance that would foster different models of development around the world. This was not so farfetched. In 1955, for example, Asian and African nations gathered in Bandung to discuss economic development, decolonization, and how these nations could, together, influence global politics. Moreover, Mexico's domestic problems with democracy seemed, to many, tame compared to other Latin American nations falling to military coups or authoritarian rule at the time. In other words, Mexico seemed poised to take on a global leadership role beyond Latin America.

Before the arrival of *the* Mexican seeds, Mexico and India had a vibrant history of plant germplasm exchange. For example, in the summer of 1964 the IARI requested, via the embassy, that the Mexican government send "sample packets of seeds, accompanied by literature giving information about the material and the necessary phyto-sanitary certificates."[36] In a reply, the National Institute of Agrarian Research (INIA) responded that it had "systematically been sending plant genetic material" to Indian institutions for domestic research and experimentation. A mere two weeks later a different application from a research institution in Dehradun requested samples of twenty-four different Mexican seed varieties including 100 grams of *Salix nigra* (black willow), 150 grams of *Pseudotsuga douglasii* (Douglas fir), 220 grams of *Arapcaria gigantea* (likely Araucaria, an evergreen coniferous tree), 800 grams of *Pinus canariensis* (Canary Island pine), and so on. Three weeks later, on October 9, 1964, the IARI in New Delhi requested six different varieties of *Medicago sativa* (alfalfa). In mid-November, when the Mexican alfalfas were ready to be shipped, an agronomist from Mexico's Ministry of Agriculture and Livestock wrote an explicit description, declaring that "in the ecological conditions of the temperate zone of central Mexico these alfalfas yield approximately 30 tons of dry material per hectare. They recover quickly after being cut and they can be cut 9 to 11 times per year."[37]

Indian institutions were not simply requesting crop plants. Queries also came for information about Mexican medicinal plants. As an October 1963 letter illustrates, the IARI requested samples and "all available information in Mexico" about producing oil from *Bursera delpechiana*, a Linaloe species that does not grow in Mexico. Undaunted, an employee from the Ministry of Agriculture and Livestock did some research and sent (one hopes in translated form) copied pages from *Useful Plants of Mexico's Flora*. This all briefly illustrates that experimental seed exchange between Mexico and India was neither new nor unusual.

Nor was the relationship between the two countries limited to matters of national flora. Mexico also served as a conduit for India into other Latin American countries, in particular Cuba. In early 1965, for example, India tried to send Cuba a copy of the film *Do Bigha Zamin* (1953), about the plight of a poor farmer who is forced to become a rickshaw driver in Calcutta to pay off his debts.[38] The film was initially sent via the United States but, because of its destination, was returned to India. Mexico, one of the few countries never to break off diplomatic relations with Cuba, often served as an intermediary between the island nation and the rest of the world. Moreover, Mexico felt, as expressed often in diplomatic correspondence, a profound affinity toward India and as such had been the first Latin American country to acknowledge India's independence and establish diplomatic relations. Finally, Mexico had hosted several goodwill museum exhibits, talks on India, and had gifted the newly independent nation a garbage truck and school houses. In short, there was a healthy exchange between the two countries.

As part of his embassy duties, Paz produced a near-daily, thorough analysis of India's political situation. But the looming food crisis began to crop up in most of his reports. For example, he wrote a multipage report concerning protests over the opening of an additional steel plant in northern India (as opposed to the underdeveloped south) and mused about demonstrations over the ban on eating beef due to not simply a question of politics or religion but, rather, because of growing hunger. The country, he concluded, was in an "agitated state." Moreover, he detailed how in Bihar and Uttar Pradesh the lack of grains had reached alarming proportions and, physically underlining his concern in the text, added that "despite the government's denial the press reports that farmers are suffering from hunger (not malnutrition but, I repeat, *hunger*)."[39]

On April 19, 1965, Octavio Paz sent a significantly different memo to the secretary of foreign relations. His superiors had inquired about the specific role that Mexico was playing in the development of wheat varieties in India. After some weeks, Paz responded. He explained that he could now confidently deliver more thorough information because Dr. Ignacio Narvaez, an agronomist with Mexico's SAG, had visited the embassy in Delhi and provided key details. Paz reported that the efforts in India were coordinated by the IARI in collaboration with the RF and the SAG. Narvaez was a young scientist working in the Cereals Department of the INIA, under the umbrella of the SAG. He had studied at the Antonio Narro Autonomous Agrarian University and was already known as a talented wheat breeder when he met and duly impressed Norman Borlaug. Shortly after their meeting, Borlaug invited Narvaez to join the team of researchers at the Office of Special Studies. Two years later, Borlaug hand-selected Narvaez to be the resident wheat adviser in Pakistan.[40] Narvaez would go on to obtain his PhD in agriculture at Purdue University in 1969. He was part of a wave of Latin American graduate students who attained doctoral degrees in agriculture from American institutions with funding from the RF.[41] In the 1974 Hearings before the Select Committee on Nutrition and Human Needs, Dr. Narvaez was described as a "native Mexican and one of Borlaug's earliest and most talented wheat apostles."[42] In Pakistan, Borlaug provided overall program guidance, but it was Narvaez, with his Pakistani counterparts, who was at the helm.[43]

The history of Mexico's incursion into India, as retold to Paz by Narvaez began in 1963 with the founding of the International Maize and Wheat Improvement Center (Centro Internacional de Mejoramiento de Maiz y Trigo, CIMMYT), created jointly by "the SAG and the Rockefeller Foundation." The work at CIMMYT, the letter continued, produced "a great number" of varieties of wheat with such abilities and "flexible adaptability to distinct environments" that these varieties were now part of more than thirty wheat improvement programs around the world.[44] Paz explained further to his superiors that Mexican seeds were being used differently in these countries: as progenitor or as a commercial variety. In India, Mexican seeds, "especially Sonora 63 and 64," were mainly progenitors to create Indian wheat varieties, though they were also used commercially in the country.

The report continued with a crucial detail.

> Mexico's participation consists of, first, the training of young scientists, mainly from the middle and far East, at the [wheat] center in Mexico (in which there are currently 31 foreign fellows) and, second, the Mexican government has granted Dr. Narváez permission to participate directly in these programs, not only in India and Pakistan, but in other countries of Asia and Africa.

Paz was referring first to the hundreds of young scientists from several dozen countries who traveled to Mexico to train with Borlaug, learn his methods, and return to their home countries where they, in turn, would train future generations of scientists. It was these so-called wheat apostles who ensured the successful spread of high-yielding hybrid wheat seeds across the world. Between 1943 and 1963, 550 interns participated in agricultural research and training at the Office of Special Studies in Mexico.[45] Of those 550, 200 received Master of Science degrees and approximately thirty received doctoral degrees. Since they trained in Mexican installations in Mexican fields with Mexican scientists led by Borlaug, Paz described it as a "Mexican" training program. Yet, Paz warned in the report that both he and Narvaez, the *experto Mexicano*, felt that "Mexico is not receiving sufficient credit in the Project."[46]

Paz's diplomatic memo made clear his concern, as he detailed: Mexico's aid had quickly become a story of an American (Borlaug) and American aid (in the guise of RF funds). At that time, Paz, or later a superior, underlined a phrase in pen to emphasize Narvaez's belief (now firmly expressed by Paz) that "*our country is not getting its due for its work*." Though Mexico's role in solving the global food crisis may be, he acknowledged, deemed "modest," it, however, "does not cease to be vital." He urged his superiors to find a way to better channel their various efforts so as to "make more noticeable Mexico's participation."[47]

Paz was gently alluding to a perennial problem of Mexican institutions: replicating efforts to accomplish a single goal. The problem was not limited to agriculture. This was the case, for example, in health care, where five different government-run health systems tended to the Mexican population, with their own separate rosters (and differing pay scales) for physicians and nurses, member-only hospitals, and even lists of separately covered pharmaceuticals—with little or no coordination between them. For that reason, Paz specifically referenced the SAG and requested that the two different ministries on the ground in India (Mexican Foreign Relations and the SAG) coordinate efforts between embassy employees and agricultural technicians. In fact, he suggested, wouldn't it be worth creating a new diplomatic title for Dr. Narvaez, who after a short trip to Mexico would return to the region for a two-year assignment: agricultural attaché?

While there is no archival record of the response to Paz's plea, only an acknowledgment that it was received, it is telling that the following seven embassy memos in the same archive folder are Indian requests for Mexican seeds. In that December 1965 memo, Ambassador Paz detailed the crisis—"a shortage of food, particularly grains"—in especially somber terms. He explained that there were three reasons for the food crisis: first, the previous year's drought; second, the suspension of plans to send more grains from the United States; and third, the inability to plant and harvest caused by the growing military conflict between India and Pakistan.[48]

A few years later the *Kabul Times* reported that after being "accustomed to news from India of famine and distress," astoundingly "India Expects Record Grain Harvest." The article continues that the bumper harvest "would not have been possible but for pioneering research work undertaken in Mexico and elsewhere," and while not elaborating on the "elsewhere," it detailed the various Mexican seed varieties and the type of research conducted in the country. Moreover, it highlighted that "when grain experts Dr. Ignacio Narvaez from Mexico and Dr. Norman E. Borlaug of the Rockefeller Foundation first went to Pakistan" they were met with "polite incredulity," but the "Mexican wheat" had proven itself.[49] With Ford Foundation aid, the All-Pakistan Wheat Research and Production Program, a project of "accelerated wheat improvement," was launched in Pakistan.[50]

Recalling the many "heroes in Pakistan's wheat revolution," a 1989 publication listed the millions of unknown farmers who "discarded old traditional methods and enthusiastically took up the new technology," as well as the minister and secretary of agriculture, before centering on the vital role of the wheat revolution's "field commanders." All but one were Pakistani: "S. A. Quereshi from the Punjab, M. A. Munshi from the Sind, M. Suleman Khan from the Northwest Frontier, and Ignacio Narvaez from Mexico."[51] The initial years of semidwarf wheat and high-yield varieties were invariably described as an "activist period," in which it fell to the "resident advisor, Dr. Ignacio Narvaez, [to be] involved in many policy matters of seed and input delivery."[52]

In hindsight, Paz's memos about the erasure of Mexico's on-the-ground contributions to end the food crisis in South Asia seem prescient. By 1966 the narrative we would all come to know about "Borlaug" seeds in India was already solidifying itself.[53]

Indian Historical Context for an "Adequate Measure"

On February 15, 1966, *The Statesman* published Indian president Sarvepalli Radhakrishnan's address to Parliament. As expected, the address focused heavily on India's "foundation of peace" and the spirit of the Tashkent Declaration, a peace agreement between India and Pakistan signed on January 10 that brought an end to the Indo-Pakistani war of 1965. Food production was, however, India's main concern that year. Addressing the "failure of monsoons" and the continued scarcity of food grains in the country, Radhakrishnan acknowledged that grain production would be down to 70–77 million tons compared to the previous year's 88 million tons. Major cities in the states of Maharashtra, Gujarat, Mysore, Rajasthan, and Andhra Pradesh were already facing "serious scarcity."

The only solution to this grave concern was, according to Radhakrishnan, science. "The present difficulties only re-emphasize the need to concert and implement measures to increase the production of foodgrains in the shortest possible time. Only by the application of modern science and technology can agricultural production increase in an adequate measure."[54]

Yield became, as it had in Mexico, the measure of success. Any concerns over the social and economic impact on poor farmers were pushed aside in order to produce more, faster. In his study of wheat in fascist Italy, historian Tiago Saraiva reminds us of the power of "mythical numbers" when it comes to wheat. In fact, Saraiva also signals how high-yield seeds became the cornerstone for transforming Italian fields under Mussolini.[55] For both Mexico and India, wheat was also the vehicle that would transform first a region and then, especially for India, the nation. The success of this transformation would initially be tethered to the adoption of hybrid seeds and production of high yields and would not focus on the well-being of farmers themselves. In other words, the initial measure of HYV seed technology would be how the seed performed in foreign ground, a scientific, quantifiable measure devoid, as it had been in Mexico, of its social context.[56]

The greatest hope for India, the Indian leader explained, was "the use of improved varieties of seeds which are particularly responsive to the application of fertilizer." This entailed planting new varieties in millions of acres (32 million by the end of the Fourth Plan), creating new fertilizer plants, and propelling irrigation projects throughout the country.[57] The three-volume hagiography *Borlaug* summed it up in the following dramatic fashion: "With Nature suppressing the food supply, with America talking of curtailing food shipments, and with famine of unprecedented enormity looming, Mexican wheats have become South Asia's last best hope."[58]

On February 16, 1966, India's agriculture minister, Chidambaram Subramaniam, invited thirty-four ambassadors and eight members of international organizations with offices in Delhi (among them the United Nations Food and Agriculture Organization [FAO], the World Health Organization [WHO], and the United Nations Children's Fund [UNICEF]) to discuss the growing hunger crisis. At the time the Indian government deemed the situation "serious" but nonetheless it had confidence in plans to funnel food to urban spaces and regions where grains were most apparently absent. The nation's major concern was the worsening shortage of rice, especially since all neighboring nations, also hit by drought, were in a similar situation. In the meeting the minister emphasized that India needed external help to overcome the crisis and was consequently relying on friendly nations and any aid they could provide.[59]

Octavio Paz took note in his memos of the virtue of Mexican seeds and their role in curtailing India's food crisis. He wrote, "It is worth mentioning that using these seeds as progenitors allowed a resolution in a relatively short time period of the indian [*sic*] problem with this grain. These varieties' yields are two or three times more than the Indian varieties. . . . In a visit that the undersigned made to the IARI I was able to witness the enthusiasm with which they are carrying out the reproduction of Mexican varieties and the technicians' optimism that [these seeds] will solve the problem of wheat production in their country."[60] Meanwhile, Indian farmers embraced Mexican seeds.

Jounti Seed Village, or Mass-Producing a *Chamatkar* (Miracle)

How did the seeds get to farmer's fields? M. S. Swaminathan explains the vital network of seed production in India in the following fashion:

> There are three important links in the seed production chain. First, the "nucleus of breeders' seed" has to be produced by the research institution where the breeder who developed the variety works. Secondly, the nucleus seeds should be sown, preferably in state or state-controlled seed farms for the production or "foundation seed." These foundation seeds should be distributed to registered growers, who are progressive farmers, for the production of certified seed. The certified seed is distributed among farmers for raising crops for consumption. Competent technical guidance and supervision are essential at all of these states in order to ensure that the cultivator gets seeds of high quality and purity and free from infection.[61]

Given the above parameters, seed multiplication was an issue. Reminiscing about this time period, Swaminathan explained that the lack of "adequate technical manpower" severely hampered seed multiplication in the first three plan periods.[62] With the stakes set so high—to avert famine—the pressing question became: How does one produce adequate quantities of seed of new varieties while ensuring high quality and keeping them free from infection? In India the solution was to transform an entire village into a "seed village" that would produce the new varieties, which would then be released across India. Scientists from the Division of Genetics of the IARI, led by Swaminathan, set off in search of a "seed-producing unit."[63] In November of 1964 they settled on the village of Jounti in Delhi state.[64] They selected Jounti because farmers in all other villages asked about payment for the seeds that they would produce and Jounti was the only village that agreed "to take to seed production on scientific lines and subject their plots to field inspection."[65]

In the winter (*rabi*) season of 1964–1965, Sonora 64 and Lerma Rojo 64 were distributed in Jounti and seventy acres were sowed with high-yield wheat seeds. The seed production was so dramatic that in May 1965 the farmers of Jounti organized themselves into a cooperative, the Jawahar Jounti Seed Cooperative Society Ltd., which could single-handedly meet the seed needs of the entire state of Delhi. In 1967 Prime Minister Indira Gandhi traveled to Jounti to celebrate farmers' success. She also inaugurated the society's seed processing building. By that year the entire village had "become skilled in seed production."[66]

As news of yields spread, the seeds were perceived to have incredible powers and as such quickly became sought after and, more importantly, farmers began to note when scientists were set to release new varieties. As a researcher noted, "when we began to talk about triple dwarfs, they expected a big jump in yield from them. Farmers in Punjab tried very hard to obtain these seeds, even to the extent of trying to steal them; some went in for multiplying the seed for sale, and others cheated farmers by passing off other varieties as triple dwarfs. We did take steps against such cheating but farmers were crazy about getting these dwarf varieties."[67]

There was an explicit lesson to be drawn in 1968. As "The Evolution and Significance of Jounti Seed Village" noted, poor farmers on small plots, "given the necessary technical guidance . . . can transform quickly a traditional and static agriculture into a scientific one."[68] In 2016 a journalist with the *Indian Express* traveled back to Jounti (written as Jaunti in his article), "where the Green Revolution started" and was able to find a handful of the original farmers who had taken part in launching the famous seed village. Yet he also found "parched fields [that] today tell a story of decline."

The seed village had enriched fertilizer and pesticide companies but the farmers, now with depleted land, were forgotten.[69]

Field demonstrations were not the only effective means to communicate the power of HYV seeds. In response to farmers' enthusiastic reaction and the need for "bridging the gap between scientific know-how and farmers' do-how," television and radio shows were created. *Krishi Darshan* (Agriculture Vision) premiered on January 26, 1967, and today is the longest-running television series in India. It continues to broadcast today to nearly eighty villages near Delhi.[70]

At the time, however, easy access to all inputs pushed farmers to embrace HYV seeds. For those who might hesitate, the government was quick to provide cash, thus furthering the idea of the need to obtain *only* HYV seeds. As Dr. O. P. Gautam, head of the Department of Agronomy at the IARI, made explicit, fast and significant subsidies were unexpected:

> The Government offered an outright input subsidy of Rs.500, to start with. That was certainly a surprise! Farmers' response was also very enthusiastic. I remember Mr Sivaraman attended one of our national demonstrations, and at the end of it he told the farmer concerned that Government was arranging credit so that he could obtain seed for the next season. The farmer, in reply, untied his turban, extracted some money from it, and offered it to Mr Sivaraman, saying: "Here is an advance for the seed; I can pay if necessary, but I must have really good seed!"[71]

Government subsidies did not just go to farmers but rather they funneled money to research centers working on HYV dwarf seeds. This unsurprisingly shaped research. As an Indian researcher recalls at the time, "before 1961–62," when researchers were "playing with the available material of tall wheats," there were eminent scientists such as Chaudhari Ramdhan Singh in Punjab, but their research was "naturally, limited."[72] He goes on to explain that while now (he was speaking in the 1990s) scientists produced thousands of crosses, in his entire career one of India's most distinguished wheat scientists, Dr. Bhatnagar (working in Rajasthan), made only one cross.[73] As in the case of Taboada and other plant scientists in Mexico, who were establishing their own research agendas reflecting national needs and concerns, RF aid for high-yielding seed research was embraced, which often derailed independent, domestically driven research. Funding, then, would push and propel the specific type of seed success narrative as well.

High-yielding seeds introduced a new level of standardization of farming that had not existed previously. As Gautam remarked, "precision and timeliness in operation became clear to everyone." HYV seeds led to short-duration wheats (the time gap between planting two crops was short), which had a quicker window of time when they could be planted, and this meant an increased need for mechanization at all stages of planting and harvesting. For example, "pumping sets in irrigation . . . mechanized harvesting and threshing; in Punjab you saw storage bins, and so on."[74] In a handful of years, entire farming practices were transformed to accommodate the needs of foreign seeds.

Early on there was so much focus on the seed and its power that programs were created to ensure that seed would be available to all who wanted it. Access soon became a problem of class and, in India, caste. As in Mexico, "big and influential farmers had the resources to obtain and grow these varieties . . . [but] they did not reach the vast number of poorer farmers."[75] The hybrid seed was deemed too "costly" and its scarcity raised the price. In addition to Jounti, the National Seed Project was launched and the formation of seed corporations ensured that seed production was "more systematic, more scientific and more broad-based."[76]

The standardization of seeds decreased the availability of local varieties yet normalization was billed as a good trait. As an Indian wheat scientist commented, "we, in the different states used to recommend only our own varieties and would not touch varieties from other states. In the national program, all the varieties which were good were being multiplied."[77] As another researcher put it, "The timing was right: we had the technology, the inputs were available, and the Government was then able to change its policy."[78] All was set in place for it to be a success.

By the summer of 1968 it was clear that Mexican "miracle" seeds (Hindu *chamatkar*, "miracle," or *chamatkari*, "miraculous") had performed better than even the best predictions. Yet a July 1968 five-page, single-spaced, confidential letter addressed to the minister of foreign relations from Octavio Paz again raised serious concerns, not about the seeds or the science but rather about how the story of this scientific innovation was being narrated. In the letter, titled "The Case of Mexican Wheat: Its Influence in Mexico-India Relations," Paz admits that Mexico is no longer associated with "Mexican" seeds.

The diplomatic memorandum begins by summarizing the leaps in production that India has achieved in a remarkably short amount of time. Though current successes are site-specific (mainly in Punjab, Haryana, Maharashtra, and Gujarat), Paz predicted that given current yield results and barring climatological shifts, India would be wheat self-sufficient. Not surprisingly, he added the Indian government had seized upon these numbers and transformed them into a matter of "national pride as well as political propaganda." The numbers were indeed astounding. When India gained independence in 1947, it was producing nearly seven million tons of wheat; barely two decades later, in 1968, it produced seventeen million tons. This leap is celebrated and often commented on, the numbers becoming evidence that "between 1964 and 1968, more wheat was added than in the previous 4000 years," as M. S. Swaminathan, the technological father of the Green Revolution, explained. This explosion of wheat production led Indira Gandhi to label it the "Wheat Revolution," and a special stamp bearing an image of the library of the IARI, as a symbol of the introduction of science into agriculture, was released to much fanfare.[79]

Octavio Paz was at the unveiling of the stamp and commented at length on the ways in which India chose to commemorate and memorialize this period in its history. In addition to the stamp, IARI researchers were lauded as heroes and given medals and diplomas for their scientific solution to India's most pressing crisis. Though all speeches were delivered in Hindi, diplomats were handed printed booklets of all the speeches in English, except for Prime Minister Gandhi's, whose impromptu speech Paz was still able to understand because his neighbor "kindly translated" the speech in real time. Paz was dismayed by there being no mention of Mexico's contribution, except briefly by the minister of agriculture. All speakers, he noted, profusely thanked the Rockefeller Foundation.

Paz felt especially slighted when M. S. Swaminathan thanked Borlaug by name and mentioned "the work done by Borlaug in Mexico," adding later extra thanks to Borlaug's "collaborators" in Mexico.[80] Paz made sure to underscore that the so-called anonymous collaborators were dozens of Mexican scientists who for several decades had been working on hybrid crops, many well before the arrival of Borlaug. Paz revealed his surprise at Mexico's omission to the Indian minister of foreign business who, in turn, responded that it was his impression that it was Indian scientists who were responsible for the seeds and the scientific work on hybrids. Specifically, Indians had "adapted both the discoveries and experiments of Japanese and American researchers, especially the work of Dr. Borlaug."[81] Mexico's role, it appeared to Paz, was effectively erased.

Paz then listed the previous occasions, going back two years, on which he had voiced this concern about the retelling of HYV seeds in India with a focus on Borlaug and not Mexico. He felt so strongly about this urgent matter that he had formally requested a detailed history of the events. It is unclear if this was ever sent to him, but he added:

> This Embassy has previously raised the issue that the Indian government does not appear to give, in our judgement, the due credit that the Mexican collaboration deserves. The undersigned suggests that the appropriate authorities take advantage of the upcoming trip of Indian functionaries to make them fully aware of the Mexican cooperation. It is worth repeating that on one occasion India obtained 200 tons of Mexican wheat without the publicity or recognition that the same operations are given in India when they are celebrating [the contribution of]

> other countries. Different conversations with various Indian functionaries reveal that they ignore the role of the Mexican government in these dealings and they give more importance and credit to the role of the Rockefeller Foundation, which also participated in these actions by providing technical and financial assistance via the International Maize and Wheat Improvement Center (Cimmyt).[82]

Paz felt so deeply about this that he believed a plan was needed:

> It may be expedient that a high-ranking official from the Ministry [of Mexico's Foreign Relations] . . . in an informal conversation with the Indian ambassador in Mexico speak in cordial and friendly terms about this, making clear our feelings of surprise and disappointment with India's official attitude [about Mexican hybrid seeds].

Paz was quick to clarify that relations between the two countries were "excellent, especially in the political and cultural realm." Despite this, he felt that the Indian government, although certainly "not Señora Gandhi," undervalued the many gestures of friendship expressed by Mexico. He then proceeded to list the many occasions on which Mexico had been jilted by India. He ends the detailed list by adding that "of course, the case of Mexican wheat seeds is incomparably more important and decisive." Because, he concluded, if India became food self-sufficient it would have resolved its most "pressing and distressing" problem. India solving its food problem would be, of course, a story that would be retold.

Exchange of germplasm between the two countries did not stop, nor was it confined to the 1960s. An especially fruitful period of exchange happened between 1981 and 1986 as part of the "Special Program of Scientific and Technical Cooperation in Food and Technology and Agricultural Research between the Government of India and the Government of Mexico." The long-named cooperation was signed in New Delhi in January 1981 and it was meant as an exchange of personnel and commodities between the two nations. For example, the year the accord was signed, a delegation of Mexican scientists visited India for ten days. The archival record does not reveal their specialty or the sites they visited but India reciprocated the exchange with two trips in 1982 and 1984. After this, the two groups concluded that the best area of mutual cooperation and exchange was agriculture, in particular exchange of plant germplasm focusing on native Mexican crops and plants. No mention was made in the accords of an existing history of seed and germplasm exchange between the two countries.

Shortly after the arrival of Mexican seeds in India there was also public unease about the power of narratives from the Indian press. In some ways it paralleled the concerns that Octavio Paz had raised and would continue to advance when he wrote about Mexican high-yielding seeds in India. For example, the author of a 1967 a commentary in the *Times of India* titled "Overlooked?" wondered why recognition of local scientists was slow in coming. Indeed, he remarked that it had been one year "since the dwarf varieties of hybrid wheat, rice, maize, bajra and jowar outgrew the laboratory stage" and yet, "there is still no sign anywhere that the Government intends acknowledging the efforts of the dedicated men who have laboured in the last few years to bring about this spectacular change in India's agricultural prospects."[83] There was no follow-up to the narrative concerns raised by the opinion writer. Experiments in both countries, as well as many of the key institutions, would be folded into one fascinating, easy-to-tell story of determination, hard work, and ingenuity portrayed in the figure of Norman Borlaug.

Though swiftly muted from the historical narrative, Mexico, however, was present in ways that framed the future and practice of wheat farming in India. Wheat farming became a game of math, of numbers, an attempt to meet the measures established in faraway Mexican fields. In so doing, Mexican soil needs and Mexican farming practices coexisted shadow-like in Punjabi and other Indian fields until experiments tailored the use of fertilizers and pesticides to regional needs.

Concluding Thoughts

In this chapter I brought to the foreground the contemporary concerns of Mexican ambassador Octavio Paz and agricultural scientist Ignacio Narvaez who, already in 1965, felt that Mexico and Mexicans were being written out of narratives about high-yielding seed technology. Why is it important to focus on these apprehensions? Centering other actors—such as Paz, Narvaez, or D. S. Athwal—allows us to question historical narratives whose power remains entrenched because, in this case, we continue to treat the Borlaugian interpretation of the Green Revolution as undisputed, historical truth.[84] Moving away from Borlaug as the pivot of the Green Revolution—not an erasure but a sideways move—complicates the "successes" of agricultural science in development aid. For example, the rubric to measure the success of high-yielding varieties (HYVs) was yield. In other words, the higher the yield the more certain researchers could claim success. But these are scientific measures of success, not measures of success for farmers on the ground, who were not seeing the profit of these increased yields. Nations were investing in seed production but cutting back on infrastructure, education, and health to cover the higher costs of fertilizer, pesticides, and irrigation needed to reach milestones set by a team of foreign researchers. HYVs were seen as a magic bullet to the detriment of progress in the countryside. Expanding the scope to include the concerns of Indian scientists and the work to transform these seeds into domestic ones also taps into the power of agricultural narratives in building a nation.

In this case, Octavio Paz's concerns about the erasure of Mexican science and scientists from the official narratives also has deeper implications. By pushing these contributions from the center and, effectively, out of the frame we also have been unable to critically analyze the *spaces* in Mexico where this knowledge was created. A focus on the Yaqui Valley in Sonora also complicates this story for it brings to the foreground that these high-yielding seeds would eventually fail. The majority of the world's farmers do not resemble the wealthy Sonoran farmers whose input, lands, funds, and participation were vital for producing the seeds that would then travel the world.

If we start our narrative in the Yaqui Valley with those wealthy farmers who benefited the most from hybrid seed research, we realize early on that this project, once applied on a global scale, would be destined to fail small-plot farmers.

Acknowledgments

A version of this chapter appeared as Gabriela Soto Laveaga, "Beyond Borlaug's Shadow: Octavio Paz, Indian Farmers, and the Challenge of Narrating the Green Revolution," *Agricultural History* 95, no. 4 (Fall 2021). I am grateful to the editors of *Agricultural History* for their permission to republish a close version here.

Notes

1. The quotation continues: "except perhaps in the rainfed areas where C 306 still dominates"; M. S. Swaminathan, ed., *Wheat Revolution: A Dialogue* (Madras: Macmillan India, 1993), 26.

2. Phone interview with Raúl Valdés, May 14, 2016. The same story is quoted in *Octavio Paz, embajador de México en India: Documentos e informes* (Mexico City: IMR-SRE, 2014), 15.

3. Interview with M. S. Swaminathan, December 16, 2019, Chennai, India.

4. During his Nobel Peace Prize acceptance speech, Borlaug alluded to the focus on him as a lone contributor to the project to end hunger via agricultural science and emphasized, "I am acutely conscious of the fact that I am but one member of that vast army" of "hunger fighters"; Norman Borlaug's Acceptance Speech, on the occasion of the award of the Nobel Peace Prize in Oslo, December 10, 1970, https://www.nobelprize.org/prizes/peace/1970/borlaug/acceptance-speech/.

5. Marci Baranski's incisive dissertation takes on the science behind the claim of wide-adaptability of hybrid seeds: "The Wide Adaptation of Green Revolution Wheat" (PhD diss., Arizona State University, May 2015).

6. Courtney Fullilove, *The Profit of the Earth: The Global Seeds of American Agriculture* (Chicago: University of Chicago Press, 2017), prologue.

7. Álvaro Obregón's personal archives, for example, describe how he conducted business across the border for just one of his crops, chickpeas, while Plutarco Elías Calles's papers show that his political clout in the region remained unchallenged even after his political exile in 1936; Fideicomiso Archivos Plutarco Elías Calles y Fernando Torreblanca, Mexico City, http://www.fapecft.org.mx/.

8. John Dwyer, *The Agrarian Dispute: The Expropriation of American-Owned Rural Land in Postrevolutionary Mexico* (Durham, NC: Duke University Press, 2008).

9. The majority of these were hydraulic technologies for the vast irrigation system in the valley.

10. The first experiment station in the Yaqui Valley was established in 1909 by the Companía Richardson, a Los Angeles–based company given the concession, initially, to extend the Sud Pacific Railroad and, later, to take over the irrigation projects of the bankrupt Sinaloa and Sonora Irrigation Company; José Rómulo Félix Gastélum, *Pancho Schwarzbeck: Campesino y empresario del Valle del Yaqui, Sonora, México* (Hermosillo: Concepto Gráfico, 2007); Cynthia Hewitt de Alcantara, *Modernizing Mexican Agriculture: Socioeconomic Implications of Technological Change, 1940–1970* (Geneva: UNRISD, 1976).

11. Interviews with Jorge Artee Elías Calles in Ciudad Obregón, Sonora, September 2016, and Karim Ammar, principal scientist and head of durum wheat and triticale breeding for CIMMYT, Mexico, April 2020.

12. Much has been written about two car rides in the 1940s that cemented the need for US technical assistance and influenced how the Mexican countryside—and its potential—would be understood by foreigners. The first was the tour by Vice President–Elect Henry Wallace, during which he saw a technology-poor countryside with low crop yields linked to the legacy of an oppressive hacienda system. The second was "the journey of the 'three ancients,'" Elvin Stakman, Paul Manglesdorf, and Richard Bradfield. Yet beginning the story of agricultural research in Mexico with these trips braids a philanthropic interpretation to the narrative of scientific pursuits. In other words, it paints domestic Mexican agricultural science and its scientists as lacking and in need of imported American expertise. See Joseph Cotter, *Troubled Harvest: Agronomy and Revolution in Mexico, 1880–2002* (Westport, CT: Praeger, 2003); Deborah Fitzgerald, "Exporting American Agriculture: The Rockefeller Foundation in Mexico, 1943–53," *Social Studies of Science* 16, no. 3 (1986): 457–483; David A. Sonnenfeld, "Mexico's 'Green Revolution,' 1940–1980: Towards an Environmental History," *Environmental History Review* 16, no. 4 (1992): 28–52; Nick Cullather, *The Hungry World: America's Cold War Battle against Poverty in*

Asia (Cambridge, MA: Harvard University Press, 2011); and, for popular audiences, Noel Vietmeyer, *Borlaug*, vol. 2, *Right off the Farm, 1914–1944* (Lorton, VA: Bracing Books, 2004). Gilberto Aboites Manrique, *Una mirada diferente de la revolución verde: Cienca, nación y compromiso social* (Mexico City: Plaza y Valdes, 2002), seeks to highlight Mexican contributions to agricultural science before the creation of the Mexican Agricultural Program; and Tore C. Olsson, *Agrarian Crossings: Reformers and the Remaking of the US and Mexican Countryside* (Princeton, NJ: Princeton University Press, 2017), though still focusing on cross-border exchange, traces some development ideas not to Mexico but to the rural South.

13. J. George Harrar, "Draft of 'Agriculture and the Rockefeller Foundation,'" *100 Years: The Rockefeller Foundation*, https://rockfound.rockarch.org/digital-library-listing/-/asset_publisher/yYxpQfeI4W8N/content/draft-of-agriculture-and-the-rockefeller-foundation-.

14. Historians have examined research agendas of crop breeders affiliated with Mexico's Institute of Agricultural Research (IIA) and the Office of Special Studies (OSS) and the differences in their interpretation of plant breeding. The two camps could be divided into those who favored "seeds for farmers" and the others who sought "seeds for agribusiness" (*empresarios*). See Aboites Manrique, *Una mirada diferente de la revolución verde*, 78; and Karin Matchett, "At Odds over Inbreeding: An Abandoned Attempt at Mexico/United States Collaboration to 'Improve' Mexican Corn, 1940–1950," *Journal of the History of Biology* 39 (2006): 345–372.

15. Aboites Manrique, *Una mirada diferente de la revolución verde*, 85.

16. Aboites Manrique, *Una mirada diferente de la revolución verde*, 85; and Ana Barahona Echeverría, Susana Pinar, and Francisco J. Ayala, *La genética en México: Institucionalización de una disciplina* (Mexico City: Universidad Nacional Autónoma de Mexico, 2003).

17. Aboites Manrique, *Una mirada diferente de la revolución verde*, 89.

18. As they are in most books, both the Yaqui Valley farmers and its experimental station are described as "in bad shape" in Jules Janick, ed., *Plant Breeding Reviews*, vol. 28 (Oxford: John Wiley, 2007), 16, 25–26.

19. Rodomiro Ortiz, David Mowbray, Christopher Dowswell, and Sanjaya Rajaram, "Dedication: Norman E. Borlaug, the Humanitarian Plant Scientist Who Changed the World," in Janick, *Plant Breeding Reviews*, 28:10.

20. Not surprisingly, many though not all of these were found in or near Mexican border states. For example, cotton research stations would later be created in the Comarca Lagunera; stations devoted to sugarcane could be found in central Mexico; and a Mexicali-based company in Baja, La Jabonera, focused on researching cotton seeds.

21. Letter from John Allen of Oakland, California, to General Alvaro Obregón, September 27, 1919, Fideicomiso Archivos Plutarco Elías Calles y Fernando Torreblanca, Mexico City, Digital Collections, content dm number: 2402.

22. *Agricultura* (Mexico: Secretaria de Agricultura y Fomento, June 1934).

23. Nicolas Ardito-Barletta, "The Costs and Social Benefits of Agricultural Research in Mexico" (PhD diss., University of Chicago, 1971), 194.

24. Ardito-Barletta, "The Costs and Social Benefits of Agricultural Research."

25. Swaminathan, *Wheat Revolution*, 20.

26. Swaminathan, *Wheat Revolution*, 20–21.

27. Swaminathan, *Wheat Revolution*, 32.

28. Janick, *Plant Breeding Reviews*, 28:25.

29. In reality there about 150 strains of wheat received from Mexico and two, PV 18 and S 227, were selected by Athwal. PV 18 had red grains but Indian farmers and consumers preferred an amber-colored grain. See "New Varieties of Miracle Wheat," *The Times of India*, February 7, 1971. For the quotation, see "India's Food: A Land of Plenty," *The Times of India*, May 16, 1967, 8.

30. "India's Food," 8.

31. "India's Food," 8.

32. "India's Food," 8.

33. M. S. Swaminathan, *50 Years of Green Revolution: An Anthology of Research Papers* (Singapore: World Scientific Publishing, 2017), 61.

34. Swaminathan, *50 Years of Green Revolution*, 7. Newspaper articles discussing Indian farmers' embrace of new seed technologies contradict both Valdés's diplomatic report (at the beginning of this chapter) and most reports of the early on-the-ground trials.

35. "Asunto: Aprehensión de los dirigentes comunistas (fracción de izquierda)," Octavio Paz to Minister of Foreign Relations, January 6, 1965, in *Octavio Paz, embajador de México en la India*, 145.

36. Secretaría de Relaciones Exteriores (SRE), Archivo Histórico Genaro Estrada (AHGE), III-2998-13, 6.

37. SRE, AHGE, III-2998-13, 21.

38. SRE, AHGE, XIV-720-17.

39. SRE, AHGE, Topográfica III-2820-11, Mexico India, in *Octavio Paz, embajador de México en la India*, section "1966–1968," 85.

40. Haldore Hanson, the Ford Foundation's representative in Pakistan, convinced the foundation to finance a wheat research and production program in the country. In 1964, with backing from Pakistan's government, the Accelerated Wheat Improvement Program was started in West Pakistan; Centro Internacional de Mejoramiento de Maiz y Trigo (CIMMYT) and Christopher Dowswell, *Wheat Research and Development in Pakistan* (Mexico: CIMMYT, 1989), 32–33.

41. Narvaez would go on to have a significant international career in sustainable agriculture, including as program director of Sasakawa-Global 2000, a 1980s agricultural project in Sudan. In that capacity he spoke widely about food supply, science, and hunger. He commented on, for example, a paper presented by economist Amartya K. Sen on hunger and economics. *Science, Ethics, and Food: Papers*

and Proceedings of a Colloquium Organized by the Smithsonian Institution, ed. Brian W. J. LeMay (Washington, DC: Smithsonian Institution Press, 1988), 71–74.

42. *Hearings before the Select Committee on Nutrition and Human Needs of the United States Senate, Ninety-Third Congress, Second Session, Part 1—Famine and the World Situation, Washington, DC, June 14, 1974* (Washington, DC: Government Printing Office, 1974), 215; quotation from Don Paarlberg, *Norman Borlaug: Hunger Fighter* (Washington, DC: Government Printing Office, 1970), 15.

43. Some of the initial findings of best seed selections not only in Pakistan but also in India were detailed in Charles F. Krull, Angel Cabrera, Norman Borlaug, and Ignacio Narvaez, "Results of the Second International Spring Wheat Yield Nursery, 1965–1966," *Research Bulletin no. 11* (Mexico: CIMMYT, August 1968).

44. SRE, AHGE, Exp. III-2998-13 (1964), April 19, 1965, pht 3468–3470.

45. Leon F. Hesser, *The Man Who Fed the World: Nobel Peace Prize Laureate Norman Borlaug and His Battle to End World Hunger* (Dallas, TX: Durban House, 2006), 64.

46. SRE, AHGE, Exp. III-2998-13 (1964), April 19, 1965, pht 3468–3470.

47. SRE, AHGE, Exp. III-2998-13 (1964), April 19, 1965, pht 3468–3470.

48. Memo of December 3, 1965, Expediente 54-0/65, SRE, AHGE.

49. "India Expects Record Grain Harvest," *Kabul Times*, October 1, 1968, 2. The first contact that Pakistani wheat scientists had with Borlaug had come eight years earlier when he toured the country as a member of a FAO-RF team that studied wheat production in South Asia, North Africa, and the Middle East. The team recommended "practical training of regional wheat scientists," and so with RF funding, young Pakistani researchers, as well as those from other regions, began to arrive as trainees in Mexico starting in 1961. It is worth noting that all trainees, not just those from Pakistan, went home "carrying packets of seeds and the knowledge of how to use them to best advantage"; CIMMYT and Dowswell, *Wheat Research and Development*, 31.

50. It was led by president Ayub Khan, who was himself a farmer. Warren C. Baum, *Partners against Hunger: The Consultative Group of International Agricultural Research* (Washington, DC: World Bank, 1986), 10.

51. CIMMYT and Dowswell, *Wheat Research and Development*, viii.

52. CIMMYT and Dowswell, *Wheat Research and Development*, 79.

53. A few weeks later, in January 1966, memos from Mexico's embassy in New Delhi continued to highlight the hunger spreading across the country. Regarding the "lack of wheat," an attaché reported that India seemed to have found a supply solution with the United States but rice supplies continued to be an issue, especially because the apparent insistence by the United States that payments for crops be made in dollars and not rupees. Other countries were intent on helping India. Brazil, for example, was willing to donate grain if India could cover the transportation costs. It, unfortunately, could not. Memo of January 3, 1966, Expediente 54-0, no. 16, SRE, AHGE.

54. Ambassador Paz's nearly verbatim summary and a newspaper clip of the entire address are in the SRE archives: SRE, AHGE, Topográfica III-2820-11, Mexico India, in *Octavio Paz, embajador de México en la India*, section "1966–1967," February 15, 1966.

55. Tiago Saraiva, *Fascist Pigs: Technoscientific Organisms and the History of Fascism* (Cambridge, MA: MIT Press, 2016), 25. Saraiva also makes the connection that this was not simply about agricultural output but that it was meant to propel the mechanization of rural areas with the use of fertilizers and machinery.

56. This chapter is part of a larger project that focuses on the sociohistorical context that shaped how science in the Yaqui Valley, in particular the push for hybrid seeds, was shaped by local farmers' needs and then exported to the world. Gabriela Soto Laveaga, *The World in a Wheat Field*, forthcoming.

57. After independence in 1947, the Indian government organized the economy in five-year plans. The planning system functioned from 1947 to 2017, when Prime Minister Narendra Modi dissolved the Planning Commission.

58. Noel Vietmeyer, *Borlaug*, vol. 3, *Bread Winner 1960–1969* (Lorton, VA: Bracing Books, 2010), 114.

59. "Rice Sought from All Friendly Nations," *Hindustan Times*, February 17, 1966.

60. During this time Octavio Paz, already internationally recognized, requested permission to accept invitations by various institutions and to travel to present his poems and essays in Europe. In his stead the embassy's chargé d'affairs, Raúl Valdés submitted the daily memo. On both February 15 and 17, 1966, Valdés writes extensively about Mexican seeds. Curiously, while listing the many Mexican varieties on the ground, he took care to emphasize how Mexico's contributions are "modest" yet significant. It is difficult not to speculate if he was aiming to characterize the role of Mexico as such while still insisting that it be acknowledged; SRE, AHGE, Memo of January 3, 1966.

61. Swaminathan, *50 Years of Green Revolution*, 30.

62. Swaminathan, *50 Years of Green Revolution*, 30.

63. Swaminathan. *50 Years of Green Revolution*, 30.

64. The name of the village is sometimes given as Jaunti.

65. Swaminathan, *50 Years of Green Revolution*, 30.

66. Swaminathan. *50 Years of Green Revolution*, 32.

67. Swaminathan, *Wheat Revolution*, 13.

68. M. S. Swaminathan, "The Evolution and Significance of Jounti Seed Village," Reprinted from *Indian Farming* (January 1968), http://59.160.153.188/library/sites/default/files/Prof%20Jounti%20seed%20village%201968.pdf.

69. The journalist inquired why Jounti was selected, and the Jounti villagers repeated that they did not ask about subsidies; other commonalities were that they owned between fifteen and twenty acres each, "which they tilled themselves"; the fields were plowed by bullocks that also treated the crops to separate the grain from the chaff and powered the *rahat* or Persian wheels to draw water from wells for

irrigation; Harish Damodaran, "After the Revolution," *The Indian Express* (Mumbai), December 6, 2016.

70. Swaminathan, *50 Years of Green Revolution*, 12.

71. Swaminathan, *Wheat Revolution*, 7.

72. Swaminathan, *Wheat Revolution*, 15.

73. A cross between Jaipur local and C591. Other notable wheat scientists were Dr. Ekbote in Madhya Pradesh and Dr. T. R. Mehta in Uttar Pradesh.

74. Swaminathan, *Wheat Revolution*, 7.

75. Swaminathan, *Wheat Revolution*, 18. Strikingly, Saraiva explains that in the Battle of Wheat a campaign initiative was the "formation of associations and consortia of farmers financed by the state with the aim of producing and distributing new high-yield seeds"; Saraiva, *Fascist Pigs*, 35.

76. Swaminathan, *Wheat Revolution*, 18.

77. Swaminathan, *Wheat Revolution*, 18.

78. Swaminathan, *Wheat Revolution*, 68.

79. Swaminathan, *50 Years of Green Revolution*, ix.

80. SRE, AHGE, Topográfica III-2820-11, Mexico India, in *Octavio Paz, embajador de México en la India*, section "1966–1968" (ver).

81. SRE, AHGE, Topográfica III-2820-11, Mexico India, in *Octavio Paz, embajador de México en la India*, section "1966–1968" (ver).

82. Letter from Octavio Paz to Minister of Foreign Relations in Mexico City, July 26, 1968. "El caso del trigo mexicano. Su influencia en las relaciones entre México y la India," SRE, AHGE, III-2998-13.

83. "Current Topics: Overlooked?," *The Times of India*, May 23, 1967, 8.

84. For example, in spring 2020, an episode of the PBS show *American Experience* on Norman Borlaug, "The Man Who Tried to Feed the World," continued to push this narrative, and though it attempted to address the negative effects of the agricultural techniques pushed as part of development projects, the analysis remained nonetheless focused on Borlaug.

Chapter Seven

Moving Coffee from the Forests of Colonial Angola to the Breakfast Tables of Main Street America, 1940–1961

MARIA GAGO

Introduction

"Coffee for Holland means blood for Angola."[1] The title of a Dutch committee report on colonial Angola released in 1972 neatly captured the opinion of the international community, three years before the end of the Portuguese Empire.[2] The siege of Portugal tightened; international pressure and sanctions to coerce decolonization increased. The report announced that a boycott was underway that sought to eliminate Angolan coffee from the Dutch market. It argued that taxes on Angolan coffee were paying for the colonial war that had broken out in 1961. But, more importantly, it insisted that the practices of forced labor, which the Portuguese authorities claimed had ended, still existed. Henrique Galvão, the well-known Portuguese colonial inspector who, in 1947, had compared forced labor to slavery, was praised on the first page. How had the empire been able, despite the obstacles and impediments put in place by the international community, to keep Angola as one of the world's largest coffee producers? The report gave some interesting clues: Portugal continued to receive foreign support from NATO and it was represented in the International Coffee Agreement as a coffee-producing state, "even though coffee is not grown in Portugal." A more interesting answer appeared in a sentence frequently overlooked: "Thanks to the Portuguese community in the United States these [Angolan] planters had managed to secure a place on the American market." What did it mean to "secure a place" on this particular competitive market?

Two strands of historiography help us understand *why* Angola became one of the largest coffee producers in the world.[3] Historians of colonial Angola demonstrated, on the one hand, that the practice of forced labor, maintained until very late in the Portuguese Empire, gave large coffee producers and capitalists in Angola the conditions they needed to thrive in the global market.[4] On the other hand, historians studying coffee as a commodity showed that new consumption trends after World War II, namely instant coffee, led to an increase in demand for robusta coffee, the type of coffee that was produced in Angola.[5] These two strands remain as yet unconnected. In other words, we know very little about how these two realities—of local labor and global markets—relate historically to each other. This chapter uses the history of science and technology, and a firm engagement with the specificities of the materiality of Angola coffee, to suggest some connections and to point in other directions too.[6] Robusta, or *Coffea canephora*, is a species indigenous to the cloud forests of northern Angola—tropical mountainous forests permanently covered with fog during the dry season—and it was here, in places where it used to grow spontaneously, that it was cultivated. These were the forests that nurtured the Portuguese Empire when, in the 1950s, Angola became the first robusta provider to the United States.[7] To make robusta grow in this particular time and space meant using an agroforest system that reproduced remote ecological relationships between plants, species, soil, and climate.[8] *How* was one to transform this deep-rooted agroforest crop into stable commodity? *How* was one to make it travel from the cloud forests of colonial Angola to the core of industrial capitalism?

This chapter addresses these questions by focusing on the work of the agricultural scientists of the Coffee Export Board, a marketing board created in the Portuguese Empire in 1940. Based on a detailed analysis of the board's knowledge-making practices, it shows that the circulation of this commodity was the result of ongoing negotiations between the Portuguese Empire and its main buyer, the United States. It reveals that the process of moving robusta coffee was embedded in the knowledge produced in these negotiations and materialized through practices of standardization, which were themselves oriented to the needs of the American market and industry. Moreover, it discloses the pressure to standardize imposed by the Portuguese Empire over coffee producers, and discusses its different effects on European fazendas and *lavras* (the colonial term used to describe African smallholdings, fields cultivated by Africans). By analyzing the engagement of agricultural scientists—namely, of Artur Rocha Medina—with the circulation of Angola robusta, the chapter seeks to unveil less visible dynamics that are also crucial if we are to understand how the Portuguese Empire, like any empire, depended on the ability to combine and shift strategies.[9] In so doing, the chapter emphasizes the importance of the history of science and technology to expose interconnections between African environmental history, the history of European colonialism, and the history of capitalism and American hegemony.

In the postwar world, Angola coffee was seen as an anachronism, "the blood" that continued to gush despite the winds of change, a reminder that fascism in Europe was still alive. Henrique Galvão contributed substantially to this perception. It is no accident that he was the first source cited in the above-mentioned Dutch report. Starting his career as an inspector in the Portuguese colonial administration and as an enthusiastic supporter of the authoritarian regime of António Oliveira Salazar, Galvão eventually became one of the most vocal critics of that regime and of Portuguese colonialism. Robusta coffee and its modes of production using forced labor contributed significantly to this disenchantment. It was while he was responsible for the inquiry into the Portuguese marketing boards in the 1940s that he became increasingly convinced of the abuses that were being done to African people. The reports he produced during this decade have been studied extensively by historians.[10] But to what extent did this emphasis on these forms of power contribute to concealing and obscuring other layers of history, other agents involved in the production and circulation of this colonial commodity?

Robusta Coffee: From Indigenous Plant to Agroforest Crop

Galvão arrived in Luanda, the capital of Angola, in February 1945, in charge of the inquiry into the three Portuguese marketing boards: the Colonial Grain Export Board, the Cotton Export Board, and the Coffee Export Board.[11] He had been entrusted with these new duties by Marcelo Caetano on September 19, 1944, two weeks after Caetano had become minister of the colonies.[12] Created in the 1930s, marketing boards emerged in a decade of contradictory trends, when development policies promoted by colonial administrations and the increase of private investment led to the growth of agricultural production, while at the same time the prices of tropical agricultural commodities in the global market fell sharply in the Great Depression and showed no signs of recovering.[13] This situation, which drove many European and African producers to despair, forced colonial powers to rethink their imperial strategy. They established these boards to serve as buyers of the entire output of major agricultural colonial crops destined for export. To the public they were presented as welfare institutions, their goal being to create stabilizing funds that could be used for the benefit of producers, improving their production systems and assuring them necessary protection in case of future recessions. In practice, as Galvão would soon find out, these funds were not distributed equally among producers.

After weeks of traveling across Angolan territory Galvão finally found the time to write to Caetano. His opinions about the marketing boards were, unfortunately, generally negative. But there was one among them whose functioning he considered "deplorable": the Coffee Export Board (hereafter the Board).[14] According to him, it was a farce to claim that the Board was performing a "great service" to small producers, "freeing them" from the merchant by buying their coffee at a fixed price. The only ones to benefit from the export system imposed by the Board were those Europeans who could export their own coffee by paying a considerable fee, namely, large coffee producers and export firms. Galvão was especially concerned with African producers. Discussing their situation in particular, he emphasized how little technical assistance had been offered by the Board. The four technical brigades that aimed to improve native production were for him "pure propaganda stunts."[15] They reached an "insignificant number" of *lavras* and were viewed badly by Africans, "who think they are only going to prepare the way for someone who will rob them of their lands." Galvão pointed out that no huskers had been introduced in these regions, where coffee was still "hulled with a pestle as in the old days of Adam and Eve." By invoking the Biblical image, he sought to emphasize the failure of the Board to improve native production systems. Yet, his image also captured something intrinsically archaic about the way of making coffee grow in Angola, something that was not exclusive to the African growers.

The first coffee fazendas were set up at the beginning of the 1830s. The initiative came from Portuguese settlers "with experience in Brazil" and members of the mestiza families of Luanda.[16] This was a time when Portugal began to recenter its imperial project on the African continent after the dissolution of the Luso-Brazilian Empire.[17] At first, these coffee growers tried to reproduce the American model used in Brazilian fazendas based on arabica coffee (*Coffea arabica*), which was cultivated in large sun-drenched plains. Later, however, they realized that this system was destined to fail and so they turned to the native species, robusta coffee.[18] They did not limit their attention to the plant itself. They engaged with the whole ecosystem where wild forms of robusta coffee used to grow, in those tropical and mountainous forests of north Angola known as cloud forests. I have argued elsewhere that by inventing an agroforest cultivation system, which reproduced existing ecological relationships, these first growers were also rooting this promising imperial cash crop in the local history of this place.[19]

These first coffee fazendas were located on the lands of African chiefs in the district of Cuanza Norte (north of Angola), in the regions of Golungo,

Zenza, Dembos, and Cazengo.[20] It is important to remember that Portugal's military and administrative occupation was very weak at the time. This was particularly true for these coffee-producing regions in the north of Angola where the economic activity of the Mbundu societies was intensive.[21] Here, the most influential rulers were not the Portuguese but the African chiefs.[22] Indeed, until 1870, Portugal's colonial authority was heavily dependent on their loyalty. This explains why the first coffee fazendas in Angola were located in the lands of the African chiefs and why, one decade later, in the 1840s, the African chiefs of Golungo and Cazengo were organizing their own fazendas.[23] As regards labor relations, all these fazendas—European, mestiza, and African—were inspired by the American model, relying on a large number of slaves and a division of labor as needed by different production activities, like picking, weeding, or pruning.

This agroforest cultivation system, deeply rooted in the ecology of the cloud forests, produced markedly atypical plantation systems.[24] They were slave-dependent but they were not, unlike sugar or cotton, a monocrop space. Robusta coffee plantations were also compatible with the calendar of food production and relatively cheap in terms of initial capital, particularly when compared with sugar fazendas.[25] These specificities of robusta as an indigenous species and an agroforest crop also help to explain why some local African coffee growers were able to resist when, in the last quarter of the nineteenth century, under increasing pressure by the colonial regimes, African chiefs and mestiza families started to lose their fazendas. At the start of the twentieth century, Angola coffee had a new composition: European fazendas and *lavras*. As scholars of Angola have noted, without however explaining exactly how, the emancipation of a "coffee peasant elite" began.[26]

When in 1945 Galvão traveled across Angola and visited the various coffee-producing regions, he was also visiting this intricate historical assemblage of cloud forests, coffee plants, African growers, and Europeans. At this point, one-quarter of the colony's total coffee production came from small African producers.[27] It is relevant to note that we still know very little about this African part of the story. Historians working on twentieth-century Angola have covered a good part of the history of coffee production, but they were mainly focused on the history of labor and on those Africans who were coerced to work in the European fazendas.[28] Many of these later Africans were not locals, but migrants, some of them coming from the Angolan Plateau, in the south of the colony, and knew nothing about forests, mountains, or robusta. The local ones, the ones who inhabited the regions of the cloud forests, not only knew robusta and how to plant it, but had already been identified by colonial scientists as important assets to the empire.[29] Moreover, in merely agronomic terms (not in terms of scale, evidently), European and African production systems were not that different.[30] Both used the same agroforest system, and both depended heavily on the cloud forests. This agro-environmental aspect, I argue, helps us understand the predicament of Angola coffee production: European fazendas, which were highly dependent on the coercion of an African workforce, produced side by side with *lavras*. It was the job of scientists of the Board to make sense of this unstable and potentially conflictual situation and extract from it the maximum advantage for the colonial authorities.

Toward Standardization

In the same year that Galvão arrived in Angola, Artur Rocha de Medina, an agricultural scientist of the Board, and João Carrolo, an Angolan *fazendeiro* with "eighteen years of experience," were leaving for Brazil.[31] Their mission was to see what was being done in the world's largest coffee-producing country. They were also excited to visit the Instituto Agronômico de Campinas, in the state of São Paulo, a leader in coffee breeding and genetics research, a project that had been underway since the 1920s and that had gathered researchers and funds from all over the world.[32] This buzzing environment of coffee production and science attracted the Portuguese to the land where they had left the imprint of language and slavery. But they did not take much time to realize that despite Campinas's advanced research, not much had been translated into the fields.[33] The old Brazilian way of planting coffee persisted: the total destruction of virgin forests, full sunshine cultivation of arabica coffee in virgin soil (an aspect that guaranteed the farmer much higher yields), rapid impoverishment of the soil (after a certain number of years of sun and soil erosion), plantations abandoned, and the cycle repeated with more destruction of more virgin forests. The environmental disadvantages of this cycle of destruction were very vivid when Medina visited Campinas in 1945 and 1946. He congratulated himself for doing a much better job in Africa: Brazilian coffee-growing practices were clearly much less "modern" than the ones adopted in Angola. In contrast, he was astonished by the level of Brazilian mechanization used in preparing coffee for market, which included the so-called coffee-processing operations, like drying, hulling, polishing, sorting, and so on.[34]

Three years after his trip to Brazil, Medina discussed these topics at the First Agronomic Colloquium in Angola, held in Nova Lisboa (Huambo).[35] He estimated that 100 percent of African farms and 75 percent of European small farms did not have mechanical installations for hulling; 25 percent of medium European farms did not have technological facilities,

50 percent were equipped for mechanical hulling, and only 25 percent had full processing facilities; finally, 100 percent of large European farms had technological facilities, but only 60 percent had full processing facilities. According to him, "the cheaper process of solving the problem of coffee-processing operations in Angola is the construction of a network of processing factories located in the producing regions at points of easy coffee inflow."[36] These processing and homogenizing factories should respond to the needs of small and medium farms, while large farms should rely on their own facilities. According to Medina these factories should be managed by "producer cooperatives," "like in several countries of South and Central America," or by traders, who then "increase their social engagement." There was also the state-funded option "which it is necessary to resort to in the absence of private investment."

The Board adopted the third management model. And in fact, this plan was soon implemented. In the same year that Medina was speaking, the first factory was being opened in Uíge, the main native coffee production zone. Within four years, there would be five more. All of these factories were located near ports. The order in which they were built from north to south reveals the Board's concern to give priority to native and small or medium European producers. After Uíge, it was the turn of Ambriz and Ambrizete (between 1949 and 1953), both of which received a great quantity of coffee coming from African *lavras,* including "an appreciable amount of Encoge coffees."[37] Then it was Luanda (in 1953), "one of the largest, most modern and best equipped in the world," which received both European and native coffee production. Finally, the last two factories (around 1954) were located at Porto Amboim and Novo Redondo, the most important recipient ports of coffee produced in large European fazendas.[38] All the machinery and associated technology that equipped these factories came from Brazil, namely the most important of the engines, the Blasi huller, a Brazilian technology used worldwide.[39]

Meanwhile, the Board was also investing in a new coffee classification system. This was an old project of the Agriculture Services of Angola, a project that gained new life with this centralizing imperial institution. The Regulation for the Classification of Portuguese Coffee (Regulamento para a Classificação dos Cafés Portugueses) was published the year Galvão was in Angola, in 1945. It stated that all coffee destined for export was the responsibility of the Board, whether it was classified in the colonies or in the metropole. It also determined that, after passing through the processing operations, export coffee should be "submitted to classification." The Board's experts were in charge of the classification process. The coffee lots were brought into the Board's facilities and were classified in the presence of the lot owner or its legal guardian. The process was based on a sampling method: first, a portion of the several bags composing the lot was selected, never less than 20 or 25 percent of the total of the lot; then, the coffee extracted from each bag was blended and three samples (450g each) were again taken. One of these samples was used to classify the whole lot, while the other two would be archived in the facilities of the Board, in case a second classification was required.[40]

According to this classification system Angola robusta was first classified in terms of its origin and biological variety. The "origin" was determined based on seven coffee-producing regions: Cazengo, Golungo, Cabinda, Encoge, Ambriz, Libolo, and Novo Redondo/Amboim. Next, coffee beans were classified in terms of their form (moka or bago chato), type (based on the number of defects of the grain), and size, a process that was based on a sequence of complicated procedures. The process ended when the coffee lots brought to the Board were classified in one of the following four grades: "first quality" (fewer than 165 defects in 450g of coffee), "second quality" (defects ranging between 165 and 250), "no choice" (more than 250 defects), and "coffee wastes" (resulting from cleaning and processing operations). The coffee lots were then packed in bags with the "official brand" of the Board. This brand consisted of a triangle, with written on each face: (1) the name of the colony, (2) the coffee-producing region, and (3) the word "coffee," followed by the type and size of the bean.[41] Finally, and very important, the Board had full responsibility for classification but shared the rights of exportation with the "authorized exporters," those who had enough resources to pay an annual fee to the Board, namely, the large European coffee producers and big export firms of Angola. All the other producers, small and medium-sized, were required to sell coffee to the Board at prices which it itself set.

Needless to say, several attempts to corrupt the functionaries of the Board in charge of this classification process were made over the course of time. The colonial archive shows this unequivocally.[42] Yet, despite the flaws of the Board as an institution there was something happening here that was much more relevant historically: the creation of a standard for Angola coffee. This aspect has escaped recent scholarship on the late Portuguese Empire, more focused on analyzing what the Board was unable to do, namely from sources produced by actors such as Galvão. The immense variety of coffee produced in this colony—by Africans and Europeans, small and large growers, using private and public homogenizing equipment—could now be reduced to one brand and a small number of grades. Robusta coffee, an indigenous plant to the cloud forests of north Angola, which for a long time had remained an agroforest crop of heterogeneous origins, was being transformed into something else. In short, robusta had become a stable commodity.

No doubt this was an uneven scheme. By creating the category of "authorized exporter," which only those with financial resources could aspire to, the empire was saying that only large European capitalistic enterprises (large fazendas and trading firms) could have a share in coffee export profits. Moreover, the pressure to standardize was conceived in order not to touch the forced-labor regimes upon which these enterprises depended. On the other hand, this scheme also reveals that African growers continued to play an important role in the coffee business and that they were considered crucial assets for the empire. As explained above, improving the processing and homogenizing procedures of coffee coming from *lavras* was at the top of the Board's priorities and actions. The art of modernizing, without interfering with entrenched production systems (both European and African) but rather managing their different needs (even if conflictual) in order to optimize the colony's total production, was the challenge that the Board efficiently met with the help of its agricultural scientists and technicians. As a stable commodity, Angola robusta started to travel. To understand its trajectory, we have to go back in time and, first of all, recognize the different dynamics at play when the Portuguese marketing boards were created.

Portuguese Fascism and the Speed of Capitalism

Marketing boards, as mentioned above, were the product of a transnational process of state economic interventions. In Tropical Africa, the first marketing boards emerged in east and central British Africa, with the creation of the Marketing Board for Maize, in Southern Rhodesia (in 1931), and the Kenya Coffee Board (in 1933).[43] Later, in 1942, and as a "wartime measure," the British Empire created the West African Control Board, which covered several colonial states of British West Africa. Between 1947 and 1949 eight more marketing boards, each devoted to a separate commodity, emerged in Sierra Leone, Gambia, Ghana, and Nigeria. In the French Empire the first attempt to launch similar institutions, the *sociétés de prévoyance*, was also in the 1930s, though it was only after World War II that these institutions were taken more seriously with the emergence of the *caisses de stabilisation*. In the case of coffee, these forms of state intervention existed beyond the colonial world: some of the best-known examples are the state agency Instituto Brasileiro do Café, in Brazil, and the semiprivate Federación Nacional de Cafeteros de Colombia (Fedecafé), in Colombia.[44]

In the Portuguese case, however, they were also intricately connected with the fascist experience. Corporatism, as a third way of organizing society, and an alternative to both socialism and liberalism, was a central dimension of Salazar's Estado Novo (Portuguese for New State).[45]

Corporatist ideas made their way into liberal democracies, but it was in the emergent European fascisms that they found the breeding ground to flourish, being used not only as a model to organize national economies but as a powerful instrument of social control. Portuguese marketing boards (of colonial grain, cotton, and coffee) were classified as "institutions of economic coordination," the same family of metropolitan corporatist institutions as the National Board of Wine, the National Institute of Bread, and the Regulatory Commission of Cod Fish. The similarities between *metropolitan* institutions of economic coordination and their *imperial* versions are obvious: they both acted as governmental agencies, controlling and regulating economic sectors, product by product, in terms of production, wages, circuits of distribution, prices, imports and exports, and so on.[46] But while the metropolitan ones were centerpieces in the corporatist machine, the same cannot be said about the imperial ones.

A good example of an imperial marketing board that was well articulated with the corporatist machine of the Portuguese fascist regime was the Cotton Export Board. The historiography of the Portuguese Empire has already demonstrated to what extent this institution worked as a heavy hand of the metropolitan state in the colonies, providing the elites of the north of Portugal working in the textile industry with the cotton they needed.[47] It is obvious that in this case imperial marketing boards strengthened the corporatist machine in the metropole, a process in which scientists played a crucial role.[48] The picture changes significantly, however, when we turn from cotton to coffee. Unlike the cotton industry, there was no coffee roasting industry in Portugal.[49] Despite the several attempts to articulate the Coffee Export Board with corporatist institutions in the metropole, like the Guild of Grocery Storers (Grémio dos Armazenistas de Mercearia), they were often in vain or with very weak results. The big capital and the well-established elites were in Angola, not in the metropole. The agency was in Luanda, not in Lisbon. If, during the 1930s, the Angolan elite (large coffee producers and trade firms) saw in the Board the solution to their problems, a way of draining the product to the Portuguese internal market, after 1945 their strategy completely changed. From one moment to the next, the demand for robusta coffee in the global market increased exponentially. Angola coffee elites ceased to be interested in the metropolitan market and shifted to the global market, namely to the United States. Much more permeable to market dynamics, Angola coffee forced the Salazar regime to look beyond its ideological limits and to experiment with new trade compositions.

This sudden demand for robusta was grounded in a technology: instant coffee, or soluble coffee. Instant coffee had been tested among consumers since World War I. An important marker was 1929, when robusta was included in the American definition of "coffee"; until then, only arabica and liberica species were recognized as such by the United States Department of Agriculture.[50] The crucial component of instant coffee is robusta coffee. Steven Topik, a historian of coffee, explains: "drinkers of instant coffee were concerned with speed and convenience, not the quality of the brew," and therefore "the small number of roasters who captured this capital-intensive market used low-priced beans, especially Robusta beans that Africa and Asia began growing."[51] For quite some time the chemical composition of robusta beans was translated, in market language, into undesirable properties—namely, bitter taste and high percentage of caffeine—preventing this coffee species from competing with the perfume and delicate taste of arabica beans. Yet, things started to change in the interwar period. And the taste revolution came after World War II, when suddenly instant coffee became a cultural trend. By the 1960s, one-third of the coffee prepared at home in the United States was instant coffee.[52]

This was the race that the Salazar regime chose to run; the race to feed the drinkers, industrialists, and politicians "concerned with speed and convenience." In Angola the impact of the trend of instant coffee was enormous. The volume of Angola coffee exports had already doubled between 1920 and the mid-1930s because of the increase in demand for robusta, sought after as a "filler" of certain blends. But after 1945 the rise was spectacular: between 1946 and 1960 coffee exports quadrupled. During the first three years, the Netherlands occupied the first position in the rank of Angola coffee importers. The United States took the lead in 1949 and held it until the end of the Portuguese Empire.[53] The Angolan elites and the Salazar regime knew how to take advantage of instant coffee, it was their window of opportunity. It remains to be seen how they won the race and became the first robusta coffee provider to the United States. The next section deals with this question, discussing the role of the Board and their scientists in "securing a place" for Angola coffee in the US market, and showing how, in the process, they were forced to revise and adjust the standardizing procedures and instruments designed for this commodity.

Coproduction: Robusta Negotiated

With the end of the Inter-American Agreement (1940), which, for five years, had closed this gigantic market to coffee-producing states outside the American continent, the transactions of Angola robusta between Portugal and the United States returned. The first selling operation consisted of 1,500 tons of Angola robusta, destined for the American army in France.[54] The beneficiaries were two large fazendas, Sociedade Remus and Companhia Angolana de Agricultura (CADA).[55] These first transactions started well but problems soon arose. According to documentation of the Board, the Americans started to question the agreed terms of the arrangements, eventually demanding that "the exporters of the colony could only use their credits after coffee exports had been accepted by the [American] Department of Agriculture."[56]

In Lisbon this was considered unacceptable. The Board fought for its own terms and ultimately won. On October 10, 1945, it announced by telegram that the Americans had dropped their demand.[57] The cautious tone of the telegram, however, shows clearly that this was a shaky beginning. For the Board, there was no doubt that Portugal's "delicate position" should be taken seriously, "because if we fail our inspection we lose future privilege."[58] There were good reasons for being reserved. Not long after, a shipment of Angola coffee was retained in New York, because it contained berry borer with "worms, alive and dead, and also excrements."[59] The US State Department responsible for food security warned the Portuguese authorities that future shipments would have to be prepared with "maximum care," otherwise they would be rejected.[60] The Board met the American threat with suspicion, and ironically:

> Fortunately, berry-borer is not a disease exclusive to Angola. It exists in every coffee in the world, to a greater or lesser extent. One can fight this disease but its effects, either in Angola, Brazil or the Dutch Indies, can be only mitigated by establishing a limit of percentage of defects on the product. These are the terms regulating exportation in every producing market. How can it be that Angola's berry-borer bears a disease other than Brazil's berry-borer when the insect that causes it is the same in both countries?[61]

Even more interesting is the internal reflection this episode generated. On January 16, 1946, the president of the Board wrote to the Minister of the Colonies in Portugal claiming that there were "strong reasons . . . for not believing in the good intentions of the American trade, especially when this country, after invoking the principles of commercial liberty and provoking a rise in export markets, intends to achieve unclear goals."[62] According to him, the coffee exported to the United States had been submitted to rigorous inspection through the Board's classification system, which, as he stressed, was based on "rules adopted internationally in this matter." For instance, "the maximum limit of 10% of berry-borer allowed by American law was not exceeded, nor even reached the limit because our legislation, by means of intelligent prudence, reduced this limit to 8%." This meant that, in their view, problems were being unjustly created. Power was behind this, not science, they believed. The solution adopted was simple: merge science and power and make concessions. In 1947, Carl Borchseniu, a high-ranking representative of the American roasting industry, visited Angola to resolve the problem of the high percentage of berry-borer. According to the agricultural scientists of the Board, not only did he solve the problem, he also played an "extremely important" role in the standardization of Angola coffee.[63]

The second time that American institutions intervened in the commodification of Angolan coffee occurred between 1954 and 1956. Central to this episode is the 1949 law that prohibited the export of "second quality" and "no choice" coffee to the United States. From that time on, the Board approved the exportation of only "first quality" coffee.[64] This high-quality policy coincided with the beginning of the war between African robustas and American arabicas. To the public the Board justified this policy with arguments of "brand identity": the goal was to produce "the best Robustas" in the world.[65] Was it? The colonial archive shows that this policy was heavily criticized by Angolan producers and exporters.[66] They complained that they also wanted to export their second-quality coffee to the American market under the same conditions as did other African states, such as the Ivory Coast, Uganda, and Belgian Congo. The Board wouldn't budge. In 1954, their complaints were so strident that the Board's technicians decided to reconsider the high-quality policy. However, before taking any decision, in August 1954, Lisbon "consulted" the Green Coffee Association, an American organization of coffee importers, brokers, and agents. The decision was then made. Nothing could be changed, the "prestige" of the commodity depended on this law.[67]

A third American intervention took place in 1956 when, after two years of further protests, the high-quality policy was finally waived.[68] It was now possible to export "second quality" coffee to the United States. This new policy implied changes to the Regulation for the Classification of Portuguese Coffee. These were, first, to increase the upper limit of the "second quality" category to 250 defects and, second, to recategorize "no choice" coffee as "third choice" coffee, and to allow it to have up to 720 defects. Moreover, some of the four regulated grades were subdivided. Likewise, a sample of Angolan coffee (450g) under inspection at the Board's facilities could be classified into six possible grades: (1) "first quality" (up to 165 defects); (2) "second quality AA" (from 166 to 250 defects); (3) "second quality BB" (from 251 to 360 defects); (4) "third quality CC" (from 351 to 550 defects); (5) "third quality DD" (from 551 to 720 defects); and finally the category "industrial purposes" (with more than 720 defects).[69] These changes in the Regulation for the Classification of Portuguese Coffee were all made with the approval of the Green Coffee Association. As an expert of the Board recalled:

> The Green Coffee Association played a leading role in clarifying the situation of exporters and importers before Ordinance nº 15.913 and the regulations that followed, giving very useful suggestions and submitting the matter to the "Standard Type Committee," and contributed greatly to the idea that the Portuguese official entities only want regularity in business and good credit for the product, thus defending interests that are bilateral, as shown by the acceptance of the subdivision of the second and third grades.[70]

To sum up, the commodity that started to circulate in the years after World War II, insistently sold by colonial propaganda as "Portuguese coffee," was, after all, coproduced with the Americans. Two institutions are key to understanding this case of coproduction: the Board, acting as the spokesperson for the Portuguese Empire, and the Green Coffee Association, representing the interests of the American market. A similar process had occurred in Brazil in the early twentieth century, when agents of the American roasting industry arrived to purchase directly from producers and take control over prices and grades; the same thing happened in other coffee-producing countries of Latin America later in the century.[71] This chapter captures the moment when this hegemonic power expanded from the American to the African continent, a trajectory that, as demonstrated, preserved the interests of capitalists on both sides of the Atlantic: on the Portuguese side, by creating an imperial institution that was designed to divide coffee profits between Lisbon and large European producers and traders in Angola; in the case of the United States, by delegating the negotiations to an institution that aimed at protecting the interests of the American brokers, importers, industrialists, and other agents in the coffee business.

Conclusion

In bringing Galvão and Medina together in this narrative, I want to stress the importance of the history of science and technology to imperial history and to transnational approaches in particular. In recent years a significant effort has been made to understand the interactions between the metropolitan and colonial processes of the late Portuguese Empire, and how these were shaped by international, transnational, and interimperial dynamics.[72] We have now a more nuanced understanding of these different scales of empire and how they intertwined. Particular attention has been given to the internationalization of the debate on forced labor and how it was instrumentalized by the Portuguese Empire as a way to legitimize its colonial possessions, a discussion that has brought Galvão into the spotlight.[73] Regardless of the major advances put forward by this scholarship, little has been written, however, about how Portuguese forms of colonial governance interrelated and networked with world market dynamics. This chapter is a contribution in this direction. Following the movement of Angola coffee through the lens of Medina and his colleagues of the Board, it identifies standardization as the missing link in the historiography, and points out the actors, connections, and dynamics that made this knowledge producing practice possible.

One simple conclusion that we can draw from this chapter is that the narrative of the Portuguese Empire as backward simply disappears when we switch from *why* to *how*. As I have shown, the success of Angolan coffee as an imperial commodity was due not only to the practices of forced labor, or to a stroke of luck in the market (instant coffee), but also to a set of reforms in colonial governance that transformed this agroforest crop into a stable and competitive commodity.[74] These reforms were framed around the standardization of coffee beans, involving the construction of state facilities for the homogenization and improvement of coffee-processing procedures and the centralization of coffee classification and exports, a development plan that has been ignored in the literature up to now. Thus, taking Medina seriously stresses the importance of strong regulatory state institutions like marketing boards to understand the process of economic integration of the late Portuguese Empire. More importantly, it invites us to reflect on the role that agricultural scientists played in imagining this integration process, and how this was done by weaving local environments, colonial contexts and market opportunities. Cotton and its scientists, as scholars have demonstrated, pushed the Portuguese Empire to the metropolitan economy, merging the imperial trajectory with the political experience of corporatism that was underway under Salazar's regime.[75] Coffee and its scientists, by contrast, with no roasting industry in Portugal to aim for, unveil an empire at variance with protectionism, corporatism, and metropolitan elites, which remained nevertheless highly state-centralized. Robusta, and its indigenousness and agroforest plantation systems, forced the empire to forge alliances with local elites (both European and African) in order to make the cloud forests politically stable production systems.

Taking standardization as a knowledge-making practice also forces us to recognize that negotiations to "secure a place" in the world coffee market occurred outside of the traditional political and diplomatic spheres. It is not possible to understand how Angola robusta traveled without going through a set of improbable negotiators (agricultural scientists), working in uncanny spaces (the Board's warehouses), on encrypted documents (Regulation for the Classification of Portuguese Coffee), published in obscure journals (like *Revista do Café Português*). Through the analysis of their practices, we understand that the circulation of this commodity was not a fluid process but a highly negotiated one, involving concessions on the Portuguese side proper to a context of global decolonization. These negotiations lie hidden in the technicalities of a standard classification system. As this chapter demonstrates, this was not a fixed and immutable legal object, but it mutated and adapted to the circumstances as they changed. The adjustments were made to meet the needs of the American roasting industry and market (represented by the Green Coffee Association) and involved a high-quality state policy that regulated exports to this country. It also shows that the same classification system determined who could, and who could not, export, in the end privileging capitalist enterprises in Angola at the expense of other producers. This story of coproduction and uneven distribution of profits is what allows us to connect the two historiographies mentioned at the beginning of this chapter, offering thus more evidence to the case of an American hegemony built on European imperialism. In doing so, it also adds one more layer of irony to the performance of Salazar's regime; the regime that strove to freeze time in colonial Africa by prolonging the practices of forced labor was the regime that promised to revolutionize time at the breakfast tables of main street America, feeding postwar working societies with no time to drink coffee.

Finally, to see the circulation of Angola coffee as a case of coproduction reminds us that asymmetric relationships do not mean total power of one side over the other. This is true for the relationship between the Portuguese and the Americans, but also for the relationship between the empire (government and enterprises) and Africans. To follow our scientists is also to realize that small African growers were not neglected in the grand imperial scheme of the years that followed World War II. Galvão shows that Portuguese investment in *lavras* was nonexistent or ineffective. Medina shows that, on top of this apparent inefficiency, infrastructures and a classification system were created, which conveniently channeled their product to the market. To say this is also to recognize that the pressure for standardization did not eliminate small African growers. This chapter aligns therefore with commodity studies scholarship that has shown that standardization (of coffee and other tropical commodities) led to a shift from large to small units of production, which eventually led to the decline of the plantation model. Such an extreme outcome did not happen in the case of colonial Angola, but one could argue that standardization ended up empowering those Africans who wanted to grow their own coffee.[76] In exploring these nuances, this chapter offers more evidence to Frederick Cooper's thesis that the cases of imperial success in Africa were the ones in which empires were able to form alliances with local authorities and create "flexible relations of production that cannot be reduced to either 'peasant' or 'capitalist.'"[77] In fact, it expands Cooper's thesis by suggesting that the local authorities to which the empire was adapting were not only political or cultural but also environmental. Through the lens of the Board's agricultural scientists, we see an empire that was identifying not only what had to be done (standardization) but also what had to be preserved: the long-term relationship between cloud forests, local Africans, and Europeans. As I mentioned before, this intricate assemblage of humans and nonhumans was deeply understood by historians who studied nineteenth-century Angola but was ignored by historians who worked on the twentieth century. This chapter illustrates that these intricacies remained when colonial rule intensified and Angola robusta became a global commodity.

Acknowledgments

This chapter is based on studies undertaken at the Institute of Social Sciences, University of Lisbon. I thank Tiago Saraiva, Staffan Müller-Wille, and John Krige, as well as my colleagues in the European University Institute, Roberta Biasillo and Andrés Vicent Fanconi, for their extremely useful comments and suggestions on previous versions of this chapter. It was at the Max Weber Programme that I found the conditions to write it, so I am also very grateful to all the personnel and scholars involved. The chapter also benefited from discussions of the research project Transcap: The Transnational Construction of Capitalism in the Long 19th Century (PGC2018-097023-B-100 del plan nacional de I+D), based in Madrid. Finally, I thank Alyson Price for her support in editing and language revision.

Notes

1. *Coffee for Holland Means Blood for Angola* (Amsterdam: Angola Committee, 1972), http://www.aluka.org/action/showMetadata?doi=10.5555/AL.SFF.DOCUMENT.nizapl026.

2. The Portuguese colonial empire ended in 1975, one year after the April 25 revolution that ended Estado Novo (1933–1974), the regime ruled by the dictator António Oliveira Salazar.

3. In 1970 Angola became the world's fourth-largest coffee producer; David Birmingham, "A Question of Coffee: Black Enterprise in Angola," *Revue Canadienne des Études Africaines / Canadian Journal of African Studies* 16, no. 2 (1982): 343–346, at 343.

4. See Gerald J. Bender, *Angola under the Portuguese: The Myth and the Reality* (Berkeley: University of California Press, 1978); Douglas Wheeler and René Pélissier, *História de Angola* (Lisbon: Tinta-da-China, 2009); Alexander Keese, "The Constraints of Late Colonial Reform Policy: Forced Labour Scandals in the Portuguese

Congo (Angola) and the Limits of Reform under Authoritarian Colonial Rule, 1955–61," *Portuguese Studies* 28, no. 2 (2012): 186–200.

5. See Steven Topik, "The Integration of the World Coffee Market," in *The Global Coffee Economy in Africa, Asia, and Latin America, 1500–1989*, ed. William Gervase Clarence-Smith and Steven Topik (Cambridge: Cambridge University Press, 2003), 21–49. It should be said, however, that the concrete case of Angolan coffee is not addressed in this magisterial work about coffee. For a connection between instant coffee and Angola see Irene S. Van Dongen, "Coffee Trade, Coffee Regions, and Coffee Ports in Angola," *Economic Geography* 37, no. 4 (1961): 320–346.

6. This chapter builds on a body of scholarship that has been using stories of crops and their movement and rootedness to rethink narratives of global circulation; see Francesca Bray, Barbara Hahn, John Bosco Lourdusamy, and Tiago Saraiva, "Cropscapes and History: Reflections on Rootedness and Mobility," *Transfers* 9, no. 1 (2009): 20–41. Examples of this approach in the case of the Portuguese Empire can be found in Tiago Saraiva, *Fascist Pigs: Technoscientific Organisms and the History of Fascism* (Cambridge, MA: MIT Press, 2016); Marta Macedo, "Coffee on the Move: Technology, Labour and Race in the Making of a Transatlantic Plantation System," *Mobilities* 16, no. 2 (2021): 262–272.

7. Van Dongen, "Coffee Trade, Coffee Regions, and Coffee Ports," 320.

8. This argument is made in Maria do Mar Gago, "Robusta Empire: Coffee, Scientists and the Making of Colonial Angola (1898–1961)" (PhD diss., Instituto de Ciências Sociais, Universidade de Lisboa, 2018). I expand the point in "Rooting Coffee: John Gossweiler, Plant Geography and African Growers (1898–1939)," forthcoming. See also Maria do Mar Gago, "How Green Was Portuguese Colonialism? Agronomists and Coffee in Interwar Angola," in *Changing Societies: Legacies and Challenges*, vol. 3, *The Diverse Worlds of Sustainability*, ed. Ana Delicado, Nuno Domingos, and Luís de Sousa (Lisbon: Imprensa de Ciências Sociais, 2018), 229–246.

9. Jane Burbank and Frederick Cooper, *Empires in World History: Power and the Politics of Difference* (Princeton, NJ: Princeton University Press, 2010).

10. See, for example, Jeremy Ball, *Angola's Colossal Lie: Forced Labor on a Sugar Plantation, 1913–1977* (Leiden: Brill, 2015); Jorge Varanda, "'A Bem da Nação.' Medical Science in a Diamond Company in Portuguese Angola" (PhD diss., University College London, Wellcome Trust Centre for the History of Medicine, 2006); Douglas Wheeler, "The Galvão Report on Forced Labor (1947) in Historical Context and Perspective: The Trouble-Shooter Who Was 'Trouble,'" *Portuguese Studies Review* 16, no. 1 (2009): 115–152; José Pedro Pinto Monteiro, *Portugal e a Questão do Trabalho Forçado* (Lisbon: Edições 70, 2018).

11. The Colonial Grain Export Board and the Cotton Export Board had been created in 1938; the Coffee Export Board was created in 1940.

12. "Relatório sobre a Junta de Exportação do Café Colonial" of the Comissão de Inquérito Parlamentar dos Elementos da Organização Corporativa, Arquivo

Nacional da Torre do Tombo (hereafter ANTT), Arquivo Oliveira Salazar, PC-17C cx. 512, pt. 4, p. 291.

13. On marketing boards see Robert H. Bates, *Markets and States in Tropical Africa: The Political Basis of Agricultural Policies* (Los Angeles: University of California Press, 1981); William O. Jones, "Food-Crop Marketing Boards in Tropical Africa," *Journal of Modern African Studies* 25, no. 3 (1987): 375–402. For the case of coffee, see Benoit Daviron and Peter Gibbon, "Global Commodity Chains and African Export Agriculture," *Journal of African History* 2, no. 2 (2002): 137–161.

14. Letter from Henrique Galvão to the Minister of the Colonies, September 30, 1944, Lisbon, File "Inquéritos de Henrique Galvão / Metrópole," subfile "Documentos singulares, no 1," ANTT, Arquivo Marcelo Caetano (hereafter AMC), cx. 8.

15. "Letter-report" from Henrique Galvão to de Minister of the Colonies, received February 18, 1945, 22–23, File "Inquéritos de Henrique Galvão / Colónias," subfile "Cartas-Relatórios, no 1," ANTT, AMC, cx. 8.

16. Jill Rosemary Dias, "Angola," in *Nova História da Expansão Portuguesa*, vol. 10, *O Império Africano 1825–1890*, ed. Valentim Alexandre and Jill Rosemary Dias (Lisbon: Editorial Estampa, 1998), 320–556, at 451–452.

17. On this turn from Brazil to Africa, see Valentim Alexandre, *Velho Brasil, Novas Áfricas: Portugal e o Império (1808–1975)* (Lisbon: Edições Afrontamento, 2000); Gabriel Paquette, *Imperial Portugal in the Age of Atlantic Revolutions: The Luso-Brazilian World, c. 1770–1850* (Cambridge: Cambridge University Press, 2013).

18. About the first experiments to cultivate coffee in Angola, see Dias, "Angola"; Aida Freudenthal, *Arimos e Fazendas: A Transição Agrária Em Angola, 1850–1880* (Luanda: Edições Chá de Caxinde, 2005); David Birmingham, "The Coffee Barons of Cazengo," *Journal of African History* 19, no. 4 (1978): 523–538.

19. See Gago, "Robusta Empire."

20. Dias, "Angola," 451–452.

21. Mbundu are the speakers of Kimbundu language.

22. Jill Rosemary Dias, "Changing Patterns of Power in the Luanda Hinterland: The Impact of Trade and Colonisation on the Mbundu, ca. 1845–1920," *Paideuma: Mitteilungen zur Kulturkunde* 32 (1986): 285–318, at 290.

23. Birmingham, "The Coffee Barons of Cazengo."

24. Historians working on the nineteenth century (Jill Dias, Aida Freudenthal, David Birmingham) are exemplary in capturing the nuances as well as the entanglements with the environment.

25. Dias, "Angola," 452–453.

26. The agricultural scientists Fernando Oliveira Baptista and José Mendes Ferrão, the geographer Mariano Feio, and the anthropologist José Redinha give important historical accounts about this peasant coffee elite. See Fernando Oliveira Baptista, *O Café em Angola: Um Panorama Socioeconómico* (Lisbon: 100 LUZ, 2012); José Eduardo Mendes Ferrão, *O Café: A Bebida Negra dos Sonhos Claros* (Lisbon:

Chaves Ferreira, 2009); Feio Mariano, *As Causas do Fracasso da Colonização de Angola* (Lisbon: Instituto de Investigação Científica Tropical, 1998); José Redinha, *Introdução ao Estudo das Sociedades e Economias Tradicionais de Angola* (Luanda: Universidade de Luanda, 1972).

27. C. A. Krug, *World Coffee Survey: A Draft of a Future FAO Agricultural Study* (Rome: Food and Agricultural Organization of the United Nations, 1959).

28. See note 4.

29. Gago, "Robusta Empire."

30. Gago, "How Green Was Portuguese Colonialism?"

31. Letter from President of the Board (Salvador de Lucena) to Minister of the Colonies, January 12, 1950, File "1º Volume," Arquivo Histórico Ultramarino (hereafter AHU), Ministério do Ultramar (hereafter MU), Gabinete do Ministro (hereafter GM), cx. 213, 1E.

32. Frederick L. Wellman, *Coffee: Botany, Cultivation, and Utilization* (London: Leonard Hill, 1961).

33. Artur Rocha Medina, *O Café No Estado de São Paulo* (Lisbon: Sociedade Astória, 1947).

34. Medina, *O Café No Estado de São Paulo.*

35. Artur Rocha Medina, "A Tecnologia do Café e o Tamanho da Propriedade," *Agronomia Angolana* 3 (1950): 3–15.

36. Medina, "A Tecnologia do Café e o Tamanho da Propriedade," 15.

37. Manuel António Matias, "Fábricas de Rebenefício e Homogeneização Do Ambrizete e Ambriz," *Revista do Café Português* 6 (1955): 37–44, at 37.

38. "Documentário da Junta: A Inauguração da Fábrica de Luanda da Junta de Exportação do Café; Alguns Dados sobre a Fábrica," *Revista do Café Português* 1 (1954): 34–36.

39. Letter from Manuel Pedro Benedito de Castro (delegate of the Board of Coffee Exports) to the Government of Angola, February 17, 1950, AHU, File "Organismos de coordenação económica, Junta de Exportação do Café Colonial, Juncafé," Governo Geral de Angola, Processo 76-B/1º (vol. 1). See also Gordon Wrigley, *Coffee* (Harlow, UK: Longman, 1988).

40. *Regulamento para a Classificação dos Cafés Portugueses*, Portaria 10,835 of January 12, 1945, 32.

41. *Regulamento para a Classificação dos Cafés Portugueses*, 30–32.

42. See, for instance, Letter from the magistrate Álvaro Rodrigues da Silva Tavares to the General Governador of Angola, December 14, 1951, File 76-B/1º, "Organismos de coordenação económica, Junta de Exportação do Café Colonial, Juncafé" (vol. 2), AHU, GANG, Governo Geral de Angola.

43. Jones, "Food-Crop Marketing Boards in Tropical Africa," 378.

44. On the valorization program implemented in Brazil in 1906 enabling the Brazilian state to intervene in the market in order to stabilize prices, see Topik, "The Integration of the World Coffee Market," 26–28; André Felipe Cândido da

Silva, "A Campanha Contra a Broca-do-Café em São Paulo (1924–1927)," *História, Ciências, Saúde-Manguinhos* 13, no. 4 (December 2006): 957–993.

45. See, for instance, Álvaro Garrido and Fernando Rosas, eds., *Corporativismo, Fascismos, Estado Novo* (Lisbon: Edições Almedina, 2012).

46. Fernando Rosas, "O Corporatismo Enquanto Regime," in *Corporativismo, Fascismos, Estado Novo*, ed. Fernando Rosas and Álvaro Garrido (Lisbon: Almedina, 2012), 17–47, at 34.

47. M. Anne Pitcher, *Politics in the Portuguese Empire: The State, Industry and Cotton, 1926–1974* (Oxford: Oxford University Press, 1993).

48. Saraiva, *Fascist Pigs*.

49. As far as I can ascertain, this industry would begin at the end of the 1950s.

50. Wrigley, *Coffee*, 57. Apparently, the Dutch lobbying, which aimed at protecting robusta production in the Dutch East Indies (Java coffee), was key to convincing the Americans. In the process of negotiation, led by the Food and Drug Administration, which organized several "discussions and hearings," the Green Coffee Association "took an active part"; see Topik, "The Integration of the World Coffee Market," 47.

51. Topik, "The Integration of the World Coffee Market," 45.

52. Steven Topik, "Coffee as a Social Drug," *Cultural Critique* 71 (2009): 81–106, at 101.

53. *AEA Statistical Yearbook, 1948* (Luanda: Imprensa Nacional, n.d.), 358–359; *AEA Statistical Yearbook, 1949–1950* (Luanda: Imprensa Nacional, n.d.), 440–441.

54. Letter of the Board (S.L./F.S.) to the Vice-President of the Conselho Técnico Corporativo, Lisbon, September 21, 1945, File "24/46 Café", AHU, MU, GM, cx.148, 1/1D (1945–1947).

55. CADA was considered by the Food and Agricultural Organization of the United Nations in 1959 as being "the biggest" fazenda in the world; Krug, *World Coffee Survey*, 64.

56. Letter from the Board (S.L./F.S) to the Vice-President of the Conselho Técnico Corporativo, Lisbon, October 1, 1945, File "24/46 Café," AHU, MU, GM, cx.148, 1/1D (1945–1947).

57. Telegram from the Board to (?), Lisbon, October 11, 1945, File "24/46 Café," AHU, MU, GM, cx.148, 1/1D (1945–1947).

58. Telegram from the Board to (?), Lisbon, October 11, 1945.

59. Doc. 8, s/d, File "24/46 Café," AHU, MU, GM, cx.148, 1/1D (1945–1947). Cf. Letter from President of the Board to the Minister of the Colonies, Lisbon, January 16, 1946, File "24/46 Café," AHU, MU, GM, cx.148, 1/1D (1945–1947). Berry-borer (*bago furado* in Portugal; *broca* in Brazil) is the name of a disease caused by an African insect, *Stephanoderes coffeae*, that bores into the berries of coffee; see Wellman, *Coffee*.

60. Doc. 8, s/d, File "24/46 Café," AHU, MU, GM, cx.148, 1/1D (1945–1947). Cf. Letter from President of the Board to the Minister of the Colonies, Lisbon, January 16, 1946, File "24/46 Café," AHU, MU, GM, cx.148, 1/1D (1945–1947).

61. Letter from President of the Board to the Minister of the Colonies, Lisbon, January 16, 1946, File "24/46 Café," AHU, MU, GM, cx.148, 1/1D (1945–1947).

62. Letter from President of the Board to the Minister of the Colonies, Lisbon, January 16, 1946, File "24/46 Café," AHU, MU, GM, cx.148, 1/1D (1945–1947).

63. Jaime Loureiro, "A Portaria 15.913, de 19 de Jullho 1956 e a Classificação Dos Cafés Portugueses," *Revista do Café Português* 11 (1956): 33–44, at 38.

64. Loureiro, "A Portaria 15.913, de 19 de Jullho 1956," 38.

65. See also Alfredo Sousa, *Ensaio de Análise Económica do Café* (Vila Nova de Famalicão: Junta de Investigações do Ultramar, 1958).

66. Documentation in File 76-B/1º, "Organismos de coordenação económica, Junta de Exportação do Café Colonial, Juncafé," AHU, GANG, Governo Geral de Angola.

67. Loureiro, "A Portaria 15.913, de 19 de Jullho 1956," 39–40.

68. Loureiro, "A Portaria 15.913, de 19 de Jullho 1956," 42.

69. "A Portaria Ministerial Nº 15.913: Balanço Dos Efeitos de Uma Medida Legislativa," *Revista do Café Português* 16 (1957): 63–65.

70. Loureiro, "A Portaria 15.913, de 19 de Jullho 1956," 44.

71. Topik, "The Integration of the World Coffee Market," 46.

72. See Miguel Bandeira Jerónimo and António Costa Pinto, "Introduction. The International and the Portuguese Imperial Endgame: Problems and Perspectives," *Portuguese Studies* 29, no. 2 (2013): 137–141; Miguel Bandeira Jeronimo and António Costa Pinto, eds., *Portugal e o Fim do Colonialismo: Dimensões Internacionais* (Lisbon: Edições 70, 2014); Miguel Bandeira Jerónimo and António Costa Pinto, eds., *The Ends of European Colonial Empires: Cases and Comparisons* (Houndmills, UK: Palgrave Macmillan, 2015).

73. For instance, Ball, *Angola's Colossal Lie*; Wheeler, "The Galvão Report on Forced Labor"; Varanda, "A Bem da Nação"; Pinto Monteiro, *Portugal e a Questão do Trabalho Forçado.*

74. On this reformist movement, see Keese, "The Constraints of Late Colonial Reform Policy," about the several attempts made by high officials of the Portuguese administration to implement reforms to abolish forced labor.

75. About cotton, see Pitcher, *Politics in the Portuguese Empire*. About the role scientists played in the case of this crop, see Saraiva, *Fascist Pigs.*

76. See Benoit Daviron, "Small Farm Production and the Standardization of Tropical Products," *Journal of Agrarian Change* 2, no. 2 (2002): 162–184; Benoit Daviron and Stefano Ponte, *The Coffee Paradox: Global Markets, Commodity Trade and the Elusive Promise of Development* (New York: Zed Books, 2005).

77. Frederick Cooper, *Africa in the World: Capitalism, Empire, Nation-State* (Cambridge, MA: Harvard University Press, 2014), 22.

Chapter Eight

Statistics and Emancipation from New Deal America to Guerrilla Warfare in Guinea-Bissau

TIAGO SARAIVA

Introduction

In 1957, Mordecai Ezekiel summarized the accomplishments of "Ten Years of FAO Statistics and Economics Training Centers."[1] Ezekiel, then the director of the Economics Division of the Food and Agriculture Organization of the United Nations (FAO), insisted that statistical training, more than teaching "merely compelling information" on world agriculture, should produce experts able to both collect reliable data and analyze them for "wiser planning." Knowing how to describe reality was inseparable from improving it, a message Ezekiel believed to be of obvious resonance in the "less developed regions of Asia, Africa, and Latin America."[2] From 1948 to 1956, the FAO organized twenty-four graduate schools of one and a half to four months across the world in cities like Baghdad, Mexico City, Ibadan, Paris, Cairo, Bangkok, New Delhi, and Buenos Aires. These were attended by more than a thousand students, the large majority among them public officials funded by their respective governments.

Ezekiel's distinguished role as economic adviser of Henry A. Wallace, the secretary of agriculture in Franklin Delano Roosevelt's administration, confirms the recent insistence in tracing back to the New Deal years many United Nations (UN) development policies of the Cold War.[3] This chapter, by exploring the relation between statistical methods and reforming American democracy, contributes to a growing revisionist historiography that takes the history of the US federal state during the New Deal as an essential element of the history of international institutions for global governance formed in the aftermath of World War II, namely the UN. In addition, it is now clear that many such institutional arrangements were also built on previous colonial infrastructures, being in some cases direct successors of imperial bureaucracies.[4] The 1953 statistical training center, for example, that took place in Ibadan (Nigeria) for ten weeks and that is of special significance for this chapter, was jointly organized by the FAO and the British Colonial Office.[5] Designed with the purpose of reaching Sub-Saharan Africa, the training was done both in English and in French, thus relying also on French colonial experts. The relation between American hegemony and European colonial structures is indeed crucial to unveiling transnational dynamics of science in the Cold War.[6]

As if the challenges of bringing together US history and colonial history were not formidable enough, this text argues for the need for considering as a major part of any narrative on science and development the practices of scientists from the Global South. Historians captured by the colonial and American archives as well as by international agencies' archives tend to be oblivious to the historical agency of actors originating from less obvious geographies and with unconventional life stories.[7] Sophisticated historical accounts have already taken us away from simplistic views of science and technology as tools of empire, suggesting instead that colonial scientists through their attention to environmental and social variables put in place a powerful critique of imperial reveries of civilizing missions.[8] Nevertheless, we still lack a proper understanding of how such critique could become the basis for political imaginations of the postcolonial era.[9] The attention given here to the scientific work of Amílcar Cabral, the leader of the guerrilla movement that started in 1960 against Portuguese colonialism in Guinea-Bissau (West Africa), points at possible ways of writing the history of science also as the history of African independence.

A fully transnational history of science and development as told through the figures of Mordecai Ezekiel and Amílcar Cabral demands from the historian simultaneous engagement with (at least) four major historiographies: the history of international institutions (e.g., the UN, the FAO, and the World Health Organization), the history of science and the American state, the history of colonial science, and histories of science from the Global South. The fourth historiography is by far the least explored in transnational histories of science of the twentieth century and it will thus deserve here closer consideration. Bringing to the forefront Cabral and his statistical practices is not meant to find missing Black scientists to make the field of history of science and technology more inclusive.[10] The gesture points at an alternative direction: attention to Black historical actors like Cabral holds the promise of making the field more historically significant.

Statistics at the USDA during the New Deal

The US Department of Agriculture (USDA), under the leadership of Henry A. Wallace from 1933 to 1940, was the source of some of the most innovative and ambitious federal policies of the New Deal era.[11] Wallace's USDA, which had the highest concentration of scientists in the entire federal government, and probably in the entire world, has been aptly described as an unprecedented case (in terms of scale) of science at the service of public policy design.[12] What has been less noticed is that the science in question was just not there waiting to be used. Wallace insisted that American democracy needed not just more science but a different kind of science.

Scientists had certainly demonstrated their ability to discover new medical cures, increase industrial production, or breed more productive animals and plants. But what the Great Depression and its crisis of overproduction demanded was science less focused on increasing economic output and more concerned with making the American economy work for the whole of the American population: "The science of production is useless when products can't be distributed." For Wallace, statistics was the solution. Not the traditional statistics of counting and amassing data, but a new statistics to which he had been an early contributor, in which descriptions of reality also provided ways to tinker with it.[13]

During World War I, as a young mathematical prodigy and member of an influential Iowan family, Wallace contributed to the federal policy of mobilizing American farmers to increase pork production. To cover the shortage in animal fats of European ally countries it was necessary to guarantee prices attractive enough for Midwestern farmers to increase the slaughtering of more hogs. Herbert Hoover, the leader of the US Food Administration and later to become president during the Great Depression, emphatically declared, "Every pound of fat is as sure of service as every bullet, and every hog is of greater value to the winning of this war than a shell."[14] Considering that corn constituted the main feed of American hogs, the corn/hog ratio, or the value of a bushel of corn divided by the value of hogs per hundredweight, was the key number determining American farmers' decisions. Wallace thus embarked on a study identifying the main factors affecting corn and hog prices to establish what would be a fair offer by the federal government for American farmers to bring more hogs to market. He produced long series of agricultural prices (from 1858 to 1914) from the data made available by the USDA and by the Chicago Board of

Trade, which, as he said, registered "prices of corn belt food staples . . . more promptly and more delicately than anywhere else in the world."[15]

Wallace produced tables showing ten-year average values of the ratio based on prices of No. 2 Chicago corn and values of Chicago hog flesh. Using simple and multiple correlation, he investigated how hog prices varied with bank clearings (a proxy for business activity), hog receipts at Chicago, as well as domestic and foreign demand. To refine his method of ratio calculating he put forward a composite corn value to account for the fact that hogs had been made out of corn at varying values and that the amount of corn consumed also varied as a function of the age of the animals. Wallace had undertaken what Mordecai Ezekiel would later characterize as "the first realistic econometric study ever published."[16] Such praise did not recognize that American social scientists were building on the work of British and German scholars.[17]

Ezekiel would make use of Wallace's studies after being hired in 1922 by the Bureau of Agricultural Economics. The bureau had been established in the USDA that same year by Henry C. Wallace, the father of Henry A. Wallace, who had been appointed secretary of agriculture by President Warren G. Harding. In 1926, Ezekiel, together with G. C. Haas, published a USDA bulletin, *Factors Affecting the Price of Hogs*, exemplifying the kind of work economists of the bureau should produce to better the situation of American farmers.[18] The aftermath of World War I had led to a steep decrease in farm produce demand, leading to a pronounced reduction in farm income. Dealing with the consequences of overproduction became the center of attention of USDA economists like Ezekiel, a tendency that would be paramount during the Great Depression. Wallace had shown the way by making use of multiple correlations to establish the main factors affecting the prices of farm products. According to Ezekiel, this enabled USDA scientists to produce Agricultural Outlook statements indicating "future developments to farmers."[19] In the specific case of hogs, historical cycles of corn-hog ratios should indicate to farmers how much corn should they bring to market and how much they should feed to their hogs: "They [the farmers] can make the best adjustment only if they can reach a sound conclusion as to the future development of the hog market."[20] What in the 1920s was mostly an information and advice service for farmers would shift to "more positive programs" and form the basis "of the much broader farm action programs of the New Deal."[21]

Before the inauguration of FDR's presidency in January 1933, farm commodity prices had reached record lows. Mortgage foreclosures, unpaid taxes, bank failures, and rapid decline in land values all expressed the farm distress of the Corn Belt. Urgent measures were necessary to "increase agriculture's share in the national income," not only to help farmers but to save "the agricultural assets supporting the national credit structure."[22] This was the justification for the Agricultural Adjustment Act (AAA), a large-scale experiment in public policy launched in May 1933 by Wallace's USDA to increase agricultural commodities prices and save American farmers from bankruptcy and with them the entire economy of the country. As Ezekiel explained in his new position as economic adviser to the secretary of agriculture directly involved in the design of the AAA, government agencies were now authorized to "experiment with administrative controls over both production and marketing," rewarding farmers through benefit payments for participating in public programs.[23]

The act was applied to seven basic agricultural commodities: wheat, cotton, corn, hogs, rice, tobacco, and milk. The urgency of increasing hog prices was first met by drastically diminishing the number of marketable hogs: pork was sold to relief agencies; low-grade hogs were transformed into tankage and soap; farmers got benefit payments for marketing light pigs. The measures translated into a dramatic slaughter, funded by the AAA, of some six million piglets. Since his early advocacy for the use of statistics in public policies, Wallace had insisted that statistical expertise should not transfer decision making from the people to the federal state, replacing democracy with a centralized planning system. And in fact, the AAA corn-hog program was from the beginning designed with such concerns in mind, taking a representative producer group into active partnership in its development. By mid-June 1933, the Iowa Federation of Farm Organizations had called a meeting of corn and hog producers from which a first state corn-hog committee was formed. By mid-July, the administration promoted a meeting in Des Moines with representatives of the ten Corn Belt states (Iowa, Illinois, Nebraska, Indiana, Missouri, Minnesota, Ohio, South Dakota, Kansas, Wisconsin), forming a national corn-hog committee of twenty-five members. It included farmers, leaders of farmers' associations, agriculture journalists, directors of marketing associations, and the president of the National Grange. When members of the AAA met with this national committee, in the presence of Wallace himself, a consensus was formed over the need to reduce simultaneously corn acreage and the number of hogs in the market as the only permanent solution to the problem.

The hog reduction campaign started with a series of educational meetings at the county level in which local agents assisted by committeemen and extension specialists explained with the help of a series of charts the aims and the expected results of the program. Two main pamphlets were produced to be used in these meetings: *The Corn-Hog Problem* and *Analysis of the Corn-Hog Situation*. This was the materialization of Wallace's idea of scientific expertise serving democracy. Community meetings were then organized to explain the contract forms to individual farmers followed by community sign-up meetings. Finally, the campaign promoted the organization of permanent "county corn-hog control associations," making the control of corn-hog production a permanent feature of the daily lives of local communities. By March 1, 1934, Iowa had already signed 144,000 contracts representing about 90 percent of the corn and hog producers in the state. What was at stake was not only reaching a final fair price for farmers but also to do it by involving hundreds of thousands of people in a democratic experience.

Ezekiel, in revisiting the reach of the AAA, exulted over its "contributions to the administrative techniques for economic planning under a democracy." He emphasized in particular how the extended network of the USDA agricultural extension embodied locally by the county agent enabled "localized proposals" that were developed "from the grass-roots up from the point of view of the possibilities and needs of each local community."[24] He was now suggesting no less than that the statistical methods used in designing the AAA could also be applied "when we come to plan broadly over many industries for a general and continuing expansion in our total production and consumption."[25]

The delineation of communities relevant to or worthy of participation in the resolution of the corn-hog problem was not unproblematic. The apparently vast number of farmers enlisted to help shape and undertake this price reform effort did not include those without their own property; neither the statisticians nor their collaborators sought the participation of sharecroppers, many of whom were African American, in the reform.[26] This echoes other New Deal efforts in which Black Americans were marginal participants at best and disadvantaged bystanders at worst.[27]

Global New Deal

Wallace and Ezekiel's painstakingly detailed discussions around corn and hogs always had more global considerations in the background. Their shared understanding of the global nature of the Great Depression justified the belief that the reduction of American agricultural production through the AAA was no more than the first stage of a more comprehensive and long-lasting response.[28] Already in his position as FDR's secretary of state, Wallace was unapologetic in pointing out the myopia of US political leaders' in failing to grasp the major unintended impact of the Depression on international relations. The Coolidge and Hoover administrations of the interwar period had been oblivious to the role of the US economy as international creditor of European countries that were highly indebted from a war effort waged in large measure through the purchase of American raw materials. US high import tariffs resulted in the decreased ability of European countries to pay their debts through exports, turning them increasingly to protectionism. The excesses of production in the American economy that led to the Great Depression were not caused just by increased capacity and productivity but also by the lack of an open international market able to buy American produce. The increased revenue of American farmers achieved through the measures of the AAA was only sustainable in the long run if they could export their hogs, corn, cotton, or wheat. According to Wallace, America had to assume its new leading position in the world and the responsibilities it entailed if new global crises were to be averted. Revisionist historians praising the democratic experimentalism associated with Wallace's USDA during the New Deal years tend to lament how the war effort curtailed such reforming endeavors. Here I would like to insist instead on how the war was instrumental in scaling up the USDA experiment from the national scale to a global scale.

Wallace's most famous speech, "The Century of the Common Man," broadcasted in May 1942 to mobilize American producers for the war effort, points at this global ambition, or more precisely at universalism.[29] He depicted the war as a battle between light and darkness, good and evil, God and Satan, democracy and dictatorship, "democracy [being] the only true political expression of Christianity." Stating that the people of the United States "have moved steadily forward in the practice of democracy," he refused nevertheless the notion of an "American Century," preferring instead a more egalitarian "Century of the Common man," in which America only "suggest[ed] the Freedoms and duties by which the common man must live." The new global order Wallace envisioned, in which "no nation will have the God-given right to exploit other nations" and in which older nations, like the United States, had the privilege to "help younger nations get started on the path of industrialization," was a millenarian one. Reproducing the Social Gospel rhetoric he had learned in his Iowan youth, he called on Americans to fulfill their duty, concluding that "the people's revolution is on the march, and the devil and all his angels can not prevail against it. They can not prevail, for on the side of the people is the Lord."

Social Gospel followers at the turn of the century, such as Wallace's grandfather, Henry Wallace, strived for the realization of the Kingdom of God on this Earth. They were harsh critics of American capitalism of the Gilded Age and they were eager to put science at the service of "God's plan for democracy in the New World."[30] Wallace now urged America to face its world role and extended such vision to the entire globe. He placed American capitalist cartels—the Satan of Social Gospel preachers—side by side with Hitler, the "Supreme Devil," urging modern science to be "released from German slavery" and from cartels "that serve American greed." Only then could science fulfill its sacred democratic duty: "Modern science, when devoted whole-heartedly to the general welfare, has in its potentialities of which we do not yet dream."

Science was thus central for Americans to realize "a just, charitable, and enduring" peace. But again, this was not just any science. This was science as the one practiced in American Land Grant Colleges and the USDA, one in which the union of mind and hand, theory and practice, served the people in their millennial march toward democracy.[31] Wallace considered statistical methods obligatory for science to serve democracy. After having proven its value in the American experiment with democracy, statistics was to become integral to the millenarian American vision for world peace, a vision in which the United States would not colonize other nations, assisting them instead to achieve the same democratic standards as the United States.[32]

The personal trajectory of Wallace has served to confirm the failure of his idealistic views in the postwar years. Truman, not Wallace, was the choice of the Democratic Party to run as vice president in FDR's reelection in 1944.[33] Truman would defeat Wallace in the 1949 presidential election when the latter ran as a Progressive Party candidate winning no more than 2.7 percent of the popular vote. Visions of world peace like those offered by Wallace, in which the Soviet Union was not considered America's enemy, had given place to Truman's Manichaeism that would characterize the Cold War era.

This said, it is hard to miss how much Truman owed to Wallace and fellow New Dealers like Ezekiel when he concretized what America had to offer to the world in opposition to communism. Truman's famous Point Four of his 1949 inaugural speech promised "a bold new program for making the benefits of our scientific advances and industrial progress available for the improvement and growth of underdeveloped areas." The Point Four initiative was to be carried out through the UN, expanding the latter's program of technical assistance to developing countries, already largely funded by the United States.[34] New Dealers like Ezekiel had found an institutional new home in the UN and in American development agencies, and defined the characteristics of their projects.[35]

In 1943, Wallace would take an active role in the plans and arrangements for the Hot Springs Conference on Food and Agriculture, which led to the founding of the FAO.[36] Ezekiel, his previous economic adviser at the USDA, served in 1945 as a member of two of the FAO's first missions to Greece and Poland. Two years after, Ezekiel was overseeing the FAO's Economic Analysis Branch, later being promoted to head of the Economics Department. He would leave the FAO only in 1961, to assume a position in the United States Agency for International Development (USAID). As Ezekiel would recall, "many of the American agricultural economists who in their younger years served under Wallace in the USDA are finishing out their professional careers in AID, seeking to help the rest of the world make the same kind of progress in agriculture that the United States made in the first two-thirds of this century."[37]

Histories of UN agencies, and of the FAO in particular, even when suggesting the need to downplay the role of American experts and highlight the historical importance of actors from different geographies, do not challenge the big picture of American hegemony. Scientists from European countries, India, or Sudan certainly made good use of UN agencies to advance their own local agendas. But it would be misleading to ignore the lasting presence of American agendas first developed in the New Deal years in the institutional life of agencies like the FAO. Its Economic and Statistics Division is an eloquent example, with four of its first five directors being of American origin. Among these were distinguished New Dealers, veterans of the AAA like Howard R. Tolley, or Mordecai Ezekiel himself, responsible for the new Economics Department formed in 1959.

Any UN sponsored project aimed at bettering the health, food, or education status of a population needed statistics to establish the welfare standard of each country.[38] A 1949 UN report on "Existing Statistical Deficiencies" complained that of thirty-two countries in Latin America, the Middle East and East Asia, only eight supplied national income figures, data on external trade, or numbers on principal types of livestock.[39] A UN statistical commission was formed to coordinate the collection of demographic and economic data from the different members. Nothing very surprising here. What is more interesting for our argument is the insistence in going beyond such collection activity to accomplish the aims of American global assistance. The uplifting of nations from poverty, according to the aid experts' view, certainly needed the systematic application of science in agriculture, medicine, and industry, but to make science contribute to the general welfare, to make science serve the people, statistical methods were considered mandatory. This was the justification for the proliferation of the FAO Statistics and Economics Training Centers mentioned in the opening of this chapter. These were the centers where officers of European colonial administrations in Africa learned how to adapt their imperial agendas to the new American hegemony.

Science and Colonial Reform in the Aftermath of World War II

Amílcar Cabral, the leader of the war of independence of the Portuguese colony of Guinea-Bissau, got his first exposure to statistical techniques while an undergraduate at the Agronomy Institute in Lisbon from 1945 to 1952.[40] Cabral was born in Guinea-Bissau of parents from Cape Verde; Cabral's father was an elementary school teacher, a mid-rank position typical of the go-between role of Cape Verdeans in Portuguese colonial administration.[41] Cabral's education granted him the status of "assimilated," placing him among the reduced cohort of people of color employed as technical experts at the service of the Portuguese Empire. In 1952, as a second-class engineer of the Portuguese colonial service, he moved to Guinea-Bissau, where he had not returned since his childhood, to take charge of the local agricultural experiment station in Pessubé.[42] As in many other similar experiment stations across European colonies, the task was to improve the productivity of major local crops such as manioc, millet, or groundnuts, while exploring the acclimatization of new ones, namely cashews.[43]

Shortly after his arrival in West Africa, Cabral was asked by the governor of Guinea to undertake an agricultural survey of the colony. From November 1953 to April 1954, Cabral oversaw a team of thirty technicians, which included his wife, Maria Helena Rodrigues (also an agronomist), that crossed the whole territory of Guinea-Bissau. Analysis of collected data was completed in the following months and by the end of the year a report was delivered to the governor. This was published in 1956 in the *Boletim Cultural da Guiné* (Guinea cultural bulletin), the official mouthpiece of all forms of colonial knowledge about Guinea.[44] The publication of the report in the pages of the bulletin, next to historical, ethnological, medical, or biological research on Guinea, already indicates what was at stake in organizing the agricultural survey. While the point about science informing Portuguese colonial administration is obvious, the bulletin was self-consciously designed in the aftermath of World War II with the ambition of unveiling the cultural and natural values of Guinea and its indigenous inhabitants.

A small colony of half a million inhabitants with scarce white settler presence, with no major plantations controlled by Europeans and no Portuguese rural settlements, in contrast with the larger imperial possessions of Angola and Mozambique, Guinea-Bissau, through the pages of the bulletin, offered the alternative vision of a benevolent Portuguese colonialism promoting the development of indigenous populations based on scientific knowledge of local realities.[45] Despite its smaller dimensions, or maybe also because of these, Guinea-Bissau became the site for a major experiment in a new type of colonialism centrally guided by a scientifically informed colonial state that promised to refrain from predatory colonial practices by private interests while improving indigenous living conditions.[46]

We don't need to investigate the true intentions of Portuguese colonial officers and their actual commitment to the welfare of African populations to draw conclusions about the historical importance of such reformist agendas. The survival of a fascist regime in Portugal after the defeat of the axis in World War II was a major constraint for the incorporation of the country in international organizations, and while the country was part of the group of the founding members of NATO in 1949 (the only dictatorship included), its initial bid to join the UN in 1946 was denied, to be approved only in 1955.[47] The request made in 1953 by the governor of Guinea for Cabral to undertake an agricultural survey of the territory was the direct result of a commitment made in 1947 by the Portuguese government to participate in a world agricultural survey by the FAO to reinforce Portugal's candidacy to the UN. This immediate justification for the survey highlights how the colonial reformist agenda expressed through the works of the bulletin was strengthened by the Portuguese state's anxiety about its international status in the new international order in the aftermath of World War II. As in the cases of French or British colonies, the reform of the empire was not perceived by colonial officers involved in it as leading to the independence of an African country.[48] The opposite was true. Scientifically guided reforms should establish imperial presence on a sounder basis relegating self-rule by indigenous people to a distant future.

Sampling and the Agricultural Survey of Guinea

Delving into the details of the survey illuminates how an initiative undertaken to strengthen the empire became an opportunity to imagine an independent country. Agricultural scientists like Cabral trained at the Agronomy Institute in Lisbon were used to conduct agricultural surveys. A high proportion of students' dissertations from the 1920s until the 1940s were in fact "social surveys" of Portuguese rural life following the tradition of Fréderic Le Play's inquiries.[49] Students produced detailed monographs of rural villages and households inquiring about crops, domesticated animals, agricultural appliances, family members, eating habits, or household budgets. Cabral's dissertation, completed in 1951, although dedicated to soil erosion (more on this below) of a county in the Alentejo region, Portugal's bread basket, also included information on property size and distribution, sharecropping arrangements, and peasants' literacy rates. The favoring of Le Play's methods was no coincidence in a country whose dictator, António Oliveira Salazar, explicitly praised the nineteenth-century French engineer for having developed a science of society based on alleged natural social units (e.g., the family, the parish, or the region) and not on abstractions (individuals, classes), thus serving "the consecration of an imagined social order."[50] Le Play's surveying methods were the perfect fit for the building of a fascist regime promoted as a third way of organizing society opposed to both liberalism and socialism.

The agricultural survey of Guinea would be a very different enterprise. The FAO asked for quantitative data on areas of cultivation and respective crops; general characteristics of the population; farm animals; and productivity of main crops.[51] In Portugal, such data were made available to the central administration by information collected by local agents of an overgrown state infrastructure that regulated agricultural production in the country, and that constituted the backbone of the corporatist fascist Estado Novo.[52] This form of data collection was impossible in a colony with such a flimsy state infrastructure as was the case of Guinea that was divided into forty-one administrative districts and whose respective officers had as their main task the collection of taxes from indigenous populations. This said, Cabral did make such districts the larger territorial divisions of his survey, thus also building it on Portuguese colonial presence.

For each district, he divided the population into subsets according to ethnicity from data produced by previous anthropological inquiries and surveys: a population census of the colony undertaken in 1950 had identified more than thirty different ethnic groups, but six groups accounted for almost the totality of cultivated area, making them the subject of the agricultural survey: Balanta, Fula, Mandinga, Manjaco, Papel, and Mancanha. Cabral's survey apparently contributed to the solidification of ethnic divisions put forward by colonial ethnographies that ignored the complicated fluid identities of the peoples of Guinea. But the distinctive point of Cabral's work that is important for the argument of this text is his use of sampling at the ethnic group level. Considering the impossibility of providing quantitative information for every village for all the different groups, the number of family holdings studied in each ethnic group (homogeneous group) was calculated as a proportion of the weight of that group in the total population of the administrative district. The choice of which agricultural family holdings to study in a village of a given district was random. Cabral and his team collected information from 2,248 indigenous family holdings (from a total of 85,478), in 356 villages spread over the forty-one administrative districts.

The few historians interested in the history of statistics in Portugal have pointed to the 1960s as the decade in which sampling techniques finally became common in the country in surveys undertaken by marketing companies.[53] They have of course overlooked the history of colonial administration and alternative transnational circuits of knowledge, a common neglect for other European national historiographies.[54] Cabral's training in statistics at the Agronomy Institute did not include sampling techniques to produce surveys. Sampling did not travel well to a fascist country in which the census was an expression of its corporatist social and economic structure and in which Le Play's style of monographs was praised as contributing to social order. Cabral learned his sampling techniques instead from the training center organized by the FAO in Ibadan in 1953, referred to above by Mordecai Ezekiel. To be able to provide the information demanded by the FAO, Cabral had to follow the methods he had learned from FAO experts, namely the application of sampling techniques deemed particularly adequate for regions with scarce infrastructure, such as those in most colonies. This portability of sampling helps explain its travels into colonial settings via the UN, but this was no automatic process, as demonstrated by its later introduction in the metropole.

Erosion, Ethnicity and Colonialism

As Ezekiel mentioned in his report on the first ten years of FAO training centers, these were also supposed to prepare students to analyze and use data for better planning. Cabral followed the lead and he would immediately start evaluating agricultural policies based on the data collected in the survey. Maybe the most compelling argument put forward by Cabral related to soil erosion and agricultural practices of the different ethnic groups of Guinea. From the survey of Fulacunda district, south of the capital Bissau, it was possible to estimate for each ethnicity the total area cultivated, area under fallow, and area burnt. These numbers, Cabral remarked, were of particular significance for an agriculture relying on slash and burn for opening new areas for cultivation and on fallow to recover soil fertility. More to the point, the comparison of burnt area and area under fallow produced an index of "agricultural dynamism," a measurement of the rhythm that soil was exposed to or protected from the action of erosion elements.[55] A large index demonstrated that the rhythm of exploitation of the land would soon lead to soil depletion, a small index that the soil was being given enough time to be cultivated again.

Maybe more unexpected and more revealing, Cabral transformed this erosion index into an index of colonization. Large indexes were produced by an acceleration of the rotation: burning, maize (or millet), groundnuts, fallow. Groundnuts constituted the major cash crop produced by Guinean farmers acquired mostly by representatives of Portuguese food-processing companies. Reducing fallow time to produce more groundnuts represented thus for Cabral a direct measure of how the agriculture of an ethnic group had been transformed by and depended on the presence of Portuguese commercial interests. Ethnicities such as the Fulas were major producers of groundnuts and were subsequently among the most colonized. In contrast, the Balantas, the most populous group in the area and in the whole of Guinea-Bissau, had lower erosion indexes and were thus less colonized.

The expansion of cultivation areas by the Balantas was actually not done by slash and burn but by transforming saline mangroves on the coast into rice paddies. Cabral recognized that talking about burnt areas for the

Balanta was problematic, which did not stop him from establishing a clear contrast between colonized Fulas cultivating groundnuts in the interior and independent Balantas cultivating rice on the coast. As part of a general program of reforming the colony by scientifically demonstrating the value of local solutions, Cabral's recommendations were clear: follow the Balantas and promote rice cultivation, refrain from the Fulas' practices of always cultivating more groundnuts. Insistence on groundnuts would serve only the interests of Portuguese industry, enriching a small elite among the Fulas while condemning most of the peasants to food insecurity in the long term and making them vulnerable, not only to market fluctuations, but also to increased soil erosion. Rice also had a market value, constituting in fact the other major export from Guinea, allowing the Balantas to actively participate in a market economy. But in addition, rice was obviously a food crop that kept Balantas cultivators immune to food scarcity produced by market fluctuations.

Cabral's discussion of rice produced by the Balantas as defense against the excesses of predatory forms of colonialism is fascinating. As a copious literature on the history of the Black Atlantic has demonstrated, rice cultivation in West Africa was crucial to the establishment of a slave society in the Carolinas in the seventeenth century that had rice as its main commodity.[56] Rice from West Africa was also found across the Americas as a major staple for marooned communities. Historians willing to recover the agency of people of African origins in the Americas have offered us many narratives around rice, in clear contrast with those based on colonial commodities such as sugar or coffee. That said, expansion of the cultivation of rice in Guinea had been a major enabler of the Atlantic slave trade since the sixteenth century, supplying slave ships crossing the ocean.[57] But crucially for the argument developed by Cabral, migrations of Balanta people to the coastal mangroves was not just a strategy for finding new areas of rice cultivation, but first and foremost a defense strategy to run away from slave raids undertaken by people like the Mandingas and Fula, who historically had been major suppliers of slaves to European merchants. Yes, Balantas cultivated rice supplying the slave trade, but the labyrinth of mangroves and rice paddies of coastal Guinea protected them from being enslaved. Balantas found in rice a crop that allowed them to profit from Atlantic trade while isolating them from its most violent dimensions.[58] This was also what had fascinated Cabral while doing his agricultural survey in the mid-twentieth century: rice was a crop that simultaneously connected African peasantry to the world and sheltered it from the predatory consequences of colonialism as materialized in soil erosion and food insecurity.

The characterization of different ethnicities as a function of their agricultural production and their relation to the land had additional important consequences. The data collected following FAO methods suggested that differences between ethnicities was a question of numbers that could be tinkered with rather than of immutable qualities. Changes in agricultural practices could ultimately lead to changes in ethnicity. The many different ethnic groups of Guinea-Bissau identified by colonial agents could indeed be understood as the result of fluid identities determined by historical conditions. This construction of ethnicity by Cabral was in stark contrast with that developed by multiple Portuguese anthropological missions which established hard ethnic identities based on religious rituals, social habits, or hereditary traits. An interpretation of ethnicity as a fluid evolving reality was indeed crucial for a project of independence that recognized differences but that proclaimed the overarching unity of a national struggle.

Conclusion

Cabral's biographers do not forget to mention the importance of his agricultural survey for the success of the guerrilla movement he started in 1963 and that would lead to the independence of the country from Portugal a decade later.[59] Knowledge of the terrain constitutes an obvious major military advantage and Cabral's crossing of all the different areas of the territory to undertake the FAO survey seemed to offer just that. The point made here is that studying the history of such knowledge matters and that generic references to the expertise of the guerrilla leader are not satisfactory. The statistical knowledge used by Cabral had a particular historical trajectory enabling the political imagination of the guerrilla leader. This was not just any kind of knowledge. This was statistics taught in FAO training centers building on the experience of the New Deal and developed to reform American democracy. While in the metropole statistics strengthened the bureaucratic structure of the Portuguese fascist state, in Guinea-Bissau the statistical techniques and sampling methods used by Cabral in the agricultural survey helped make plausible a political project of national independence.

In 1968, a new Portuguese military governor arrived in Guinea-Bissau to counter the successful guerrilla movement put in place by Cabral. General António de Spínola added to his previous experience of war in Angola a deep knowledge of counterinsurgency literature produced by the British, French, and American militaries, as well as familiarity with the guerilla manuals written by Mao Zedong and Che Guevara. Judging the previous use of napalm and other violent methods by the Portuguese military as too crude and inefficient to defeat the guerrillas, Spínola launched a major colonial development effort, which he baptized "For a better Guinea."

Hearts and minds of local populations were to be conquered by the usual array of investments of the development era in public infrastructure and public health. In addition, general Spínola organized a Congress of the Peoples of Guinea, offering political representation to the different ethnicities of the colony.

The Congress hardened differences that Cabral understood as fluid. What seemed a colonial reform empowering local populations accentuated differences, empowered traditional forms of leadership, and stagnated social change. Political representation was designed as a function of the ethnic categories identified by anthropologists as immutable realities, ignoring the livelihood conditions of the populations as characterized by the FAO's survey, led by Cabral. While ethnicity was the criterion for organizing the Congress, Cabral insisted on knowledge of agricultural realities to constitute a political community. Directly inspired by the techniques he used to train the auxiliaries for the agriculture survey, he urged his guerrilla fighters to learn how to talk and listen to the local populations they wished to convince to join the fight. Here is how Cabral advised his comrades on how to start a conversation with Guinea-Bissau villagers:

> It was very rare that both chicken and rice would be served. . . . I would say, "Papa, why aren't you giving me anything but rice?" . . . "I am poor, I don't have any chicken." . . . "Why?" "Why are you asking me these questions son? Yes, I used to have cattle but the white man took everything away with that tax." . . . By that point I had had a chance to size up the old man. . . . "Papa, tell me, if something happened to come along that would make it possible for you to have a better life tomorrow, would you go along with it?" "I would go along with it."[60]

For Cabral, inquiring about rice and chickens was a necessary first step to constituting an independent political community. While this might seem a rough and ready method to make a country, it is good to keep in mind that it was by inquiring about corn and hogs that scientists embarked on one of the most consequential reforms of American democracy.

Notes

1. Mordecai Ezekiel, "Ten Years of FAO Statistics and Economics Training Centers," *Journal of Farm Economics* 39, no. 2 (1957): 221–234.

2. Ezekiel, "Ten Years of FAO Statistics and Economics Training Centers."

3. There is a growing body of historical scholarship linking New Deal policies and American development initiatives after World War II. I find the following references particularly useful: Daniel Immerwahr, *Thinking Small: The United States*

and the Lure of Community Development (Cambridge, MA: Harvard University Press, 2015); David Ekbladh, *The Great American Mission: Modernization and the Construction of an American World Order* (Princeton, NJ: Princeton University Press, 2010); Seth Garfield, *In Search of the Amazon: Brazil, the United States and the Nature of a Region* (Durham, NC: Duke University Press, 2013); Tore C. Olsson, *Agrarian Crossings: Reformers and the Remaking of the US and Mexican Countryside* (Princeton, NJ: Princeton University Press, 2017); Elizabeth Borgwardt, *A New Deal for the World: America's Vision for Human Rights* (Cambridge, MA: Harvard University Press, 2007).

4. Mark Mazower, *No Enchanted Palace: The End of Empire and the Ideological Origins of the United Nations* (Princeton, NJ: Princeton University Press, 2009); Joseph Morgan Hodge, *Triumph of the Expert: Agrarian Doctrines of Development and the Legacies of British Colonialism* (Athens: Ohio University Press, 2007); Saul Dubow, "Smuts, the United Nations and the Rhetoric of Race and Rights," *Journal of Contemporary History* 43, no. 1 (2008): 43–72.

5. *Report of the African Training Centre in Agricultural Statistics, Ibadan, Nigeria, July–September 1953* (Rome: Food and Agricultural Organization of the United Nations, 1953).

6. Axel Jansen, John Krige, and Jessica Wang, "Empires of Knowledge: Introduction," in "Empires of Knowledge," ed. Axel Jansen, John Krige, and Jessica Wang, special issue, *History and Technology* 35, no. 3 (2019): 195–202; Jessica Wang, "Plants, Insects, and the Biological Management of American Empire: Tropical Agriculture in Early Twentieth-Century Hawai'i," in Jansen, Krige, and Wang, "Empires of Knowledge," *History and Technology* 35, no. 3 (2019): 203–236.

7. For a powerful critique of such tendencies, see Clapperton Chakanetsa Mavhunga, *The Mobile Workshop: The Tsetse Fly and African Knowledge Production* (Cambridge, MA: MIT Press, 2018); Pablo F. Gómez, *The Experiential Caribbean: Creating Knowledge and Healing in the Early Modern Atlantic* (Chapel Hill: University of North Carolina Press, 2017); Gabriela Soto Laveaga, "Largo Dislocare: Connecting Microhistories to Remap and Recenter Histories of Science," *History and Technology* 34, no. 1 (2018): 21–23.

8. Helen Tilley, *Africa as a Living Laboratory: Empire, Development, and the Problem of Scientific Knowledge, 1870–1950* (Chicago: University of Chicago Press, 2011); Maria do Mar Gago, "Robusta Empire: Coffee, Scientists and the Making of Colonial Angola (1898–1961)" (PhD diss., Instituto de Ciências Sociais, Universidade de Lisboa), 2018.

9. Historiography on India is a major important exception. See particularly Prakash Kumar, "'Modernization' and Agrarian Development in India, 1912–52," *Journal of Asian Studies* 79, no. 3 (2020): 633–658; Projit Bihari Mukharji, "Profiling the Profiloscope: Facialization of Race Technologies and the Rise of Biometric Nationalism in Inter-war British India," *History and Technology* 31, no. 4 (2015): 376–396.

10. For an insightful critique of such historiography, see Amy E. Slaton, *Race, Rigor, and Selectivity in US Engineering: The History of an Occupational Color Line* (Cambridge, MA: Harvard University Press, 2010).

11. An extended version of the argument of this section may be found in Tiago Saraiva and Amy E. Slaton, "Statistics as Service to Democracy: Experimental Design and the Dutiful American Scientist," in *Technology and Globalisation: Networks of Experts in World History*, ed. David Pretel and Lino Camprubí (Cham, Switzerland: Springer, 2018), 217–255. On agricultural policy in the New Deal, see Jess Gilbert, *Planning Democracy: Agrarian Intellectuals and the Intended New Deal* (New Haven, CT: Yale University Press, 2015); Olsson, *Agrarian Crossings*; Andrew Jewett, "The Social Sciences, Philosophy, and the Cultural Turn in the 1930s USDA," *Journal of the History of the Behavioral Sciences* 49, no. 4 (2013): 396–427.

12. Theda Skocpol and Kenneth Finegold, "State Capacity and Economic Intervention in the Early New Deal," *Political Science Quarterly* 9, no. 2 (1982): 255–278; Daniel T. Rodgers, *Atlantic Crossings: Social Politics in a Progressive Age* (Cambridge, MA: Harvard University Press, 2000).

13. See Saraiva and Slaton, "Statistics as Service."

14. For the general context of American economy mobilization for World War I, see Adam Tooze, *The Deluge: The Great War and the Remaking of the Global Order, 1916–1931* (New York: Viking, 2015). For food production, see Lizzie Collingham, *The Taste of War: World War II and the Battle for Food* (New York: Penguin, 2012); Nick Cullather, "The Foreign Policy of the Calorie," *American Historical Review* 112, no. 2 (2007): 337–364. For Hoover and his food policies, see George H. Nash, *The Life of Herbert Hoover: Master of Emergencies, 1917–1918* (New York: Norton, 1996); Robert D. Cuff, "The Dilemmas of Voluntarism: Hoover and the Pork-Packing Agreement of 1917–1919," *Agricultural History* 53, no. 4 (1979): 727–747.

15. Henry A. Wallace, *Agricultural Prices* (Des Moines, IA: Wallace, 1920), 7.

16. Mordecai Ezekiel, "Henry A. Wallace, Agricultural Economist," *Journal of Farm Economics* 48, no. 4 (1966): 789–802.

17. There is an important body of work in history of science dealing with the historical emergence of these statistical methods: Theodore M. Porter, "Statistics and Statistical Methods," in *The Cambridge History of Science*, vol. 7, *The Modern Social Sciences*, ed. Theodore M. Porter and Dorothy Ross (Cambridge: Cambridge University Press, 2003), 238–250; Greg Gigerenzer, Zeno Swijtink, Theodore Porter, Lorraine Daston, John Beatty, and Lorenz Krüger, *The Empire of Chance: How Probability Changed Science and Everyday Life* (Cambridge: Cambridge University Press, 1989); Ian Hacking, "Telepathy: Origins of Randomization in Experimental Design," *Isis* 79, no. 3 (1988): 427–451; Nancy S. Hall, "R. A. Fisher and His Advocacy of Randomization," *Journal of the History of Biology* 40, no. 2 (2007): 295–325; Trudy Dehue, "Establishing the Experimenting Society: The Historical Origin of Social Experimentation According to the Randomized Controlled Design," *American Journal of Psychology* 114, no. 2 (2001): 283–302; Giuditta Parolini, "The Emergence

of Modern Statistics in Agricultural Science: Analysis of Variance, Experimental Design and the Reshaping of Research at Rothamsted Experimental Station, 1919–1933," *Journal of the History of Biology* 48, no. 2 (2015): 301–335.

18. G. C. Haas and Mordecai Ezekiel, *Factors Affecting the Price of Hogs* (Washington, DC: US Department of Agriculture, 1926).

19. Mordecai Ezekiel, "The Shift in Agricultural Policy toward Human Welfare," *Journal of Farm Economics* 24, no. 2 (1942): 463–476, quotation at 464.

20. Haas and Ezekiel, *Factors Affecting the Price of Hogs*, 53.

21. Ezekiel, "The Shift in Agricultural Policy," 464.

22. Mordecai Ezekiel and Louis H. Bean, *Economic Bases for the Agricultural Adjustment Act* (Washington, DC: Government Printing Office, 1933).

23. Ezekiel and Bean, *Economic Bases for the Agricultural Adjustment Act*.

24. Ezekiel, "The Shift in Agricultural Policy"; Edwin G. Nourse, Joseph S. Davis, and John D. Black, *Three Years of the Agricultural Adjustment Administration* (Washington, DC: Brookings Institution, 1937).

25. Ezekiel, "The Shift in Agricultural Policy," 467.

26. Theodore Saloutos, "New Deal Agricultural Policy: An Evaluation," *Journal of American History* 61, no. 2 (1974): 394–416; David Eugene Conrad, *Forgotten Farmers: The Story of Sharecroppers in the New Deal* (Urbana: University of Illinois Press, 1965); Neil Foley, *The White Scourge: Mexicans, Blacks, and Poor Whites in Texas Cotton Culture* (Berkeley: University of California Press, 1997).

27. In the 1930s, there was no louder voice in the Senate supporting the Jim Crow racist system of the American South than Theodore Bilbo from Mississippi. Making his case against antilynching legislation in 1938, he did not refrain from quoting Hitler's *Mein Kampf*, assuring his audience that "one drop of Negro blood placed in the veins of the purest Caucasian destroys the inventive genius of his mind and strikes palsied his creative faculty." Bilbo was the champion of poor rural whites, building his political career in Mississippi as a Southern populist attacking not only Northern capitalists but also the traditional planters' elite. Bilbo was nevertheless a key political ally for FDR, guaranteeing the support of the block formed by Southern politicians on whom depended the passing in Congress of New Deal policies to get the country out of the Great Depression. Among these was the Agricultural Adjustment Act, which compensated farmers for producing less in order to bring prices up for the country's major agricultural commodities. After the act was found unconstitutional, New Deal policymakers transformed it into a soil conservancy program, generously paying farmers to divert from production eroded parts of their properties, which then became the object of soil recuperation procedures. Bilbo, always concerned that progressive policies of the New Deal risked undermining racial hierarchy in the South, made sure that federal aid would not reach the majority of the Black peasantry in the region. Only landowners were entitled to participate in the soil conservation program put in place by the USDA, thus excluding from its benefits the large masses of Black sharecroppers who did not own the land they

toiled. Ira Katznelson, *Fear Itself: The New Deal and the Origins of Our Time* (New York: Liveright, 2013), 86.

28. "J. Samuel Walker, "Henry A. Wallace as Agrarian Isolationist, 1921–1930," *Agricultural History* 49, no. 3 (1975): 532–548.

29. Henry A. Wallace, *The Century of the Common Man* (New York: Reynal and Hitchcock, 1943).

30. Robert M. Crunden, *Ministers of Reform: The Progressives' Achievement in American Civilization, 1889–1920* (Urbana-Champaign: University of Illinois Press, 1985); Bradley W. Bateman, "Clearing the Ground: The Demise of the Social Gospel Movement and the Rise of Neoclassicism in American Economics," *History of Political Economy* 30, suppl. (1998): 29–52.

31. Alan I. Marcus, ed., *Service as Mandate: How American Land-Grant Universities Shaped the Modern World*, 2nd ed. (Tuscaloosa: University of Alabama Press, 2015).

32. For an insightful critique of this idea of American assistance, see Gisela Mateos and Edna Suárez-Díaz, "Development Interventions: Science, Technology and Technical Assistance," *History and Technology* 36, no. 3–4 (2020): 293–309.

33. Mark L. Kleinman, *A World of Hope, a World of Fear: Henry A Wallace, Reinhold Niebuhr, and American Liberalism* (Columbus: Ohio State University Press, 2000).

34. Olsson, *Agrarian Crossings*; Borgwardt, *A New Deal for the World*. For a more general discussion of the consequences of Point Four for American global hegemony, see John Krige, *American Hegemony and the Postwar Reconstruction of Science in Europe* (Cambridge, MA: MIT Press, 2006).

35. In *Thinking Small*, Immerwahr has convincingly demonstrated that only through this nexus can one understand the insistence of American experts on community development in India and the Philippines.

36. Ralph Wesley Phillips, *FAO: Its Origins, Formation, and Evolution, 1945–1981* (Rome: Food and Agricultural Organization of the United Nations, 1981).

37. Ezekiel, "Henry A. Wallace, Agricultural Economist," 801.

38. "Statistical Development in Certain Countries and Possible Remedial Measures," Statistical Commission, United Nations Economic Council, March 9, 1945.

39. United Nations, *Existing Statistical Deficiencies* (n.p., 1949).

40. For biographic details, see Patrick Chabal, *Amílcar Cabral: Revolutionary Leadership and People's War* (Trenton, NJ: Africa World Press, 2003); Ronald H. Chilcote, *Amílcar Cabral's Revolutionary Theory and Practice: A Critical Guide* (Boulder, CO: Lynne Rienner, 1991); Peter Karibe Mendy, *Amílcar Cabral: Nationalist and Pan-Africanist Revolutionary* (Athens: Ohio University Press, 2019); António Tomás, *O fazedor de utopias: Uma biografia de Amílcar Cabral* (Lisbon: Tinta-da-China, 2007).

41. The identification of Cape Verdeans with Portuguese imperial rule was particularly salient in Guinea-Bissau since this West Africa territory had been administrated from the Cape Verde islands until 1879.

42. For Cabral's trajectory as an agronomist, see Filipa César, "Meteorisations: Reading Amílcar's Cabral Agronomy of Liberation," *Third Text* 32, no. 2–3 (2018): 254–272; Carlos Schwartz da Silva, "Amílcar Cabral, um agrónomo antes do seu tempo," *Buala*, May 21, 2012, https://www.buala.org/pt/a-ler/amilcar-cabral-um-agronomo-antes-do-seu-tempo; José Neves, "Ideologia, ciência e povo em Amílcar Cabral," *História, Ciências, Saúde-Manguinhos* 24, no. 2 (2017): 333–347; Sónia Vaz Borges, "Amílcar Cabral: Estratégias políticas e culturais para independência da Guiné e Cabo-Verde" (master's thesis, University of Lisbon, 2009).

43. For an overview of the work of colonial experimental stations in Africa, see Christophe Bonneuil, "Development as Experiment: Science and State Building in Late Colonial and Postcolonial Africa, 1930–1970," in *Osiris*, vol. 15, *Nature and Empire: Science and the Colonial Enterprise*, ed. Roy McLeod (Chicago: University of Chicago Press, 2000), 258–281. For the Portuguese Empire and its relation to other fascist imperial undertakings, see Tiago Saraiva, *Fascist Pigs: Technoscientific Organisms and the History of Fascism* (Cambridge, MA: MIT Press, 2016).

44. Frederico Ágoas,. "Social Sciences, Modernization, and Late Colonialism: The Centro de Estudos da Guiné Portuguesa," *Journal of the History of the Behavioral Sciences* 56, no. 4 (2020): 278–297; Diogo Ramada Curto, *O Colonialismo Português em África: De Livingstone a Luandino* (Lisbon: Edições 70, 2020), 149–160.

45. For an overview of sciences and Portuguese colonialism in Africa, see Cláudia Castelo, "Investigação científica e política colonial portuguesa: Evolução e articulações, 1936–1974," *Manguinhos* 19, no. 2 (2012): 391–408. See also António E. Duarte Silva, "Sarmento Rodrigues, a Guiné e o luso-tropicalismo," *Cultura* 25 (2008): 31–55.

46. Peter Karibe Mendy, "Portugal's Civilizing Mission in Colonial Guinea-Bissau: Rhetoric and Reality," *International Journal of African Historical Studies* 36, no. 1 (2003): 35–58.

47. For this international context, see Valentim Alexandre, *Contra o Vento: Portugal, o Império e a Maré Anticolonial* (Lisbon: Temas e Debates, 2017).

48. Frederick Cooper, *Africa since 1940: The Past of the Present* (Cambridge: Cambridge University Press, 2002); Hodge, *Triumph of the Expert*.

49. Frederico Ágoas, "Povo, população e sociedade na investigação económica-agrária do início do século XX," in *Como Se Faz um Povo: Ensaios em História Contemporâneo de Portugal*, ed. José Neves (Lisbon: Tinta-da-China, 2010), 293–310. On Le Play, see Theodore M. Porter, "Reforming Vision: The Engineer Le Play Learns to Observe Society Sagely," in *Histories of Scientific Observation*, ed. Lorraine Daston and Elizabeth Lunbeck (Chicago: University of Chicago Press, 2011), 281–302.

50. Porter, "Reforming Vision," 282;

51. Maria H. Cabral and Amílcar Cabral, "Breves notas acerca da razão de ser, objetivos e processo de execução do recenseamento agrícola da Guiné," *Boletim Cultural da Guiné Portuguesa* 9, no. 33 (1954): 195–202.

52. Nuno Luís Madureira, *As ideias e os números: Ciência, administração e estatística em Portugal* (Lisbon: Livros Horizonte, 2006).

53. Madureira, *As ideias e os números.*

54. This is also true for France.

55. Amílcar Cabral, "Queimadas e Pousios na Circunscrição de Fulacunda em 1953," *Boletim Cultural da Guiné,* 9, no. 35 (1954): 627–643, at 630.

56. Judith A. Carney, *Black Rice: The African Origins of Rice Cultivation in the Americas* (Cambridge, MA: Harvard University Press, 2001).

57. Toby Green, *A Fistful of Shells: West Africa from the Rise of the Slave Trade to the Age of Revolution* (Chicago: University of Chicago Press, 2019).

58. Walter Hawthorne, "Nourishing a Stateless Society during the Slave Trade: The Rise of Balanta Paddy-Rice Production in Guinea-Bissau," *Journal of African History* 42, no. 1 (2001): 1–24.

59. Chabal, *Amílcar Cabral*; Mendy, *Amílcar Cabral.*

60. Chabal, *Amílcar Cabral*, 71.

Chapter Nine

Security versus Sovereignty in a Palestinian Seed Bank

COURTNEY FULLILOVE

On the drive into Hebron from Jerusalem, roadside vendors peddle grapes. A lone Israeli Defense Force (IDF) solider mans a bus stop adjacent the Israeli settlement of Migdal Oz, a satellite of the larger settlement of Gush Etzion, which has been continuously occupied since the early 1970s. A few stone finishing shops abut junk yards cluttered with car husks and a series of pottery stands. On the outskirts of Hebron, IDF vehicles line the road. On the hilltop above, smoke rises, and several heavily armed Israeli soldiers scuttle toward the disturbance. These disturbances are frequent: often acts of protest by Palestinian teens who set tires ablaze.[1]

The presence of Israeli soldiers in a West Bank city requires explanation. According to the terms of the Oslo Accords of 1993 and 1995, Hebron should have been under the full control of the Palestinian Authority (PA) for both civilian and security matters. But political realities shifted radically after the assassination of Israeli prime minister Yitzhak Rabin and the hasty defeat of his temporary successor, Shimon Peres. The new prime minister, Benjamin Netanyahu, who ran a campaign attacking the concessionary nature of the Oslo Accords, declined to withdraw from Hebron because of the presence of some 450 Jewish settlers in the city center, citing principles of security and reciprocity to which Palestinians must accede. In 1997, Hebron was cleaved in two by a protocol signed by Netanyahu and PA president Yasser Arafat, circumscribing 20 percent of the city under the authority of Israeli security. The Hebron Agreement signaled the failure of the Oslo Accords by sanctioning the redeployment of Israeli forces into West Bank cities, increasing already stringent restrictions on Palestinian mobility, and effecting the further fragmentation of Palestinian territories according to the Israeli "principle of separation."[2] Today, some 850 Jewish settlers inhabit the old quarter of the city. Altercations between settlers and Palestinians are common, as are escalations involving IDF soldiers who aim to shield settlers from attacks.

This chapter is not about overt clashes between settlers and Palestinians, but rather about a seed bank in the city of Hebron serving the surrounding Palestinian agricultural lands. It is nevertheless impossible to characterize cultivation in the South Hebron Hills, or the West Bank more broadly, without reference to the territorial divisions that structure land use and settlement, and to the local institutions that support agriculture as a means of land control. The drive south from Jerusalem provides a case in point. The stony hills around the roadway into Hebron are lined with terraces planted with fruit trees. Almond, olive, and grape vines are visible on some newly planted terraces. Others appear to be crumbling.

Terracing is an ancient agricultural technology well suited to dry and mountainous areas with little soil. But these terraces are not ancient. Some appear newly planted. In fact, the oldest date to the 1990s, in the aftermath of the First Intifada, a multiyear civil uprising by Palestinians against Israeli occupation. These roadside terraces were planted with the support of the Union of Agricultural Work Committees (UAWC), a Palestinian nongovernmental organization (NGO) serving farmers in the southern West Bank/Jordan Valley and Gaza.

The UAWC is one of multiple agricultural relief committees founded in the context of the First Intifada. In the midst of mass uprising, Palestinian leadership realized the extent to which their political agendas must make economic independence central to a project of liberation, and its parties courted civil society organizations accordingly.[3] As a result, the agricultural relief committees founded in the 1980s became linked to the underground parties of the Palestine Liberation Organization (PLO).[4] Founded in 1986, the UAWC was loosely affiliated with the Popular Front for the Liberation of Palestine (PFLP), an organization which has embraced political violence as a path to national liberation.

This relationship has rendered it, along with many other Palestinian NGOs, vulnerable to charges of abetting terrorism. Most recently, in July 2021, the Israel Defense Forces (IDF), citing security reasons, raided the UAWC headquarters in Al-Bireh, seizing computers and hard drives and ordering the office closed for six months.[5] In September 2019, Israeli authorities arrested two employees of the UAWC for suspected involvement with an PFLP-sponsored bombing the previous month near an

Israeli settlement in the West Bank. (At the time of publication, the trial is ongoing.)[6] In response to subsequent pressure by pro-Israeli groups, the Dutch government suspended payments to the UAWC and launched a review of the UAWC's possible links to the PFLP.[7] The UAWC objected to Israeli charges as a smear campaign and held that the review would open the door to "escalating incitement . . . and manipulative interference" with the UAWC and other Palestinian NGOs working on contested land.[8]

Independent of the political parties that sought their sponsorship, the agricultural committees born of the 1980s have provided essential social services to West Bank residents. Headquartered in Nablus, the UAWC supports land development projects encompassing the construction of agricultural roads, retaining walls, land rehabilitation, and water systems. Through its Community Seed Bank in Hebron, founded in 2010, the UAWC serves farmers in the southern West Bank and Gaza. It targets crops suited to rainfed agriculture, which it regards as a traditional and technologically appropriate solution to climate changes and lack of access to water resources.

The seed bank's objectives are simultaneously broad and pragmatic, oriented toward both global food politics and local conditions. The organization has benefited from foreign aid received in the decades since the Oslo Accords, receiving substantial aid from international agencies and European governments. It houses an array of seed preservation equipment standard in international genebanks, much of which has been donated by European seed companies. In 2014, the UAWC became the first Arab member of Via Campesina, an international movement dedicated to peasants' rights. By its own description, the UAWC "seeks to enable the Palestinian farmers socially and economically to strengthen their resistance on the Palestinian land and to protect the environment, biodiversity, and agricultural heritage." In its determination that Palestinian farmers "have sovereignty over their food," it combines commitments to environmental protection with the prerequisites to control territory and resources.[9]

In its simultaneous entanglement with international technical assistance, peasants' rights movements, and a politics of national liberation, the UAWC pivots between a praxis of international biodiversity preservation (food security) and local control (food sovereignty) with uncertain relations to international politics. Food security and food sovereignty are each novel categories of global knowledge, oriented, respectively, toward state or community control of seeds and other resources for food and agriculture systems. Ex situ seed-banking practices support state-centered visions of food security. Seed banks initially supported projects of international modernization, subsequently applied to the futures of postcolonial nation-states. Collections supplied breeders in developed countries with access to a diverse pool of global germplasm for the production of new varieties of staple grains, generally for wide release and large-scale cultivation.[10] By contrast, the concept of food sovereignty is a product of 1990s antiglobalization movements critical of corporate and multinational control of food systems.[11] In contrast to state-centered approaches organized around the modernization of postcolonial nation-states, advocates of food sovereignty prioritize community control of resources for food and agriculture.

In a globalized food system, the principles of food security and food sovereignty have taken shape as transnational ventures, and it is in the movement across borders that they come to grief or fruition, satisfying or thwarting hopes for local, national, or international control of resources. In the occupied West Bank, however, neither knowledge nor goods move freely, as the recent shuttering of the UAWC headquarters suggests. The region presents instead a jagged circuit of geographic, infrastructural, and political barriers. A 2015 United Nations Conference on Trade and Development (UNCTAD) survey of the Palestinian agricultural sector named these conditions as byproducts of the occupation: "restrictions on access to land, water and markets; loss of land to settlements and the separation barrier; demolition of infrastructure and the uprooting of trees; restrictions on access to agricultural inputs; shortage of credit for agricultural production; [and the] flooding of Palestinian markets with agricultural imports from Israel and settlements."[12] These conditions hamper agricultural productivity, as they also thwart local livelihoods and representation.

In the twenty-first century, the Palestinian agriculture sector is unique for the same reasons it is hampered. It is a troubling irony that barriers to commerce and development may preserve knowledge and technologies eroded by the progressive commercialization and industrialization of agriculture. Imposed isolation and underdevelopment make Palestine a refuge of locally adapted seed varieties and modes of cultivation not fitted to large-scale production. Rainfed agriculture utilizing terraces and swales constitutes a distinctive perennial agroecosystem, which has survived only in the absence of access to water for irrigation and other inputs for the expansion of intensive agriculture.[13] The UAWC seed bank aims to support these agricultural practices by collecting and preserving seeds adapted to local conditions. On the face of it, these practices share little with high-modern genebanks and the apparatus of international research and development (R&D) they support, and yet the UAWC draws heavily on international capital and technological practices to support its seed bank, often in ways that trouble its commitments to self-determination.

Historians of science have labored to explain how natural science figured as the handmaiden of European empire. The quest for useful plants provided the machinery of imperialism and colonization through European botanical gardens. Forged against fears of colonial degeneracy and the pursuit of valuable natural resources, these scientific projects provided the foundation for nineteenth- and twentieth-century models of development rooted in concepts of social evolution and economic growth.[14]

Science functions more ambiguously in an early twenty-first-century settler colony marked by occupation and incursions of settlements illegal under international law. The post-Oslo landscape of international and local NGOs masks the persistent influence of European nation-states and their settler colonies (including the United States) in organizing natural resources at a global level. In the context of international R&D, agricultural scientists target drylands as reservoirs of plant genetic material to breed crops resistant to the conditions of a warming world. Meanwhile, development schemes remain funded by international finance institutions, donors, and NGOs intervening in occupied territory, leading to an array of disjointed programs focused on technical assistance and capacity building. International organizations thus mediate the transmission of knowledge alleged to be ancestral and local, at times leading to bloat and redundancy among organizations dedicated to Palestinian development. Misplaced donor priorities and funds imperil the vitality and stability of the society they wish to support. Even so, local organizations devise practices and vocabularies that aim to circumvent international control.

Seed saving, as any technological practice, derives meaning from the social, political, and economic systems in which it is embedded. The UAWC promotes ethics of preservation prioritizing local sustainability and sovereignty over food systems. Yet the neocolonial power relationships in which ex situ seed banking is embedded dictate dependence on international capital and imperatives of global sustainability that mitigate local control. This chapter explores the extent to which high-tech ex situ seed banking is compatible with local commitments to food sovereignty rooted in the support of rainfed agriculture and how those commitments relate to the broader pursuit of Palestinian national sovereignty. As a study of seed banking, it locates a twenty-first-century UAWC seed bank in Hebron within a broader history of dryland agricultural research in Palestine and explores the extent to which the UAWC's recent membership in the Via Campesina alliance reinforces or challenges the universalizing tendencies of international politics.

The UAWC Seed Bank

The UAWC began to prioritize the collection of local seeds in 2003, founding its seed bank in 2010. The Community Seed Bank, located in Hebron, serves the farmers of Hebron, the southern West Bank, and North Bethlehem. The seed bank supports rainfed agriculture exclusively, focusing on drought-resistant summer crops, especially tomato, squash, and cucumber. Its multiplication unit serves some 1,000 farmers, purifying, regenerating, and multiplying varieties collected from farmers in the region. In 2017–2018, 3,000 dunums were newly planted with rainfed crops. The UAWC positions rainfed agriculture as a solution for the region in the face of climate change. Water shortages have been increasing in the last ten years, especially in the southern area of the West Bank. UAWC representatives see local seeds, traditionally cultivated, as a solution to farming in these conditions.

The seed bank currently focuses on forty-five local crops from twelve plant families, focusing on vegetables unique to local markets. These are widely recognized and have established market demand. The top summer crops are tomato, serpent cucumber, squash, okra, eggplant, and pumpkin. Winter crops of primary interest are carrot, radish, turnip, broad bean, and prickly alkanet (from the wild). In the *Solanaceae* family, tomato, pepper, and eggplant are the primary crops. Each crop has special characteristics associated with Palestinian cuisine. Tomato varieties vary from village to village and have irregular shape, a sour flavor, and a juicy body that makes them well suited to processing. They are typically used for juicing and sauce, and dried tomatoes for winter use. In the *Leguminosae* family, pea, chickpea, and especially cowpea form parts of traditional Palestinian breakfast. In the *Cucurbitaceae* family, priority collection is of (small, green) pumpkin, squash (cultivar), watermelon, and serpent cucumber, for which there is especially high demand. These are prized for drought and temperature tolerance and have morphological features that distinguish them from other varieties. In the *Brassicaceae* family, cauliflower, yellow radishes, cabbages, turnips, and carrots are targets of collection, especially, in recent seasons, the black carrot, indigenous to Palestine, and prized for its anticarcinogenic properties.

In the last two years, the UAWC has also begun collecting wild relatives of vegetables and medicinal plants. Many of the medicinal plants collected double as culinary herbs, reflecting the extent to which community practices may trouble the boundary between cultivated and wild. While field crops have predictable harvest times, sample size and harvest time vary enormously for medicinal plants in the UAWC collection, which include chamomile, anise, coriander, sage, thyme, rosemary, basil, and alkanet. While a focus on culinary and medicinal herbs targets local use values, the UAWC's interest in wild relatives of cultivated crops is consistent with a shift in international conservation strategies from cultivated plants to their genetic relatives. Researchers value this material for crop diversity in breeding. These multiple logics of collection indicate the complex orientation of the UAWC seed bank to local and international priorities.

FIGURE 9.1 Union of Agricultural Work Committees LSB cleaning room (author's personal collection).

The seed bank operates with the benefit of donated technology. The seed cleaning room in the Hebron seed bank (figure 9.1) is bafflingly free of the chaff and debris that typically litter the floor of such a facility. This clinical aura is due in part to the presence of a state-of-the-art Dutch seed blower gifted by Dutch agribusiness, a wonder of steel and transparent tubes that separates the chaff from the seed and banishes the former via an amber cyclone encased in glass. Three staff members, all young women master's students at local universities, manually clean or use the machine to separate seed from chaff. In rows on the table are bins of garlic, pumpkin, squash, and spinach seed. The women sort beans into different varieties by color and shape.

The seed blower was manufactured by Seed Processing, the second-largest producer of agricultural equipment in the Netherlands, behind the Switzerland-based giant Syngenta. A grading machine, also donated by Seed Processing, sorts a parsley sample by size (figure 9.2). Funding from the Netherlands also supported the upgrading of the laboratory and seed storage facilities, along with a nursery and upgraded breeding and reproduction facilities. Since 2013, the UAWC has received 17.5 million euros

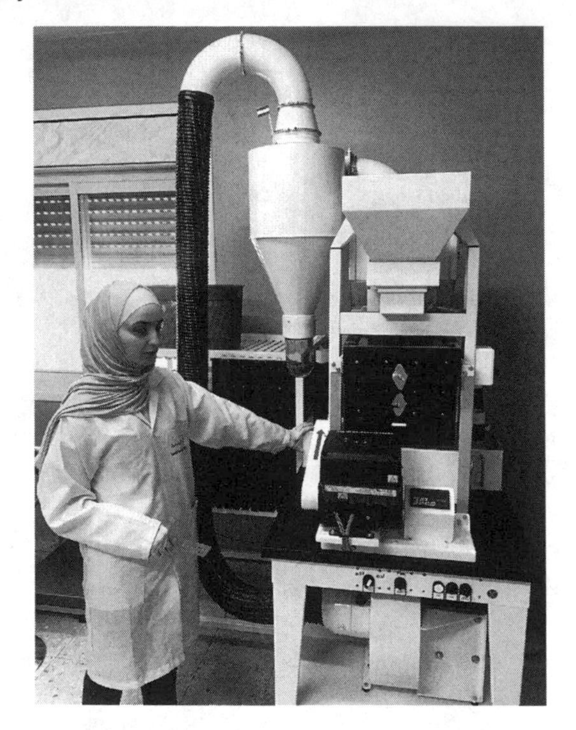

FIGURE 9.2. Local seed bank grading room (author's personal collection).

in funding from the Netherlands, through the Land and Water Resource Management Project, which is slated to end in 2021.[15]

While the Netherlands has supplied the bulk of its funding, the UAWC has significant support from other European governments and NGOs. UAWC funding from Oxfam Solidarité (Belgium) and the Belgium government ended in 2010. The UAWC has also received 1.4 million euros from Norwegian People's Aid (Norway) and additional funding from the Agence Française de Développement (France), Organizzazione per lo Sviluppo Globale di Comunita' in Paesi Extraeuropei Onlus (Italy), Associazione di Cooperazione e Solidarietà (Italy), and Solidaridad Internacional Andalucia (Spain). Oxfam Novib and Médecins du Monde France have supplied additional funding to relief programs in Gaza. Canada has also contributed significant funding through Oxfam Italia, the United Nations Office for the Coordination of Humanitarian Affairs, and CARE International.[16] The recent suspension of funding from the Netherlands suggests the vulnerabilities of reliance on international aid.

Yet it is also more than a little ironic that the bulk of aid to the UAWC, with its commitment to small-scale, rainfed agriculture, has come from an agricultural giant. The Netherlands is the second-largest agricultural exporter in the world, trailing only the United States. It is stunning that a country which is less than twice the size of New Jersey could attain this position, and it was not accomplished, needless to say, with a commitment to low-input agriculture such as that practiced in the West Bank. The bulk of the production in the Netherlands is in the southern province of Zeeland, which is less than 3,000 square kilometers.

In practice, international capital and technology inevitably alter both the practice and the product of traditional agricultural systems. Ex situ seed-banking practices require specific technologies for the viable preservation of seed. UAWC staff developed these over time through technical assistance and capacity-building programs sponsored by an array of institutions. These institutions and trainings establish a hierarchy of knowledge in the practice of seed saving. For example, although the UAWC's genetic fingerprinting initiatives with the Palestinian Polytechnic University are limited by costs, a UAWC project manager nevertheless characterizes genetics as the "most trusted" method of plant identification, taking precedence over morphological characterization and cultural knowledge.

The UAWC's collecting programs conform to international standards for ex situ seed banking. There are three levels of storage: short-term, medium-term, and long-term. Consistent with international conservation strategy, the UAWC prioritizes ex situ extraction, multiplication, improvement, and purification. In addition to funding from Oxfam and the Netherlands, in 2009 the UAWC engaged in capacity-building training for best practices in seed preservation with the International Center for Agricultural Research in Dry Areas (ICARDA). No training has taken place since the war in Syria. The UAWC sampled and distributed multiplied seeds from local varieties beginning in 2003, but only in 2010, after the ICARDA training, did it prioritize purification and improvement. New facilities supported these practices.

The majority of seeds are in short-term storage for maximum distribution, aimed at increasing the amount of land farmed (figures 9.3–9.5). Packs of 100–150 grams of each variety are entered into medium-term storage, available in case of loss or deterioration of working collections. Vacuum packed in aluminum bags to control moisture content, 50 grams or less of each sample are preserved for up to fifty years at twenty degrees Celsius in long-term storage, intended to serve as protection against natural disaster, war, or other kinds of total loss. These seeds are tagged in the database, along with lab data, date of planting, and the name of the farmer overseeing field reproduction. The UAWC also houses a small laboratory, which conducts germination, moisture, and viability tests. Farmers are supplied general germination percentage for every sample in the bank. The viability

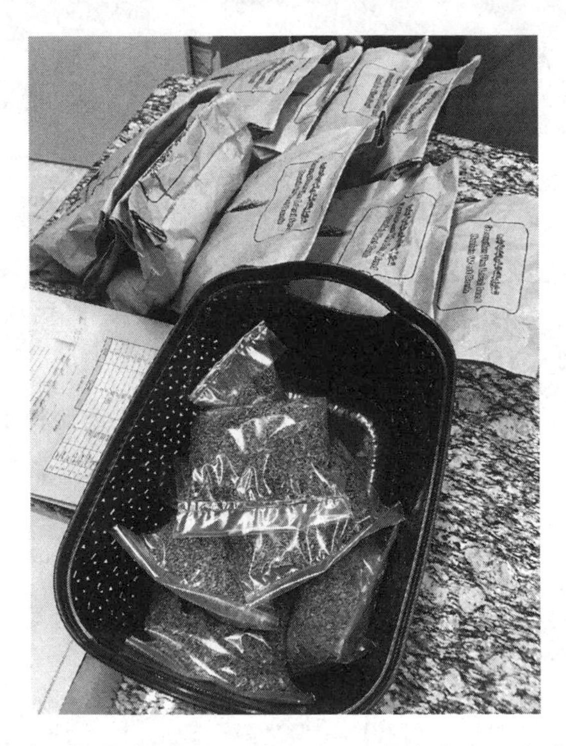

FIGURE 9.3. Seed packets for distribution to Gazan farmers (author's personal collection).

FIGURE 9.4. Working collection for distribution (author's personal collection).

FIGURE 9.5. Medium-term storage (cooled room) (author's personal collection).

test applies tetrazolium to assess the potential to germinate. Based on the result of lab tests, the sample is entered into the drying unit or extracted from the fruit and dried.

In these respects, the UAWC's looks much like other seed banks, with row upon row of frozen seed. But knowledge does not freeze; nor would traditional agricultural systems remain static in the absence of these ultramodern technologies. The UAWC selectively adapts introduced technologies to its own ends. Staff utilize equipment from Dutch agribusiness companies, including the seed blower and grading machine, but they eschew chemical methods of seed cleaning and extraction in favor of organic and heat-driven methods. They advocate the use of manure in preference to chemical fertilizers or pesticides, regarding organic production less as an ethical or political commitment than as a technology appropriate to the seeds they distribute and to the economic and infrastructural constraints of Palestinian agriculture. The seed bank makes its seed cleaning and grading machines available to local farmers. (Farmers also replicate these technologies in creative ways, using fans, for example, to reproduce the effects of the seed blower.) UAWC staff serially interview farmers in local villages about best practices, organizing meetings and distributing questionnaires through a network of key farmers dispersed through the villages. These efforts to center local practice are by no means exceptional in international agricultural development.[17] They nevertheless suggest the ways the UAWC constantly renegotiates its relationship to international networks of science and capital in drylands agriculture, and the extent to which these networks have structured local land use over time.

Dryland Agricultural Systems in Palestine as Targets of Development

Pejorative characterizations of drylands have a long history.[18] Biblical rhetoric cast dry regions in the Ancient Near East as barren, desolate places of trial and suffering. In turn, the promise to "make the desert bloom" has been a point of mission and pride for Israeli settlers.[19] While agricultural innovations have supported forestation, water conservation, and drip irrigation techniques, the intention of making the desert bloom can miscast the characteristics of arid ecologies not well suited to agriculture and those that support styles of subsistence agriculture long adapted to local communities. As Omar Tesdell has argued, cultivation is both an abstract and concrete claim to land, historically refigured through colonial classification and adjudication. Land struggles between Palestinian Arabs and Zionists groups in the early twentieth century provide but one set of entangled claims to control over supposedly cultivated or uncultivated land, mediated by the emerging profession of drylands agricultural science.[20] International wheat breeding initiatives, and the focus on Palestine as a site of domestication, helped remake drylands as targets of colonization.[21]

These efforts were part and parcel of a nation-building exercise consistent with those of other settler colonies. European colonial policy and Israeli occupation facilitated governance by circumscribing and cataloging practices of cultivation. Omar Tesdell has characterized the practice of agricultural science in Palestine as "the result of interaction across modern European settler colonies, whose defining characteristic is an attempt to consolidate control over land."[22] Early Zionists, American students of biblical archaeology, and agriculturalists with interests in dry farming met in the Holy Land, securing access to resources through the machinery of the British Mandate government.

The depiction of arid lands as degraded and barren was a misinterpretation, based initially on evolutionary theories of human history posted by late nineteenth-century ethnologists and later on the modernization theories prevalent in the post–World War II period. Modernization theorists such as Walt Whitman Rostow posited that all civilizations proceeded through one path of development and looked, at the end, eerily like the United States.[23] Industrialized countries led the charge to modernize agriculture. From the 1950s to the 1970s, the United States and Europe competed to establish themselves as dominant exporters of food, then of agricultural inputs, based on a model of input-intensive industrial agriculture. The export of high-yielding seeds and agricultural methods developed by American agronomists aimed to usher in a "Green Revolution," averting the Red alternative by increasing rural prosperity.[24]

Global conservation strategies developed to match these agendas, oriented at first toward state control of natural resources, then toward an international order that recognized the sovereignty of UN member states over others. Aiming to build on the alleged successes of the Green Revolution, the Food and Agriculture Organization (FAO) of the United Nations supported programs of agricultural modernization and the free exchange of germplasm between countries for the use of breeders. When international agricultural research centers turned their attention to biodiversity loss, it was to argue that public and private breeders should have access to global plant genetic resources: moving seed stocks out of the field and into banks from which they could circulate to countries with the capital to pursue research.[25] Today, fifteen international agricultural research organizations managed by the Consultative Group for International Agricultural Research (CGIAR) effectively govern the free transfer of global germplasm through standard material transfer agreements defined by the International Treaty on Plant Genetic Resources (2001). The Middle East and North Africa are served by the ICARDA, located in Aleppo, Syria, until 2010 and now dispersed through Lebanon, Jordan, and Morocco.

As a nascent state, Israel followed a similar trajectory to other colonized territories in the wake of World War II. Policy discourses about local land use mythologized some agricultural practices and degraded others. Both British Mandate and Israeli state governments selectively adapted existing legal and scientific machinery to justify interventions into Palestinian land and labor.[26] A primary theme was that Palestinian agriculture was degraded, backward, and primitive, and that the landscape was wasted and barren. The Ottoman-Israeli legal apparatus was used to mark lands as uncultivated, thereby claiming them for the new state of Israel.[27] These fictions facilitated occupation, governance, and the cultivation of dependency.

In June 1967, after brief but decisive conflicts with the surrounding Arab states, Israel occupied the West Bank, along with Gaza, Sinai, and the Golan Heights. In the West Bank, Israel supported policies of agricultural modernization intended to bind Palestinian farmers to the Israeli state technical apparatus. As Omar Tesdell summarizes: "developmentalist policies of the early occupation were meant not only to create dependency of Palestinians on Israeli government and companies for technology and technical assistance, but also to increase the effectiveness of modes of sovereign power such as tree uprootings, limitations on exports and trade, control of crops and planning, and draconian water restrictions."[28] Palestine effectively became "a captive market for finished Israeli goods."[29]

While Israeli occupation took on distinctive forms, it shares features with the neoliberal, globalized food system derived from European imperial geopolitics: specifically, as Philip Salzmann has characterized it, land grabbing, or "accumulation by dispossession . . . within the corporate food regime." Israeli pretexts for land dispossession resembled those used in other settler colonies: displacing current inhabitants, characterizing territory as uncultivated, and casting peasant agricultural practices as primitive and unproductive. While the market replaced the state as the "primary guarantor of food security" after the 1970s, it remained sponsored and enabled by dominant states.[30]

The signing of the Oslo Accords in 1993 left Palestine under the twin control of Israel and international finance institutions, marking a moment of neoliberal restructuring and defeat for a nationalist project of liberation.[31] Specifically, the division of the West Bank into Areas A, B, and C, with Area C under full Israeli administrative control, normalized dispossession of Palestinian territory. This reordering paved the way for incursions of Israeli agribusiness and further contributed to the marginalization of rural communities based on peasant agriculture.[32] The World Trade Organization and the International Monetary Fund further institutionalized asymmetries in power between states inherited from their imperial pasts, dictating loan conditions to the governing PA.[33] The PA's rural policy, outlined in the 2008 Palestinian Reform and Development Plan, adopted a market vision for agriculture supported by international lenders. Among its proposals was a plan for the Jordan Valley (Area C) that would convert peasant land into large-scale production for export using day labor.[34] The PA's acquiescence to neoliberal structural adjustment policies also hobbled community development initiatives and economies of resistance that had flourished during the First Intifada (1987–1993).[35]

Oslo muted both the sensibility of acute conflict and immediate aspirations for national liberation. The depoliticized development practice that took shape in its wake catered to donors rather than to communities. Palestine has received some $24 billion in assistance since 1993.[36] In 2008, the Agricultural Project Information System, managed by the Palestinian Ministry of Agriculture with assistance from the FAO, included some 170 international nongovernmental, local nongovernmental, and community-based organizations, UN agencies, and donors that represent the agricultural sector of West Bank and Gaza Strip.[37] Hobbled by financial dependence on international donors, the PA remains subservient to the priorities of international actors.[38]

In this climate, international organizations took up the mantle of European colonial governments in shaping institutions and regulations to organize natural resources in occupied territory. International development agencies prioritize the market potential for an expanded agricultural sector liberated from the impediments of occupation. UNCTAD emphasizes that Palestinian agricultural yields are 43 percent of Israel's and half of Jordan's. It recommends support required to develop Palestinian agricultural infrastructure, assist farmer cooperatives, and stabilize production and transportation costs. The implicit goal is to increase the productivity of the Palestinian agricultural sector for the purposes of trade and development.[39]

Institutions dedicated to scientific research interface somewhat differently with Palestinian agriculture, even as they remain oriented toward market agriculture. For researchers within the international complex of food security, the utilitarian or functional benefits of locally adapted crops take precedence. That is, crops are important for their potential adaptations to climate changes, not because they have immediate local potential or because they represent cultural systems with a long history. Drylands are a focus of twenty-first-century breeding research because they host plants and crop varieties adapted to drought, salinity, and high temperatures. Seed samples also form the basis of new research into drought-resistant wheat varieties.[40]

In recent decades, biodiversity preservation advocates have emphasized that ex situ conservation of seeds in genebanks must be complemented by in situ conservation of traditional farming systems, often confined to drylands and mountainous areas not extensively cultivated for commercial purposes. While strategies for in situ preservation have been drafted by research funded by the Global Environment Facility and United Nations Development Programme, these programs have retained a primarily development-oriented perspective.[41] Such research emphasizes the adaptability of landraces to harsh conditions and low-input agriculture, and it recommends the pursuit of increased yields through participatory breeding, water harvesting, conservation agriculture, and integrated pest management. Although contemporary R&D often relies on the knowledge required to sustain locally adapted seeds, generally it remains for the purposes of its transformation for commodity production and progressive integration into world markets.

Occupied Territory

Even as it figures as a laboratory for drought-resistant crops, the landscape of the West Bank bears the marks of conflict: IDF artillery ranges, the snaking separation wall, and the copious infrastructure of occupation. Blockades and armaments are obvious elements of such an infrastructure. Perhaps less so are the warring trees: pine "peace forests" planted by the Jewish National Fund face off against olive groves symbolic of Palestinian persistence on the land. Shaul Cohen has described this confrontation as "competitive tree planting" for control of Jerusalem's borders.[42] It also reflects contrasting views of what constitutes a healthy landscape. Zionists committed to David Ben Gurion's project to "make the desert bloom" used irrigation to engineer lush greenery, in stark contrast to the dry, scrubby appearance of rainfed farming. This replanted landscape masks demolished villages, obscuring the material origins of Israeli agricultural practice on newly cleared lands after 1948. In the wake of violent expulsions of Palestinians (*al-Nakba*), the newly formed Villages Section of the Israeli Custodian of Absentee Property of the state of Israel scrambled to take up the olive and citrus groves in time for harvest.[43]

Seventy percent of the population in the West Bank is rural, constituting some 400 villages, the majority of which practice dry farming. The Israeli government restricts the importation of fertilizers under the banned dual-use substances policy, which targets materials that can be used to produce explosives. The barrier wall constructed along the Green Line, marking the 1949 Armistice border, runs entirely through Palestinian land and restricts movement of farmers and products to a handful of border checkpoints. The expansion of Israeli settlements in the West Bank further erodes land control and presents competition for Palestinian farmers. The Israeli government provides incentives to settlement in priority areas, including 70 dunums of land, tax breaks, housing subsidies, mortgage plans, and discounts on utilities including water.[44]

Moreover, Area C, designated by the Oslo Accords and occupied since 1967, constitutes 62 percent of the West Bank and includes most of the fertile land, natural resources, and water. That amounts to 60,000 dunums of land unavailable to Palestinians for cultivation. Ninety-five percent of the fertile Jordan Valley, primarily characterized by irrigated agriculture, is in Area C, under full Israeli control, its abundant water resources and underground reserves inaccessible to Palestinians. Indeed, rainfed agriculture on arid lands at higher elevations is one adaptation to Israeli control of water resources and fertile land. Checkpoints inhibit the movement of agricultural products, which suffer high spoilage rates even if allowed to pass. Palestinian master plans for development frequently remain blocked or stalled in Israel bureaucracy. IDF soldiers frequently destroy water pumps and other water infrastructure constructed for agricultural projects. Palestinian farmers charge that the Israeli government disproportionally prevents them from digging wells, even as it allows them in illegal settlements.[45] Groundwater and soil are often contaminated with sewage and toxins from illegal dumping and drainage from uphill settlements.[46]

In spite of this litany of obstructions posed by occupation, in many respects the Palestinian agricultural landscape is quite typical. The agricultural sector is marked by disinvestment and market laissez-faire. Israeli produce floods Palestinian markets. Most is loaded with nitrate and chemical residues. These intensive farming practices have contributed to elevated nitrate levels and salinity of the soil. The date palm industry, which was an adaptation to dry and salty soil in the Jordan Valley, now contributes significantly to the problem. It has also hastened a shift from sharecropping to seasonal labor that compromises the livelihoods of local farmers.[47]

Agriculture is a declining share of the Palestinian gross domestic (GDP), dropping from around 19 percent in 1987 to 13.3 percent in 1994 to 5.7 percent in 2008. By 2013, the agricultural sector contributed 4.1 percent of the Palestinian GDP and 3.4 percent of the West Bank's, and it has hovered around these figures since.[48] This number omits food processing, wholesale, and retail markets, which are grouped with other sectors. However, the bigger issue with these figures is that they focus on who is enriched by agriculture rather than the number and type of people who rely on it for their livelihoods or sustenance.[49] As of 2010, the sector employed 11.5 percent of Palestinian workers, of whom a disproportionate number are women.

Many issues with specific features in Palestine nevertheless remain problems in many economies. These include "problems with land registration, the decline in arable land, and the parceling of land into small plots, as well as urban expansion at the expense of farmland and outmigration of youth from the countryside."[50] Many charge, with some justification, that the PA doesn't prioritize agriculture and that it applies standard market models geared toward commodity export and growth. Less than 0.1 percent of the PA's overall budget is allocated to the agricultural sector, leaving farmers dependent on the support of NGOs. Donor aid skews priorities away from the needs of farmers. Israeli agribusiness further threatens the livelihoods of Palestinian farmers.[51]

There are other challenges, which Palestinians share with the rest of the world: few farmers want to farm using backbreaking, labor-intensive, minimally productive techniques required for rainfed farming. There is no market for vegetables undercut by cheap Israeli imports. In light of this competition, many farmers pursue intensive agriculture with introduced seeds that see short-term gains and long-term losses in productivity. Intensive agriculture contributes to vulnerability to pests and disease and to soil salinity, which jeopardizes the viability of the land for cultivation. Along with the spread of unsuitable varieties, the closure of extension services imperils knowledge required to farm. Outmigration to cities among youth contributes to knowledge loss, which further threatens agrobiodiversity. Finally, many agricultural assistance programs have responded to conflict, leading to an emergency relief mentality. Yet as FAO Palestine director Azzam Saleh has observed, Palestine's is not the classical emergency context: it is not a hurricane, but rather a "slow releasing stress" with "recurrent shocks."[52]

In contradiction to a neoliberal development model stands an array of local institutions committed to community prosperity through local stewardship of natural resources. At the university level, nearly every Palestinian university has departments and centers dedicated to water resource management and sustainable land use. At the community level, Um Selim farm, the Bethlehem Farmers' Association, Bustana Community Supported Agriculture, and the Heirloom Seed Library in Battir, among others, prioritize the preservation of local seed varieties.

Palestinian NGOs, however, have led the charge in organizing farmers. These organizations have declared their intention to support rural livelihoods, self-sufficiency, and, as Glenn Robinson puts it, "to create a network of people throughout the occupied territories who shared the idea that, in effect, vegetables and politics were closely intertwined."[53] Palestinian liberation movements in the 1970s and 1980s drew on Third World anti-imperial struggles. The ideology of *sumud* (steadfastness) *muqawim* (resistance) gave these movements shape, situating resistance politics internationally while offering a rebuke of the "passive *sumud*" peddled by the PLO after 1967.[54] Grassroots organizations such as the agricultural committees of the First Intifada brought socially transformative concepts of "equality, social justice, women's liberation, respect for manual labour, alternative development, appropriate technology, and voluntary work" into the national liberation movement. This idea of "development for liberation" provided alternatives to prevailing market dogmas.[55]

Now, nearly forty years on, the politics of Palestinian agricultural NGOs bear faint resemblance to their founding moments, especially as international aid has altered the profile of each organization. Yet some features remain. The agricultural NGOs also focus on control of natural resources, including land and water, as a means to achieve both self-sufficiency and territorial sovereignty. They pursue the cultivation of land, both to diminish dependence on Israeli goods and to forestall Israeli confiscation of uninhabited lands. Cultivating land makes it inaccessible for settlement or protective restrictions applied to nature preserves. Many also support agricultural modernization to enable competition with Israeli producers.[56] The UAWC and the Palestinian American Research Center each organize farmers' committees, including women's cooperatives, throughout the West Bank.

Via Campesina and Food Sovereignty

The founding of the seed bank in 2010 marked a strategic shift in the UAWC's orientation from food security to food sovereignty. Palestinians "already have resources," notes seed bank manager Do'a Zayed, but they "cannot access them." These resources include abundant groundwater, which Palestinians are forbidden to tap by digging wells. In these conditions, control, rather than quantity, is the appropriate emphasis of any conservation strategy.[57]

In 2014, the UAWC became the first Arab member of the Via Campesina.[58] The UAWC's membership in the Via Campesina followed a Spanish delegation to Palestine targeting human rights violations in the countryside. Thereafter, Via Campesina took a more active role in promoting Palestinian causes and institutions, including the Boycott, Divestment, and Sanctions (BDS) campaign targeting the state of Israel.[59] Hiba Al-Jibeihi, the UAWC's international outreach coordinator, characterizes advocacy as working on both local and international levels, in the latter case working "to enlighten the international community on Palestinian issues and to increase the number of people who support Palestine and the rights of its people." She points to the UAWC's awareness of "the impacts of capitalism as regards the destruction of cooperative principles" and of nature itself.[60] In support of Palestinian food sovereignty, she exhorts the international community "to fight Israeli agribusiness through supporting BDS, and to put pressure on governments via effective demonstrations to end Israel`s constant violations of human rights."[61] The UAWC's embrace of Via Campesina's anti-agribusiness ethos thus complements its critique of Israeli occupation.

Founded in 1993 by farmers' organizations from Europe, Latin America, Asia, North America, Central America, and Africa, Via Campesina embraces, and has done much to popularize, the concept of food sovereignty. First suggested at the World Food Summit in 1996, the concept of food sovereignty was elaborated by Via Campesina as an antiglobalization movement prioritizing peasant control over natural resources. While definitions of food security postulated within a national security framework stressed the availability of sufficient quantities of staple grains, the Via Campesina's Nyéléni Declaration of 2007 defined food sovereignty as "the right of peoples to healthy and culturally appropriate food produced through ecologically sound and sustainable methods, and their right to define their own food and agriculture systems." The organization opposes the commodification of food in favor of community rights of self-determination of production, distribution, and consumption. In the language of the declaration: "It puts those who produce, distribute and consume food at the heart of food systems and policies rather than the demands of markets and corporations."[62]

Food sovereignty is an international movement, with 164 member organizations in seventy-three countries. UN definitions of food security have taken on board many aspects of Via Campesina's formulations, shifting its definition to stress that a society is secure if its supply of food is not simply sufficient in quantity but also "culturally appropriate."[63] Even the lexicon of food security has widened to encompass the security of families and communities rather than national stocks. Yet major distinctions between international projects of food security and food sovereignty remain.

The UAWC shares Via Campesina's commitment to preserving local heritage and the knowledge related to the seed. It identifies outmigration, knowledge loss in cities, and a generational gap between elders and youth as threats to the vitality of local agriculture. It sponsors both community-level data collection and capacity-building programs focusing on youth and children, including community awareness, school programs, education for women, and the training of agronomists at universities. The UAWC also prioritizes empowering women farmers. In Palestine, as in many parts of the world, men control marketing while women do the work of planting and production. Women also traditionally serve as household managers and control the diets of children. Do'a Zayed wryly repeats the adage: "When you teach women you teach the whole family; when you teach men, you teach just the man."

Apart from these commitments to local vitality, the UAWC exploits Via Campesina's global rhetoric to support Palestinian national sovereignty. In the past decade, Palestinian NGOs initially affiliated with political parties began to shift their political allegiances to international movements with anticolonial politics. The UAWC's decision to join Via Campesina is part of a strategic international outreach campaign aligning anti-agribusiness movements with the BDS movement. Advocates of food sovereignty see this as a legitimate attempt to secure peasant control of agriculture by pursuing solidarity between local political movements. This broader political mandate of the UAWC community directs the seed bank toward not simply food sovereignty, but Palestinian national sovereignty.

In practice, territorial sovereignty and food sovereignty are indivisible projects. While the UAWC targets vegetable crops for local markets, it collects grains (barley, wheat, sorghum) less for market potential than as a landholding strategy. No one expects these field crops to turn a profit. Rather, farms are barriers to Israeli state control of uninhabited lands. The farmer takes on special symbolic status in the UAWC's appropriation of rhetoric from transnational peasant movements. In addition to being victims of dispossession, farmers are symbols of the suffering and injustice Palestinians attribute to the occupation. In adopting Via Campesina's resolutions and voice, the UAWC also applies the vocabulary of indigeneity to Palestinian residents of the West Bank, situating Bedouin and Arab national minorities in the context of international social movements, collective action, and minority nationalisms.[64]

The centrality of territorial sovereignty is underscored by the UAWC's commitment to Gaza. The UAWC operates simultaneously in the West Bank and Gaza, with five locations in each, totaling 100–120 staff in the West Bank, and 40–50 in Gaza. Travel restrictions thwart staff attempts to meet in Jordan, so all communication between offices in the West Bank and Gaza is via video conference. "It's hard to manage a project you can't see," notes Do'a Zayed of the UAWC in Hebron, but "we want to stay one body." This intention suggests the broader political mandate of the UAWC Community Seed Bank toward not simply food sovereignty, but Palestinian national sovereignty.

The UAWC's demands for Palestinian national sovereignty may appear radical in contrast to its situation within a complex of international aid agencies and European agribusiness. If the need to justify projects to international donors has subjugated the seed bank to Oslo mandates to demonstrate Palestinian readiness for statehood, managers nevertheless combine this acquiescence with a persistent commitment to a project of Palestinian liberation. This ambiguous orientation is a deliberate strategy to appeal to multiple constituencies at a global level, declining to speak to the contradictory politics of international aid and local self-determination. It nevertheless exposes the sinews of international capital and settler colonial enterprise that continue to structure the allocation of rights and resources at a global level.

International Organizations and the Politics of Local Food Systems

By a 2013 World Bank estimate, the region could produce a billion dollars in revenue per year if restrictions were eased, which is greater than the sum of all aid to Palestine ($750 million per year). These projections aim to measure Palestine against the same yardstick of high-yield production used to crown the Netherlands an agricultural giant. But it is not clear that Palestine should try to be the next Zeeland even if it could. This model of world food production has enormous ecological and human costs, and it is arguably doomed to obsolescence. After a century of evaluating Palestinian agriculture against the template of high-yield exports, the tables have turned. The climate is changing, and farmers are scrambling to figure out how their practice will have to change as a result. At its most radical, adaptation means looking to agricultural techniques that may seem old-fashioned, preindustrial, or oriented toward low but stable yields under a variety of climatic conditions. Moreover, one-size-fits-all models have produced low-quality food at low prices, but it is no longer clear that this is a sustainable or desirable way to eat. Like many other places, the Netherlands, especially the agricultural province of Zeeland, is trying to figure out why its food is bad, by which it means tasteless, uniform, and cheap.[65]

State-centered visions of food security, and the complex of international research and development they support, have too often failed local communities, recapitulating centuries of colonial control over local resources and livelihoods. International seed banks exploit conditions of dependency for the purposes of research and development oriented toward global food security. They are an end-run around conflicts over territorial sovereignty, simultaneously deriving benefit from them and rendering them invisible. In spite of its claims to safeguard against risks posed by conflict, seed banking is a hegemonic practice assembled from conflict: over territory, over natural resources, and over the knowledge required to steward them.

As an alternative to food security, food sovereignty provides another global stage for rehearsing conflict, mediated by the global advocacy of Via Campesina. While Via Campesina's programs support farmers' rights and the livelihoods of indigenous people, they may also be commoditized in the local-food movements of highly capitalized agricultural systems: witness the aisles of any Whole Foods. There as elsewhere, to be local is the new ideal: and so "local" has become a global signifier for ill-defined notions of authenticity and resistance to big food. This celebration of the local, ironically now international in scope, represents a quest for something that cannot be domesticated to a universalizing global agenda.

In practice, international organizations mediate the transmission of knowledge alleged to be ancestral and local, whether they are organizations dedicated to programs of security or sovereignty. Yet they also allow us to conceive of alternative modes of preservation, which target knowledge as a prerequisite to cultivation rather than simply harvesting raw materials for reinterpretation according to the norms of twenty-first-century biotechnology.

The roadside terraces on the outskirts of Hebron signal a path forward. Cultivated since the First Intifada, they preserve an ancient agricultural system. And yet, no technological system can stand still. On these sites, agronomists guide farmers in intercropping with multiple vegetable crops so that they can harvest in the several seasons before newly planted fruit trees reach maturity. As the decayed and renewed terraces suggest, seeds are not static elements of a continuous agricultural system. They are made and remade by varied communities of practice and they can be made to serve many masters. Palestine's rainfed agriculture provides the grid for an alternative system of food production powered by indigenous foodways. The physical and intellectual resources artificially contained by occupation can provide the basis of novel production strategies favoring sociocultural vitality.

A postscript: I finished writing this chapter in Jerusalem on May 22, 2021, the UN's annual day of biodiversity, instituted in commemoration of the 1992 signing of the Convention of Biological Diversity. Palestine acceded to the convention in 2015 as "State of Palestine," a performative declaration of national sovereignty expressed through the media of multilateral agreements in protection of the global environment. In 2021, May 22 was a Saturday, the day after a cease-fire ending an eleven-day conflict between Israel and Hamas. Over ten days, Hamas fired some 1,300 rockets on Israel, which, while mostly intercepted by Israel's "Iron Dome" defense system, killed thirteen people. Meanwhile, Israel mounted an extensive aerial bombardment of military targets in the densely populated Gaza strip, with hundreds of civilian casualties. According to the UN Office for the Coordination of Humanitarian Affairs, some 75,000 people have been displaced as a result of the bombardment.[66] During the nights I listened to convoys of Israeli jets on their way to Gaza to drop bombs. As the cease-fire took effect at 2:00 a.m. on Friday, fireworks in the east continued for hours, punctuated by police sirens: defiant celebrations of a rickety and provisional pause in an ongoing conflict, this most recent instance of which also saw the greatest violence in mixed Palestinian-Jewish cities since the founding of the state of Israel.

The immediate context of this conflict echoes the broader histories of which it is part.[67] And as there was nothing symmetrical in the destructive toll of the exchange between the IDF and Hamas, so there is nothing symmetrical in the power relations that undergird Palestinian development characterized by Israeli occupation, neocolonial power relations, and governmental intransigence. It has been a source of some frustration to me that I cannot find a way to write about the logics of agricultural development in Palestine without simply reiterating the terms of "the conflict": a shorthand which itself obscures more than it explains. Yet I hope that in continuing to probe the meticulous, persistent, and everyday labor of making food for people, we can see alternatives to a long-running script of occupation and resistance that disregards more expansive imaginations of individual and communal well-being.

Notes

1. Research for this chapter was conducted over an extended period of visits to the Union of Agricultural Work Committees (UAWC) Local Seed Bank in Hebron

while based in Jerusalem between January 2019 and June 2021 and was facilitated by a Palestinian American Research Center (PARC) National Endowment for the Humanities fellowship. I would like to thank PARC fellow Emily McKee for her feedback on a draft of this chapter.

2. For a summary of the Hebron Agreement, see Ian Black, *Enemies and Neighbours: Arabs and Jews in Palestine and Israel, 1917–2017* (New York: Grove Atlantic, 2017), 350–352.

3. Eileen Kuttab, "Alternative Development: A Response to Neo-Liberal De-development from a Gender Perspective," *Journal für Entwicklungspolitik* 34, no. 1 (2018): 72.

4. The first of these was the Palestinian Agricultural Relief Committee (PARC), founded in 1983 by a group of agricultural engineers near Jericho offering free extension services for small farmers, and rapidly in the next decade to some eighty employees and seven centers. The NGO was affiliated with the People's Communist Party. Its success inspired Fatah-affiliated parties to found their own agricultural committees. The Applied Research Institute–Jerusalem followed in 1990. See Samer Alatout, "Towards a Bio-Territorial Conception of Power: Territory, Population, and Environmental Narratives in Palestine and Israel," *Political Geography* 25, no. 6 (2006): 601–621; Ismail Daiq and Shawkat Sarsour, "Agricultural Development and the Preservation of Indigenous Knowledge," in *Ecological Education in Everyday Life: ALPHA 2000*, ed. Jean-Paul Hautecoeur (Toronto: University of Toronto Press, 2016), 227–242; The Applied Research Institute–Jerusalem/Society, "Historical Background," https://www.arij.org/about-arij/background/historical-background.html, accessed February 15, 2020.

5. "Statement by UAWC after Israeli Military Raid on Our Headquarters in Al-Bireh," July 7, 2021, https://www.uawc-pal.org/news.php?n=3779&lang=2.

6. Yonah Jeremy Bob, "Shin Bet Cleared of 'Torture' of Palestinian Accused in Rina Shnerb Murder," *Jerusalem Post*, January 24, 2021, https://www.jpost.com/israel-news/shin-bet-cleared-of-torture-of-palestinian-accused-in-rina-shnerb-murder-656509.

7. "Dutch-Funded Expansion of Palestinian Control in Area C of the West Bank," NGO Monitor, December 16, 2020, https://www.ngo-monitor.org/reports/dutch-funded-palestinian-control-area-c/; Caroline Turner, Director, UK Lawyers for Israel, "Re: Donations by the State of the Netherlands to the Union of Agricultural Work Committees," June 18, 2020, https://www.rijksoverheid.nl/binaries/rijksoverheid/documenten/brieven/2020/06/18/brief-uklfi-18-juni-2020/brief-ukfli-dd-18-juni-2020.pdf; "Statement by UAWC after Dutch Government Announces Review," UAWC, July 22, 2020, https://uawc-pal.org/news.php?n=3624&lang=2.

8. The Palestinian Non-Governmental Organizations Network, Letter to Ms. Sigrid Kaag, Minister for International Trade and Development Corporation, Re: Review and Suspension of Dutch Payments for UAWC, July 27, 2020.

9. Union of Agricultural Work Committees, http://uawc-pal.org/un.php, accessed February 15, 2020.

10. Tiago Saraiva, "Breeding Europe: Crop Diversity, Gene Banks, and Commoners," in *Cosmopolitan Commons: Sharing Resources and Risks across Borders*, ed. Nil Disco and Eda Kranakis (Cambridge, MA: MIT Press, 2013), 185–212; Marianna Fenzi and Christophe Bonneuil, "From 'Genetic Resources' to 'Ecosystems Services': A Century of Science and Global Policies for Crop Diversity Conservation," *Culture, Agriculture, Food and Environment* 38, no. 2 (2016): 72–83; Michael Flitner, "Genetic Geographics: A Historical Comparison of Agrarian Modernization and Eugenic Thought in Germany, the Soviet Union, and the United States," *Geoforum* 34, no. 2 (2003): 175–185.

11. On the historical opposition and evolution of these terms, see Raj Patel, "What Does Food Sovereignty Look Like?," *Journal of Peasant Studies* 36, no. 3 (2009): 663–673; Philip McMichael, "A Food Regime Genealogy," *Journal of Peasant Studies* 36, no. 1 (2009): 139–169; Eric Holt-Giménez and Annie Shattuck, "Food Crises, Food Regimes and Food Movements: Rumblings of Reform or Tides of Transformation?," *Journal of Peasant Studies* 38, no. 1 (2011): 109–144.

12. United Nations Conference on Trade and Development (UNCTAD), *The Besieged Palestinian Agricultural Sector* (New York: United Nations, 2015), https://unctad.org/system/files/official-document/gdsapp2015d1_en.pdf.

13. Omar Tesdell, Yusra Othman, Yara Dowani, Samir Khraishi, May Deeik, Fouad Muaddi, Brandon Schlautman, Aubrey Streit Krug, and David Van Tassel, "Envisioning Perennial Agroecosystems in Palestine," *Journal of Arid Environments* 175 (April 1, 2020): 104085, https://doi.org/10.1016/j.jaridenv.2019.104085; Omar Tesdell, Yusra Othman, and Saher Alkhoury, "Rainfed Agroecosystem Resilience in the Palestinian West Bank, 1918–2017," *Agroecology and Sustainable Food Systems* 43, no. 1 (2019): 21–39.

14. On European botanic gardens and tropical agriculture, see Richard Harry Drayton, *Nature's Government: Science, Imperial Britain, and the "Improvement" of the World* (New Haven, CT: Yale University Press, 2000); E. C Spary, *Utopia's Garden: French Natural History from Old Regime to Revolution* (Chicago: University of Chicago Press, 2000); Richard H. Grove, *Green Imperialism: Colonial Expansion, Tropical Island Edens and the Origins of Environmentalism, 1600–1860* (Cambridge: Cambridge University Press, 1995). On practices and institutions of botanical collection, see Harold John Cook, *Matters of Exchange Commerce, Medicine, and Science in the Dutch Golden Age* (New Haven, CT: Yale University Press, 2007); Paula Findlen, *Possessing Nature Museums, Collecting, and Scientific Culture in Early Modern Italy* (Berkeley: University of California Press, 1994); Londa L. Schiebinger, *Plants and Empire Colonial Bioprospecting in the Atlantic World* (Cambridge, MA: Harvard University Press, 2004); Jim Endersby, *Imperial Nature: Joseph Hooker and the Practices of Victorian Science* (Chicago: University of Chicago Press, 2008).

15. Development Aid Portal, Ministry of Foreign Affairs of the Netherlands, https://www.nlontwikkelingssamenwerking.nl/en, accessed September 11, 2021.

16. Audit by NGO Monitor, *Union of Agricultural Work Committees' Ties to the PFLP Terror Group* (Jerusalem: NGO Monitor, 2020), http://ngo-monitor.org/pdf/UAWC_0120.pdf. It should be noted that the stated purpose of this audit, conducted by a right-wing Israeli organization, was partly to refute the UAWC's claim to "reject normalization and political conditional funding" by establishing the extent of donations from foreign governments and international aid organizations.

17. The prioritization of local knowledge, indigenous and traditional knowledge, community development, and participatory plant breeding are common to many twenty-first-century development projects. Each has a complex history. Two notable historical studies of community development and participatory plant breeding are Daniel Immerwahr, *Thinking Small: The United States and the Lure of Community Development* (Cambridge, MA: Harvard University Press, 2015), and Jonathan Harwood, *Europe's Green Revolution and Others Since: The Rise and Fall of Peasant-Friendly Plant Breeding* (London: Routledge, 2012).

18. Diana Davis, *Arid Lands: History, Power, Knowledge* (Cambridge, MA: MIT Press, 2016).

19. Alon Tal, "To Make a Desert Bloom: The Israeli Agricultural Adventure and the Quest for Sustainability," *Agricultural History* 81, no. 2 (2007): 228–257.

20. Omar Loren Tesdell, "Shadow Spaces: Territory, Sovereignty, and the Question of Palestinian Cultivation" (PhD diss., University of Minnesota, 2013).

21. Omar Tesdell, "Wild Wheat to Productive Drylands: Global Scientific Practice and the Agroecological Remaking of Palestine," *Geoforum* 78 (2017): 43–51.

22. Tesdell, "Shadow Spaces," 53.

23. For example, Lewis Henry Morgan's thirty-five years of study on Indian material culture and kinship culminated in the publication of *Ancient Society* (New York, 1877). Morgan linked phases of civilization to technological development and social institutions such as property, drawing on Henry James Sumner Maine's theories of law as developing "from status to contract." Henry James Sumner Maine, *Ancient Law: Its Connection with the Early History of Society, and Its Relation to Modern Ideas* (London, 1861). On theories of social evolution, see George Stocking, *Victorian Anthropology* (New York: Free Press, 1991); Curtis Hinsley, *Savages and Scientists: The Smithsonian Institution and the Development of American Anthropology, 1846–1910* (Washington, DC: Smithsonian Books, 1981); Robert Bieder, *Science Encounters the Indian, 1820–1880: The Early Years of American Ethnology* (Norman: University of Oklahoma Press, 1989); Alan Joyce, *Shaping of American Ethnography: The Wilkes Exploring Expedition, 1838–1842* (Lincoln: University of Nebraska Press, 2001). On modernization and take-off, for example, see W. W. Rostow, *The Stages of Economic Growth: A Non-Communist Manifesto* (Cambridge: Cambridge University Press, 1960), chap. 2, "The Five Stages-of-Growth: A Summary," 4–16;

Mark Mazower, *Governing the World: The History of an Idea* (New York: Penguin, 2012); Nick Cullather, *The Hungry World: America's Cold War Battle against Poverty in Asia* (Cambridge, MA: Harvard University Press, 2010); David C. Engerman, *Staging Growth: Modernization, Development, and the Global Cold War* (Amherst: University of Massachusetts Press, 2003); Michael E. Latham, *Modernization as Ideology: American Social Science and "Nation Building" in the Kennedy Era* (Chapel Hill: University of North Carolina Press, 2000); Michele Alacevich, *The Political Economy of the World Bank: The Early Years* (Stanford, CA: Stanford Economics and Finance, Stanford University Press, 2009).

24. William Gaud coined the term in 1968 as a counterpoint to the red revolutions in the Asian countryside, arguing that the rural development activity of the last two decades would mean bumper harvests and happy farmers. Caroline Abu-Sada, "Cultivating Dependence: Palestinian Agriculture under the Israeli Occupation," in *The Power of Inclusive Exclusion: Anatomy of Israeli Rule in the Occupied Palestinian Territories*, ed. Adi Ophir, Micael Givoni, and Sari Hanafi (New York: Zone Books, 2009), 413–434.

25. Saraiva, "Breeding Europe"; Fenzi and Bonneuil, "From 'Genetic Resources' to 'Ecosystems Services'"; Flitner, "Genetic Geographies."

26. Tesdell, "Shadow Spaces," 79.

27. Tesdell, "Shadow Spaces," 84; Warwick P. N Tyler, *State Lands and Rural Development in Mandatory Palestine, 1920–1948* (Brighton: Sussex Academic Press, 2014); Shaul Ephraim Cohen, *The Politics of Planting: Israeli-Palestinian Competition for Control of Land in the Jerusalem Periphery* (Chicago: University of Chicago Press, 1993); Hussein Abu Hussein and Fiona McKay, *Access Denied: Palestinian Land Rights in Israel* (New York: Zed Books, 2003).

28. Tesdell, "Shadow Spaces," 86.

29. Abu-Sada, "Cultivating Dependence."

30. Philipp Salzmann, "A Food Regime's Perspective on Palestine: Neoliberalism and the Question of Land and Food Sovereignty within the Context of Occupation," *Journal für Entwicklungspolitik* 34, no. 1 (2018): 14–34.

31. Adel Samara, "Globalization, the Palestinian Economy, and the 'Peace Process,'" *Journal of Palestine Studies* 29, no. 2 (2000): 20–34; Salzmann, "A Food Regime's Perspective on Palestine."

32. Linda Tabar, "People's Power: Lessons from the First Intifada," unpublished manuscript, 2013, 42, https://www.rosalux.ps/wp-content/uploads/2015/03/Linda-Tabar.pdf.

33. Salzmann, "A Food Regime's Perspective on Palestine"; Holt-Giménez and Shattuck, "Food Crises, Food Regimes and Food Movements"; McMichael, "A Food Regime Genealogy."

34. Salzmann, "A Food Regime's Perspective on Palestine."

35. Kuttab, "Alternative Development."

36. Kuttab, "Alternative Development," 76.

37. United Nations, *Agricultural Projects in the West Bank and Gaza Strip, 2008*, Agricultural Projects Information System (APIS) Report, January–December 2008, https://www.un.org/unispal/document/auto-insert-203615/.

38. For a succinct synopsis of this condition, see Muriel Asseburg, "The Palestinian Authority and the Hamas Government: Accessories to the Occupation?," in *Actors in the Israeli-Palestinian Conflict: Interests, Narratives and the Reciprocal Effects of the Occupation*, ed. Peter Lintl, Stiftung Wissenschaft und Politik German Institute for International and Security Affairs, SWP Research Paper 3 (June 2018), https://www.swp-berlin.org/publications/products/research_papers/2018RP03_ltl.pdf.

39. UNCTAD, *The Besieged Palestinian Agricultural Sector*.

40. For example, Maria Buerstmayr, Karin Huber, Johannes Heckmann, Barbara Steiner, James C. Nelson, and Hermann Buerstmayr, "Mapping of QTL for Fusarium Head Blight Resistance and Morphological and Developmental Traits in Three Backcross Populations Derived from Triticum dicoccum × Triticum durum," *Theoretical and Applied Genetics* 125, no. 8 (2012): 1751–1765.

41. Joan Freeman, Sawsan Mehdi, and Mahmud Duwayri, *Conservation and Sustainable Use of Dryland Agrobiodiversity: Jordan/Lebanon/Syria/Palestinian Authority, Terminal Evaluation Final Report*, United Nations Development Program Evaluation Resource Center, December 2005, https://erc.undp.org/evaluation/documents/download/849.

42. Cohen, *The Politics of Planting*.

43. Tesdell, "Shadow Spaces," 84.

44. UNCTAD, *The Besieged Palestinian Agricultural Sector*; Samer Abdelnour, Alaa Tartir, and Rami Zurayk, "Farming Palestine for Freedom," *Al-Shabaka* (blog), July 2, 2012, https://al-shabaka.org/briefs/farming-palestine-freedom/; Abu-Sada, "Cultivating Dependence."

45. UNCTAD, *The Besieged Palestinian Agricultural Sector*; Abdelnour, Tartir, and Zurayk, "Farming Palestine for Freedom"; Abu-Sada, "Cultivating Dependence."

46. Sophia Stamatopoulou-Robbins, *Waste Siege: The Life of Infrastructure in Palestine* (Stanford, CA: Stanford University Press, 2019).

47. Julie Trottier, Nelly Leblond, and Yaakov Garb, "The Political Role of Date Palm Trees in the Jordan Valley: The Transformation of Palestinian Land and Water Tenure in Agriculture Made Invisible by Epistemic Violence," *Environment and Planning E: Nature and Space* 3, no. 1 (2020): 114–140.

48. The most up to date statistics are provided in Food and Agriculture Organization of the United Nations (FAO), *Context Analysis for the Country Programming Framework for Palestine 2018–2022* (Jerusalem: FAO, 2019), http://www.fao.org/3/CA0627EN/ca0627en.pdf.

49. Interview with Azzam Saleh, FAO Palestine director, Ramallah, October 25, 2017.

50. Abdelnour, Tartir, and Zurayk, "Farming Palestine for Freedom"; Abu-Sada, "Cultivating Dependence."

51. Salzmann, "A Food Regime's Perspective on Palestine"; Abu-Sada, "Cultivating Dependence"; Abdelnour, Tartir, and Zurayk, "Farming Palestine for Freedom."

52. Interview with Azzam Saleh, FAO Palestine director, Ramallah, October 25, 2017.

53. Glenn E. Robinson, *Building a Palestinian State: The Incomplete Revolution* (Bloomington: Indiana University Press, 1997), 53–55.

54. Salzmann, "A Food Regime's Perspective on Palestine"; Tabar, "People's Power"; Kuttab, "Alternative Development."

55. Kuttab, "Alternative Development," 73.

56. Abu-Sada, "Cultivating Dependence," 416.

57. The following section draws on an interview with Do'a Zayed, seed bank project manager, UAWC Hebron, conducted on November 11, 2019, at the Hebron Community Seed Bank.

58. The UAWC is represented at international food sovereignty conferences, including the most recent one in October 2019. Its website links to United Nations General Assembly, Resolution 73/165, United Nations Declaration on the Rights of Peasants and Other People Working in Rural Areas, United Nations Declaration on the Rights of Peasant and other People Working in Rural Areas, adopted on December 17, 2018, A/RES73/165 (January 21, 2019), http://uawc-pal.org/un.php. The declaration reiterates the principles stated in the Charter of the United Nations to recognize the "inherent dignity and worth and the equal and inalienable rights of all members of the human family as the foundation of freedom, justice and peace in the world," and applies these and successive conventions on development, antiracism, and indigenous rights to peasants and people working in rural areas. The UAWC exhorts the Palestinian government, Ministry of Agriculture, agencies and organizations of the United Nations System in Palestine, and Palestinian intergovernmental and civil society organizations to promote the implementation of the declaration, further stating its commitment to the stated principles.

59. La Via Campesina, "Israeli Violations against the Palestinian Farmers, 2015," July 1, 2015, https://viacampesina.org/en/israeli-violations-against-the-palestinian-farmers-2015/; Salzmann, "A Food Regime's Perspective on Palestine."

60. Hiba Al-Jibeihi, "Protecting Our Lands and Supporting Our Farmers" (interview with Philipp Salzmann), *Journal für Entwicklungspolitik* 34, no. 1 (2018): 109.

61. Al-Jibeihi, 'Protecting Our Lands," 110.

62. "Declaration of Nyéléni," *Chain Reaction* 100 (2007): 16.

63. Patel, "What Does Food Sovereignty Look Like?"

64. Ismael Abu-Saad, "Spatial Transformation and Indigenous Resistance: The Urbanization of the Palestinian Bedouin in Southern Israel," *American Behavioral Scientist* 51, no. 12 (2008): 1713–1754; Amal Jamal, *Arab Minority Nationalism in Israel: The Politics of Indigeneity* (New York: Routledge, 2011).

65. Roosevelt Institute for American Studies, "Conference: Futures of Food," September 6, 2019, https://www.roosevelt.nl/app/uploads/2021/07/The-Roosevelt-2019.pdf; Peter M. Burns and Marina Novelli, eds., *Tourism and Social Identities: Global Frameworks and Local Realities* (Amsterdam: Elsevier, 2006).

66. "Occupied Palestinian Territory (oPt): Flash Update #9—Escalation in the West Bank, the Gaza Strip and Israel (as of 12:00 18 May–12:00 19 May)," OCHA Situation Report, Reliefweb, May 19, 2021, https://reliefweb.int/report/occupied-palestinian-territory/occupied-palestinian-territory-opt-flash-update-9-escalation.

67. On May 10, Hamas, the de facto governing party of Gaza, fired a volley of long-range rockets on Jerusalem. In a statement, Hamas characterized the attack as a response to scurrilous Israeli police incursions into Jerusalem's Al-Aqsa Mosque during Ramadan and the pending evictions of Palestinians in the East Jerusalem neighborhood of Sheik-Jarrah to the benefit of Zionist settlers. The action by Hamas took place in the context of ongoing contest with the Fatah-governed Palestinian Authority over control of the West Bank. On April 29, the PA canceled legislative elections scheduled for May 22, in which Hamas was anticipated to make great gains. The PA also indefinitely postponed the presidential election scheduled for July 3, leaving President Mahmoud Abbas in power for the seventeenth year of a four-year term. (The last elections held in the West Bank were on January 25, 2006, for the Palestinian Legislative Council.) In the midst of Hamas's bombardment of Israel, Abbas directed his criticism toward Israeli occupation, including the recent incursions on the Al-Aqsa Mosque and the evictions in Sheik Jarrah.

Chapter Ten

How Data Cross Borders

Globalizing Plant Knowledge through Transnational Data Management and Its Epistemic Economy

SABINA LEONELLI

Introduction: Plant Data Journeys and Why They Matter

Plant phenotype data—and related metadata about plant genealogy and growth patterns—are material traces of human interactions with (and knowledge of) plants, which document various aspects of plant morphology, development, and function.[1] They comprise items as varied as photographs of leaves and measurements of root width, and are increasingly available in digital form, which makes it technically possible to copy them and disseminate them over the internet—just like any other digital, nonrivalrous good.[2] Plant phenotype data crucially inform innovation and policy relating to food security around the world. They are highly valued by scientists as key empirical grounding for new knowledge about plant development, biodiversity, health, and ecology, which could in turn elicit environmentally sustainable strategies to increase agricultural yield and resilience in the face of climate change and population growth—and more specifically, may inform the choice and deployment of specific crop varieties. There is a strong incentive to "open" these data and ensure they can be disseminated globally and consulted as a single body of evidence, no matter what their provenance is. There is also an expectation that the sharing of data can be greatly facilitated by current communication technologies, especially since a large proportion of plant data is digital rather than material and travels unmoored from actual samples of germplasm.[3]

A strong disincentive to such mobilization comes from the interests and value attached to plant phenotype data, which take a variety of different forms and are steep in long-running scientific frictions and conflictual political-economic regimes. These data typically comprise genomic information, field observations, results of field trials and experimental insights produced by publicly funded research or corporate R&D *as well as* information about specific strains and their environments generated by farmers and breeder communities. These data are often subject to specific intellectual property regimes and regulations around material transfer, some of which rigidly oppose any form of sharing (e.g., data produced by agrotech businesses), while others are more ambiguous in the extent to which data can be appropriated and interpreted by diverse stakeholders (e.g., sequencing data produced by academic laboratories). For many data sources, the framework within which exchanges and dissemination could and should happen is unclear, and sometimes downright absent. In particular, there is little explicit agreement on goals, rewards, responsibilities, and rights relating to the generation, circulation, and use of digital data. How to define data production is itself a matter of contestation: is it the result of growing the plant specimens, selecting particular strains, designing specific field trials, adopting novel measurement tools, or designing systems for data collection? The answer to this question determines who is viewed as legitimate owner of the data and who has control over their use, and yet all of these activities have a legitimate claim to being part of data production.

The enormous uncertainty around scientific, legal, and administrative views of data work is making space for new forms of bioprospecting, whereby richer stakeholders in global agriculture stand to gain from the systematic exploitation of the resources held by low-income countries. The predatory practices that have long plagued the exchange of plant materials[4] are now extending to the digital realm—with alarm growing around the emergence of a new "digital feudalism," within which control of the land is replaced by control of information as the linchpin of power struggles around food production and environmental management.[5] These challenges are exacerbated by the diversity of goals which globalized data collections are supposed to serve, which range from the comparison and selection of crop varieties to fit changing environmental conditions to the fight against fast-spreading blights that threaten staple crops such as banana, coffee, rice, maize, cocoa, and cassava.

The position of the nation-state in all of this is complex and multifaceted. On the one hand, national interests play a major role in determining plant data movements, and the crossing of borders is fraught with challenges. The geographical and geopolitical characteristics of the territories in which plant research plays out remain significant. The degree to which specific crop variants are entrenched in food cultures and local economies, the socioeconomic threat posed by pathogens that target those variants, and the extent to which specific regions are threatened by climate change (and whether they have the resources and willingness to tackle that threat with the related need to consider whether to abandon heritage breeds) are also significant. Beyond their immediate use for plant research, plant data platforms are among the sources that inform indicators such as those associated with the United Nations' Sustainable Development Goals (especially 2, zero hunger; but also 3, good health and well-being; 10, reduced inequalities; and 15, life on land), where the use of national borders to mark data provenance and the scope of data analysis plays a crucial role.[6] The trade of information around plant life has a strong colonial heritage, which is itself grounded in national ambitions and related exploitation.[7] Indeed, data about plants and crops have long been a diplomatic and political asset, with the standardization of crop measurements, and the resulting ability to disseminate and compare such data, fostering the development of a centralized, neoliberal food market in the second half of the nineteenth century.[8] Many plants are tightly linked to specific territories and ways of life, and thus regarded as national resources and forms of heritage which other nations are not welcome to plunder—a situation monitored by ad hoc international committees such as the International Committee for the Protection of New Varieties of Plants (UPOV).[9] When it comes to plant materials, there is a long history of national governments attempting to identify and control their use as well as long-standing efforts to secure fairness in germplasm exchanges.[10] International agencies such as the Food and Agriculture Organization of the United Nations (FAO) and the Convention on Biological Diversity have fostered regimes of international law that facilitate the sharing of plant materials without losing sight of intellectual property and heritage claims attached to them—for instance through material transfer agreements, the International Treaty for Plant Genetic Resources for Food and Agriculture, and the Nagoya protocol.[11]

On the other hand, national interests and related commercialization strategies are often effaced from scientific discourse around the problems of moving and linking plant data, which typically focus on the epistemic and technical challenges of making data machine-readable and interoperable to enable the search and analysis of global plant data as a single body of evidence. Many contemporary plant data infrastructures are not obviously "national," even when they are funded by governmental agencies such as the National Science Foundation in the United States or the Biotechnology and Biological Sciences Research Council in the United Kingdom. Plant data infrastructures are facilitated by the internet and often structured as international research networks or consortia,[12] with stated goals typically being to acquire more users (including more "data providers" willing to share their resources) and to underpin agrotechnical innovation no matter which countries it may affect. Many such infrastructures have relatively little strategic regard for how knowledge may eventually be translated into commercial products, a situation that is reflected in the stated mission of the Interest Group on Agricultural Data (IGAD), a working group of the Research Data Alliance that was founded in 2013 to bring together and coordinate national and international data management initiatives in this domain.[13] The four key objectives of this group are as follows:

> To promote good practices in the research domain: data sharing policies, data management plan, data interoperability; to provide a platform for networking and cross-fertilization of research ideas in data management and interoperability; to solicit and promote interactions and projects among the major international institutions and groups worldwide which work on agricultural research and innovation; to achieve data interoperability.[14]

As these goals exemplify, the discourse underpinning efforts to develop reliable data infrastructures centers on integrating morphological, genetic, ecological and developmental data to inform global innovation and policy on food security & precision agriculture. The underlying assumption is that such global pooling of resources will be used collaboratively to deliver widely applicable solutions and ultimately touch the lives of all humans on the planet.[15] Questions about the downstream commercialization and distribution of knowledge and technologies developed through data analysis are not often asked within these scientific initiatives: the focus is firmly on improving the reach, comprehensiveness, and interoperability of global data resources. Some of the key infrastructures are funded by international agencies such as the FAO and the World Bank, global institutions such as the Consultative Group for International Agricultural Research (CGIAR), or private funders such as the Gates Foundation and the Wellcome Trust, none of whom are explicitly aligned with national logics and who prefer instead to present their efforts as contributing to food security worldwide.[16]

In this chapter, I argue that this transnational vision for data management has fueled a specific epistemic economy for plant research, which presents global plant knowledge production as a common good to be harnessed for the continuing survival of our species and of the planet. This epistemic economy stands in a complex relationship with attempts to control or sanction access to plant data resources in ways that reflect national and/or private interests. To illustrate this, I consider the history of the Crop Ontology, a data linkage tool set up by a network of plant scientists and breeders spread across the world (and particularly in the Global South) in the mid-2000s. At a moment when scientific debate over the importance of "bringing data together" glossed over socioeconomic tensions in data management strategies and ownership claims, the Crop Ontology attempted to acknowledge the fraught politics of crop data at the local and regional levels and devise data semantic and linkage systems that can help to address them. Based on a series of interviews with data curators and users which I carried out between 2016 and 2018, as well as field visits to relevant sites and a review of related scientific publications and online resources, I trace some of the ways in which the Crop Ontology identified and navigated tensions in transnational plant data management.[17] I then briefly consider two applications of this resource to the study of the cassava root: Cassavabase, an open database for cassava field trials that was chiefly developed at the International Institute for Tropical Agriculture (IITA) in Ibadan, Nigeria; and Nuru, a smartphone app facilitating the diagnosis of plant disease which was developed as part of the Plant Village open resource project at Penn State. Finally, I reflect on how national and transnational goals intersect when in establishing strategies, infrastructures and norms for the movement of plant phenotypic data and the creation of a global knowledge infrastructure around cassava, using this a window into the politics of data trade, knowledge governance and international collaboration within agricultural science and biotechnology.

The Making of the Crop Ontology as a Transnational Data Broker

The aim of the Crop Ontology is to help classify and manage environmental and biological data about plant phenotypes collected from plant science experiments and crop field trials around the world, and particularly low-income countries hosting some of the main field trials for staple crops such as potato, maize, banana, and cassava. To this aim, the Crop Ontology involves the creation of a controlled vocabulary for plant traits, which provides researchers and breeders around the world with a common semantic platform to collect and analyze data. In other words, a controlled vocabulary brings clarity about the targets of research—what it is that data are being collected about. For instance, the Crop Ontology provides an explicit definition for the term "root" and formalizes its relations to other aspects of a crop system (e.g., "stem," "ease of harvest," "yield," "fiber content," and so forth, as illustrated in figure 10.1), thus effectively providing a list of phenomena to which data can be attached as potential evidence.[18] A shared controlled vocabulary such as the Crop Ontology is crucial to researchers' efforts to compare data across varieties or species, assess new observations, and study crucial biological elements for agriculture such as biodiversity and resistance to pests, climate change, and other environmental stressors.[19]

FIGURE 10.1. Cassava traits as specified by Crop Ontology.

This deceptively simple exercise involves tremendous technical challenges, including the development of standardized software and semantic systems to link large, heterogeneous datasets.[20] This goes well beyond the more traditional tasks of taxonomy, which are largely centered on the determination of boundaries between species, with less regard for the characteristics of individual specimens and their multiple roles in both the ecosystem and the food chain.[21] Even defining terms as basic as "root" and "stem" involves confronting differences in the ways in which farming communities and researchers around the world have measured and described these parts of a plant. For instance, defining a root means establishing parameters for where the root starts—not an obvious issue given its continuity with the stem and the fact that some roots grow outside the soil. Indeed, different field workers—including those employed by field trials within research facilities—have different ways to cut roots away from the rest of the plant and take measurements of those samples, which are based on different beliefs about where the root starts and what a "normal" root should look like. This in turns determines differences in other measurements, such as root weight. Capturing the granular, highly fragmented, and embodied nature of these judgments and habits is part of the mission of the Crop Ontology. It demands sensitivity to the specific contexts of data collection and inclusivity in considering a wide variety of perspectives on each term and related approaches to measurement, which is difficult to reconcile with the quest to make measurements—and the resulting data—comparable and interoperable across fields and locations. Moreover, the challenge of including multiple forms of diversity of data collection is exponentially enlarged by the need to coordinate the semantic structure and content of the Crop Ontology with those employed by the dozens of other large data infrastructures collecting crop data around the globe.[22] To be effective, the Crop Ontology thus needs to work as a data broker both at the local and at the international levels, mediating between different groups of data collectors and analysts as well as those responsible for implementing common standards across countries and linguistic, cultural, and legal borders.

The Crop Ontology was started as part of the Generation Challenge Programme of the CGIAR, launched in 2002 in order to stimulate technical advancement in plant and agricultural research of relevance to the Global South. The tagline of the program, which lasted until 2013 and was funded largely by the World Bank and the European Commission, was "cultivating plant diversity for the resource poor."[23] Effective data collection and linkage were key remits for the program, given the importance of comparing varieties and assessing biodiversity across locations as well as the necessity to improve collaboration and exchanges among the CGIAR member institutes, which are spread across four continents and often in very deprived regions.[24] The quest for computational solutions to standardize and facilitate effective data circulation was hampered by the lack of sophistication of existing plant trait descriptors, which had largely been developed to suit specific crops, without a strong emphasis on cross-species consistency.

From 2008 the Crop Ontology was spearheaded by French scientist Elizabeth Arnaud. From a background in animal biology, Arnaud switched her interest to tropical crops and bioinformatics early in her career. As a master's student she did an internship at the Centre National de la Recherche Scientifique (CNRS), where she was asked to help with computational renditions of plant taxonomies. She subsequently worked for the International Rice Research Institute on rice pathogens and then for the Banana Improvement Network, where she took charge of developing a genetic resources database. In her words: "the idea was how can we develop a system that GenBank [a genome sequencing database] data managers could install everywhere and have an exchange of data into a central database, so we could have a very global vision of what type of banana varieties and rice are conserved worldwide" (PI_14_A). These experiences sensitized Arnaud to the difficulties encountered in implementing computational systems that are developed in the Global North within low-resourced research environments and field stations. She was also alerted from the start to the importance of involving local research communities—and thus a wide diversity of cultural backgrounds, assumptions, and interests—in the development of controlled vocabularies. Her emphasis was on achieving consensus through the identification of the relevant stakeholders and the establishment of platforms for dialogue:

> What we did was first consult that community to develop the controlled vocabulary, so they all agreed about what should be in the database. Then we developed the tools and we organized workshops where the collections were; so we did one from Latin America, Caribs, one for Pacific Islands, South Asia, one for East Africa, one for West Africa. Then we called the GenBank data managers together and we did a training workshop for each of the regions. The idea was we wanted to show them how you can go from the field observation to the database, locally, and from the database send your data to a collaborative platform. So the workshop was starting by using the controlled vocabulary as a fieldbook. (PI_14_A)

This early work exemplifies the important role played by data experts such as Arnaud in brokering global data trade by mediating dialogue over standards and infrastructures among the key technical providers involved—in this case, GenBank and the stewards of local data collections dotted around the Global South. Note the peculiar use of the term "community" in the quote above, which is meant to capture a wide and heterogeneous array of data workers—ranging from local breeders to field trial managers, database curators, and managers of germplasm stocks—with an active interest in how crop data are classified and integrated. Arnaud's understanding of what expertise is relevant to the building of global data resources is broad and aimed at encompassing forms of local knowledge that do not typically feature in standard plant science textbooks, and yet are crucial to capturing the variety of plant properties and approaches to measurement characterizing different parts of the world:

> It was expert knowledge. We had a selection of names we knew from our collaborators from the scientists working with us, because they have been collaborating into projects since years. So they said, "Okay, this lady is a specialist of the description of the banana in India. This one is for African bananas." . . . Because according to where the banana grows it looks different, and the varieties and races are different. So we had an expert more or less of each region, making sure that at least that person could consult with the expert of the taxonomic group they were looking at. Then we coordinated the feedback about those descriptors. (PI_14_A)

A central preoccupation of this approach to expertise was the inclusion of local botanical knowledge, or ethnobotany; much less central were concerns with intellectual property and national legislation around plant data ownership and licensing. This was partly due to the widespread absence of clear legislation on those issues at the time, and partly to the transnational vision of this data gathering and linkage, which explicitly circumvented national logics with the expectation that a globalized data resource yields higher returns for all (including the contributing nations). Arnaud brought this understanding of what constitutes an expert community with her when taking up the leadership of the Crop Ontology. As she noted in her discussions with me, her appointment was motivated by her ability to work with researchers on the ground and her awareness of the challenges underpinning the highly localized semantics of crop morphology:

> Because of my involvement in the banana germplasm information system, the descriptors and production of variety cards, I was proposed to take the lead of the crop ontology for the Generation Change program. Before me, the project was led by a bioinformatician. He encountered a lot of difficulties to make breeder scientists understand what it was for, because his approach was too informatics-based. . . . I contributed a lot to the project for the banana ontology, and then the leader said, "Okay, I would like to hand over the project to you because I think your approach is closer to the scientists." So in 2008 I became the PI [Principal Investigator] of that project. (PI_14_A)

Arnaud managed to get the project on track with rapid progress made on the development and implementation of the system. One of her operating principles, shared with many of her collaborators, was that the Crop Ontology needed to encompass the peculiarities of each crop, to take account of their enormously different biological characteristics as well as their cultivation and processing as staple foods. The Crop Ontology thus developed a separate ontology for each crop in dialogue with the relevant experts, including breeding communities and farmers, as well as plant researchers from various locations. The resulting ontologies—including a Cassava Ontology, a Banana Ontology, a Rice Ontology, a Mildew Ontology, a Yam Ontology and so forth—needed to be interoperable (and thus employ compatible standards and terms), while retaining their own specificities. In the words of a user I interviewed, the Crop Ontology acted as a "regulatory body" in mediating differences, negotiating visualizations and standards that could work for all stakeholders involved, and producing *just enough* consensus on procedures as to enable the linking and cross-consultation of databases.[25]

This approach has come under considerable pressure by bioinformaticians and data scientists who viewed the complexity of the system as unnecessarily confusing and would rather have supported a more logically structured, top-down approach in which interoperability standards were fixed once and for all. However, it was arguably the bottom-up quality of the Crop Ontology that made it successful among its users. In 2008 the Crop Ontology comprised a core team of two: Arnaud and one computer scientist, both based at Bioversity in Montpellier. The content management was distributed across contributors from eight centers, including Mexico, India, Nigeria, and the Philippines. As the project blossomed, so did its network of partnerships, which expanded to include more and more contributors as well as collaborations with other data services focusing on genomic or breeding data. Arnaud facilitated the maintenance and expansion of collaborations by organizing in-person meetings every two years. Initially called Crop Ontology workshops, these meetings have evolved into conferences with hundreds of participants.[26]

The Crop Ontology has become the center of a large and complex landscape of interacting services, public and private innovators (e.g., Bayer), and data resources.[27] Three resources in particular have developed alongside and in coordination with the Crop Ontology (whose core staff remains minimal despite the growth in responsibilities): the Minimal Information About Plant Phenotyping Experiments (MIAPPE), the Collaborative Open Plant Omics (COPO), and the Breeding API (BrAPI). MIAPPE includes the Crop Ontology as its key standard and is in turn compatible with the Multi-Crop Passport Descriptors (MCPD) used to track the cross-national movement of material samples of germplasm. It also collaborates with the Plant Phenotype Experiment Ontology (PPEO) and is interoperable with BrAPI, itself a large resource sponsored by over twenty major institutes and foundations around the globe. MIAPPE shares with the Crop Ontology an emphasis on minimal standards that exclude as little as possible, thus enabling biologists and breeders to input data acquired through their everyday experiences of the crops.[28] Together, these resources provide a specific platform for crop-related research, which focuses on facilitating as much data movement as possible beyond national borders and making those data usable as evidence base for agricultural R&D, plant science as well as smallhold farmers and breeders.

This vision for plant data movement is clearly aligned with the broader Open Data movement promoted by the European Commission and enacted by ELIXIR (the European life-sciences infrastructure for biological information), FAO, and the Gates Foundation with regard to biological data.[29] This vision is also concretely supported by the CGIAR, where a centralized Platform for Big Data was instituted in 2016 to facilitate seamless data travel across the fifteen CGIAR institutes, each of which based in a different country across four continents. Significantly, the Crop Ontology was formally adopted as a common standard by the CGIAR Platform for Big Data in 2018, which underscores its importance as a reference tool not only for the CGIAR institutes, but also for the hundreds of data initiatives related to their work at both local and international levels, which effectively places the Crop Ontology at the center of worldwide efforts to enable plant data movements across borders. Starting as a relatively marginal tool, the Crop Ontology has thus become a key resource for data management worldwide and an obligatory passage point to secure plant data collaborations across nations.

Data Value beyond Borders: Using the Crop Ontology to Study Cassava

One of the reasons cassava databases are working is because of the ontology work, the attempt to have a standardized system for collecting data.
(Informer 32)

The ability of the Crop Ontology to mediate between many different concerns and stakeholders has been a key factor in its success as a transnational broker of data movements and exchanges over the last decade. Crop Ontology contributors have long recognized that negotiations around technical issues like interoperability and machine-readability require considering issues of highly contested cultural, social, political, and economic significance, such as: which plant varieties to prioritize and valorize; how and when to extract plant data, for which purposes, and from whom; how to compare data acquired from different territories and knowledge systems; which types of knowledge to highlight and incorporate (for example, how to include input from farmers and breeders); how to track and reward the provenance of information; what kind of access to grant to such information, to whom, and for which purposes; and what kind of economic and political support to seek while developing a vision for tackling these challenges.

To better understand the development and implications of this sophisticated understanding of the interplay between data standards and political economy, I now briefly examine the relation between the Crop Ontology and CassavaBase, a database dedicated to the collection and analysis of phenotypic data resulting from field trials on the cassava root. CassavaBase originated from a wider international (and American-led) collaboration called the Solanaceae Genomics Network, focused on building data resources for research on major crops including tomato, potato, eggplant, and various species of pepper.[30] The bulk of CassavaBase was developed primarily at the IITA, in collaboration with research partners at Cornell in the United States and other CG centers, through funding provided by the Gates Foundation via its NextGen Cassava program.[31] It is used across central Africa, Asia (especially Thailand), and South America.[32] Cassava is the main source of nutrition for 500 million people around the globe, and its resilience in the face of climate change (unique among the major staple crops) makes it attractive for future food security worldwide.[33] Data collection on cassava genomics, field trials, and wildtypes has thus blossomed over the last decade. At the same time, cassava distribution has largely been limited to the Global South and to exchange economies among small farming communities. Hence cassava has not (yet) been commodified in the same way as banana, rice, and its fellow tuber, the potato. This has important repercussions for research on this crop, much of which is not yet constrained by the kind of iron-clad intellectual property regimes binding R&D on the likes of wheat. To reflect this, CassavaBase has been conceived as an Open Data source from its inception and is endowed with the ability to release field trial data as online, freely accessible goods straight from their collection in the field via a fieldbook app which can be downloaded on technicians' smartphones.

CassavaBase curators have worked closely with the Crop Ontology to develop participatory practices to facilitate and support phenotypic data collection and use among local cassava breeders. For a start, farmers and breeders have been consulted over the choice of which cassava varieties to trial, which is crucial since any one trial takes between six and seven years and it is obviously not possible to shift variety halfway through a trial. This choice is immediately complicated by social factors such as gendered preferences in the choice of varieties and related plant traits. As it turns out, cassava breeding is largely conducted by female breeders, and yet the choice of varieties is often regarded as a community issue and discussed among the largely male leadership of each tribe. This social arrangement matters to data collection within the database, since CG-sponsored research has uncovered significant differences between the preferences of female and male breeders—with women favoring varieties that have the best chance of securing high-quality food for their families, while men prefer varieties with high yield and market value.[34] Related to these discrepancies are traits that taxonomists would not usually consider in relation to the biological classification of cassava, and yet are significant for local breeders, such as the flavor and consistency of cassava flour when cooked (a key factor in Nigeria where cassava is used to make essential meal components such as gari and fufu). In response to these factors and in an explicit attempt to secure representation for female breeders and their preferred traits, CassavaBase curators have introduced new terms in the Crop Ontology which describe biological traits of sociocultural and commercial value (e.g., "color of boiled roots," "gari content," "marketable root weight"). Data curators across CG centers have also worked toward the development of community protocols to accompany and support the choice of which varieties to test, so that local women are explicitly consulted and their expertise has as much of a chance to shape the collection and use of cassava data as the expertise of men.

Another matter on which stakeholders are consulted is the description of methods used to measure cassava traits. This is particularly important in the case of nonquantitative traits like color, but also in relation to quantitative measurements that depend on local judgments about which specimens to measure (for instance as a representative of a whole field) and at which stage in the development and processing of the crop to measure. For instance, CG research and breeders' feedback evidenced differences in the time, specimens, and types of measurements chosen by different breeders to exemplify flowering, harvest, and postharvest stages of data collection.[35] To reflect this, the Crop Ontology has provided ways to record each method used for data collection, and triangulate that with the specific skills employed by the data collector in question when measuring a specific trait. The system is meant to capture nuances in methods and field experience, while at the same time making it possible for multiple methods, enacted differently, to produce data associated with the same trait.

CassavaBase curators have strived to maintain dialogue with farmers and breeders through regular farm visits and the organization of "cassava breeder meetings" hosted at the IITA, during which feedback and insights are collected and decisions are made about which functionalities and related terms or data would be of most use to local farmers and their families. In the course of this work, larger clashes between different worldviews have become apparent, for instance concerning plant description and classification. Genetic classification of cassava into distinct varieties, for instance, has not always matched the classification used by farmers, which is grounded on differences in the plants' morphology and market value (e.g., consistency of the pulp and the outer skin). Again, in such cases CassavaBase curators worked with the Crop Ontology to garner evidence about each approach to classification and to document the methods by which different determinations have been made, thus preserving as much diversity and inclusivity as possible in the data being collected.

This emphasis on diversity has immediate political, social, and economic resonances. It enabled the Crop Ontology to keep a relatively neutral stance on different ways of conceptualizing and evaluating plant data, which are in turn connected to different motivations and interests by the researchers involved—arguably grounded in their respective locations and the socioeconomic expectations attached to their work. For example, genetic methods (and related biomarkers) tend to be favored by plant researchers in the United States and Europe, who have the instruments to implement them and tend to view them as epistemically superior taxonomic tools—particularly for research aimed at genetic manipulation.[36] The Gates Foundation is a key supporter of this approach, with a strong emphasis on increasing the rate of so-called genetic gains, thus focusing the modernization of breeding programs in countries such as India, Bangladesh, and Ethiopia on the application of genomic innovation.[37]

Morphological indicators tend instead to be favored by breeders in Nigeria, Ghana, and Brazil, who have direct experience of cassava as a local food source and acknowledge the value of nonmolecular, low-resource, noncommodified approaches to improving varieties. The Crop Ontology has been set up to incorporate such differences in ways that are granular and modular, and foster (at least in theory) the assemblage of data coming from very different sources without losing sight of what diversity in provenance may signify.[38]

To this end, the Crop Ontology also facilitated dialogue between specialized structures such as CassavaBase and international initiatives in plant data management, such as for instance the large project of plant ontology integration called Planteome, of which the Crop Ontology became a formal component in 2017. Such dialogue is not always easy and it is hard to assess its overall effectiveness (a point to which I return in the next section), but at the very least it does foster regular moments of intersection between the development of high-level standards meant for adoption across species and around the globe and the development of local tools of relevance to specific crops.

One remarkable outcome of such work has been the use of cassava data globally available online to tackle plant diseases that threaten the local economy. A recent instance is Nuru, a smartphone application that uses data collected initially from cassava (and later extended to a vast variety of crops) to facilitate the diagnosis of plant disease. Nuru was developed from a pilot project grant of CGIAR awarded in 2017 and it went on to feature as runner-up for the 2019 CGIAR Prize for best data science innovation of the year. It works by asking users to point their smartphone camera at a diseased plant. Using machine-learning algorithms trained on existing data, the app then interprets the image into a diagnosis of the disease and provides helpful tips about the characteristics of the disease and possible routes to tackle it. Crucially for countries without extensive or reliable bandwidth, this can work offline: users can download the Nuru database (though not its source code, which remains proprietary), which includes datasets and classification tools from the Crop Ontology and CassavaBase, and get their application to work without being connected to the internet. The creation of Nuru was facilitated by several contributors, including CassavaBase curators, the nonprofit research group at Penn State working under the name of Plant Village, CGIAR, and, more recently, Google. As documented in a promotional video released in 2017:

> More than 200,000 images of diseased crops have already been collected on farms at RTB cassava field sites in coastal Tanzania and western Kenya using cameras, spectrophotometers, and drones. These images will be used to train

> AI algorithms. The team recently developed an AI algorithm with TensorFlow that can automatically classify five cassava diseases building on published work covering twenty-six diseases in twelve crops. Through a collaboration with Google, a TensorFlow smartphone app has been developed and field-tested in Tanzania by IITA.[39]

The project expects to radically transform pest and disease monitoring by using artificial intelligence (AI), advanced sensor technology, and crowdsourcing capable of connecting the global agricultural community to help smallholder farmers. "Nuru" means "light" in Swahili, and while the key developers of the app are English speakers, the app has been set up from the start to include translations in local languages (such as Swahili, Twi, and Hindi). In this and many other ways, an app like Nuru exemplifies the potential payoffs attached to the vision of globalized plant knowledge that rules the push for open, coordinated travel of data. This is the kind of technology that motivates the Crop Ontology to try to negotiate the tensions arising from attempts to link locally acquired digital information to global networks, and the related effort to regulate the movement of data and related materials across national borders.

The Epistemic Economy of Transnational Knowledge and the Political Economy of the Nation-State

Considering the Crop Ontology and its complex interrelations with other data platforms and their patrons also provides a glimpse of the tensions, obstacles, and visions underpinning the supposedly smooth flow of data across national borders, epistemic communities, cultural landscapes, and socioeconomic systems. Plant data infrastructures are required to serve different forms of value, which in turn spell diverse allegiances and forms of alignment with the state and its political economy.[40] Examples of such diverse forms of value are: the *epistemic* value of data within diverse scientific perspectives (phenomic, agronomic, molecular); the *cultural* value of specific legacies that data are seen to represent and/or embody (e.g., local breeding traditions); the *social* value of specific market conditions for human flourishing and for the resolution of global challenges (e.g., the commitment to the commodification of plant varieties as a resolution to food security); the *technological* and *political* value of specific modes of action and intervention (e.g., genetic engineering of crops or pathogens, and/or the remediation and conservation of wild variants); and the *financial* value of data as inexhaustible, nonrivalrous sources of information that can however be controlled, owned, and speculated upon as commercial goods (e.g., data collection as digital bioprospecting).[41]

National interests and borders play an ambiguous, multifaceted role with respect to these different forms of valuing. As emphasized by Mezzadra and Neilson, national borders should be conceptualized not as simple geographical conventions but as complex social institutions: "borders, far from serving simply to block or obstruct global flows, have become essential devices for their articulation"; and "the multiple (legal and cultural, social and economic) components of the concept and institution of the border tend to tear apart from the magnetic line corresponding to the geopolitical line of separation between nation-states."[42] In other words, borders are "sites in which the turbulence and conflictual intensity of global capitalistic dynamics are particularly apparent."[43] Taking this into account, the frequent absence of borders from plant data linkage discourse and tools is in itself very significant and conveys a specific epistemic perspective.[44] It is not that national interests have disappeared or have become irrelevant to transnational efforts of plant knowledge production. Rather, there is a deliberate sidelining of political and social considerations when it comes to the circulation of plant data (and, arguably, of digital data in general, especially when not conceptualized as "personal data"), in favor of a scientific focus on the technical challenges and opportunities linked to data sharing. It is clear, for instance, that farmers' choice of crop variety within specific countries is rarely just a matter of consulting evidence, with national strategies around the regulation, distribution, and commercialization of specific varieties typically shaping selection by farmers and breeders. To infiltrate and inform the choice of crop varieties on the ground, plant linkage infrastructures need to simultaneously inform many different types of stakeholders, including research labs, agrotech businesses, governmental agencies, breeder groups, and smallhold farming.

To facilitate this, the stakeholders that this work is accountable for are a heterogeneous group typically defined in terms of their expertise and use of plant knowledge rather than their location, national affiliation, or strategic interests in the crop market. This is effectively illustrated by an infographic produced by the Research Data Alliance (RDA) Working Group in Semantics in 2017 (which included Crop Ontology developers), reproduced here as figure 10.2. The infographic identifies data semantics itself as a way to "reconcile points of view and data," where the "points of view" are identified as those of biologists, farmers and breeders—or nutritionists, chefs, food manufacturers, traders, information managers, and sociologists. They do not include governments, land managers, or other spokespersons for territory and related interests and borders, which positions these actors outside the core group of intended plant data users.

More generally within the contemporary big data landscape, control over data acquisition, visualization, aggregation, and retrieval tools means

SEMANTICS - THE WAY TO RECONCILE POINTS OF VIEW AND DATA

THE EXAMPLE OF "RICE"

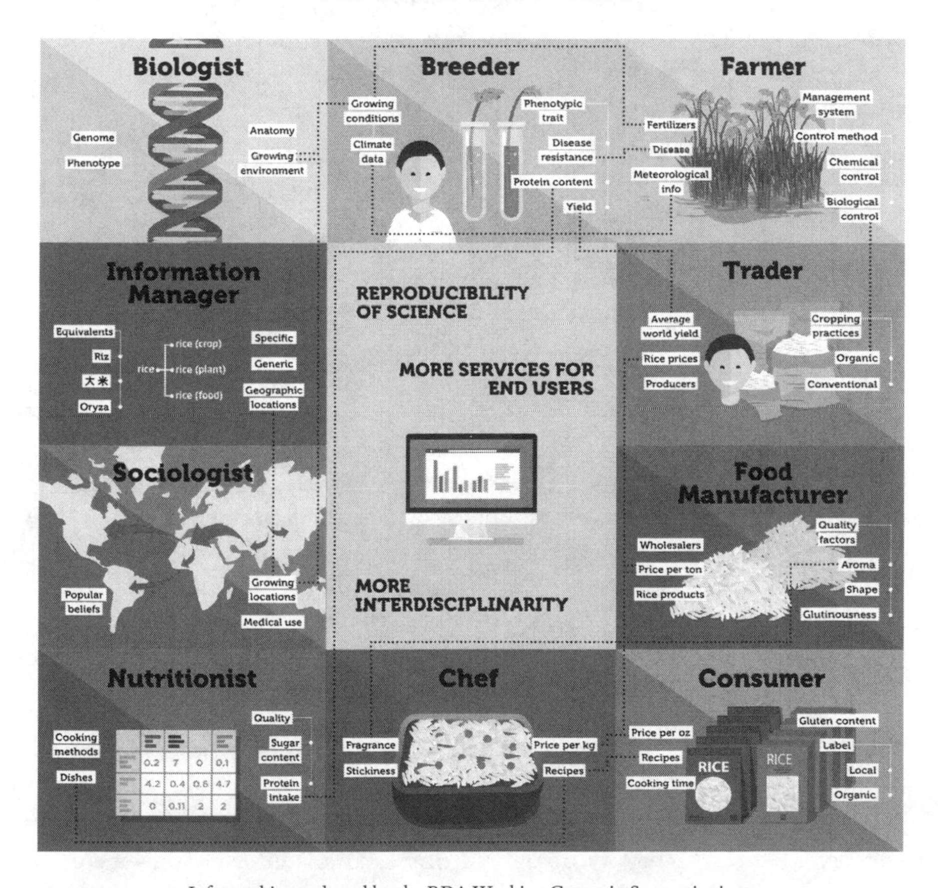

FIGURE 10.2. Infographic produced by the RDA Working Group in Semantics in 2017.

control over means of knowledge production, and thus in turn over what counts as legitimate and reliable data sources, who can participate in research, and what kind of knowledge is envisaged as output of data analysis. Analytic approaches focused on tracing networks, concerned as they are with mapping links between nodes of power and agency, are at risk of hiding away these dynamics and related transformations in knowledge landscapes. A qualitative and historicized analysis of the travels of data facilitates an exploration of the different forms of valuing data that are instigated and supported by initiatives such as the Crop Ontology, which are neither political bodies nor international agencies, and yet quietly affect which value-judgments are challenged or channeled through research communities and related institutions around the world.

The form of transnational political agency that is manifested in the making of data infrastructures produces knowledge landscapes tailored toward specific forms of value, social configurations, and related scientific regimes. One way to think about such knowledge landscapes is as *epistemic economies*: that is, knowledge systems with the capacity to encompass and support the intersection between a particular vision for the conduct of research, the technical procedures and tools through which such vision can be realized, and its social and economic motivations and functions.

The epistemic economy characterizing plant data infrastructures like the Crop Ontology emphasizes the transnational nature of plant knowledge as an essential platform for more local insights and applications. This is meant to open the way toward improving understanding of diverse situations of inquiry, including changing problem agendas, stakes, and communities of reference in plant and agricultural science. This epistemic economy operates at different levels than national politics and in ways that do not often clash directly with national borders and political economies. Indeed, the Crop Ontology functions by focusing dialogue between international data standards (and related infrastructures) and local practices of data collection and reuse, the latter typically more circumscribed than national borders or escaping them altogether (for instance, where the same crop variety is being cultivated across borders in two different states).

A key insight guiding this work and providing a way to field its complexity is the idea that tracking the history of particular data points and clusters is more significant than standardizing per se. As a crop database curator put it to me, "If you get an accession, you should trace its history, get its attributes, in which trials it has been used and its performance in the trials at every level. Quality, agrobiotics, stresses. All the information should be linked." This is a way to incorporate contextual elements of relevance to data analysis—such as the social and environmental conditions under which a field trial was designed and conducted—without compromising the chance to search across the datasets and compare results with a very different history. Whether such careful contextualization is properly acknowledged and carried across other data-sharing platforms remains to be determined, particularly given that other data resources place more emphasis on, for instance, combining genetic and phenotypic traits than on the analysis of data provenance and its significance for understanding gene-environment interactions. For the purposes of this chapter, it is useful to signpost how difficult it is, from the methodological point of view, to evaluate and measure the degree to which the specific versions of inclusiveness and regard for epistemic diversity endorsed by the Crop Ontology are preserved once the data shepherded by that system are linked, repackaged, and redistributed by other transnational data infrastructures—for instance, as part of the Proteome project (which is primarily funded by the US National Science Foundation and led by US-based scientists) and the CGIAR Platform for Big Data.

Even bigger questions can be posed around how long it will take for national logics to catch up to the economic value of plant data and what role plant data histories could play in this respect. As concluded by Krige in his study of US-UK collaboration about nuclear weapons, "technological collaboration involves making difficult trade-offs. It provides an avenue for knowledge circulation to mutual advantage, but it also opens research facilities to external scrutiny. It can keep lines of communication open until the knowledge concerned is deemed sensitive, whereupon collaborators can become competitors and sharing can give way to denial."[45] There are important lessons here about the dynamic nature of any collaboration and the fragility of data-sharing agreements: relations between countries can evolve as quickly as the potential of the technologies at hand, with expectations around the goals and usefulness of data analysis shifting accordingly. There are also significant differences between Krige's case and the transnational sharing of plant data I consider in this chapter. First, the plant data circulated through platforms such as the Crop Ontology do not, per se, constitute knowledge: they need to be complemented by a large array of tools and metadata in order to be employed as evidence for knowledge claims and interventions—which in turn requires complex division of labor and sharing/dissemination agreements and infrastructure that are arguably best obtained through large international consortia rather than targeted, centralized operations. Relatedly, it is technically difficult and financially inconvenient to focus on only a few types of data use, thus constraining and specifying the concrete goals of data linkage. The promise of countless, underdetermined possibilities for data analysis and interpretation (typically associated with big data) is a much better fit for speculation and strategic reasoning than a narrowing of the potential value of data accumulation. In turn, the strategic vagueness and multiplicity of goals associated with plant data linkage make it hard to identify who are the relevant parties and beneficiaries in these scientific collaborations and particularly to single out economic returns at the national level. This situation may well account for some of the enthusiasm underpinning commitments to "open data" over the last decade—and may explain why, as soon as specific data applications are identified with well-characterized scientific and economic returns, data work tends to become restricted to a small number of named partners within the boundaries of a specific project.

The difficulties in managing the indeterminacy of value of scientific resources at the national level are apparent when considering the battles over genomic data and their relation to sovereignty that took place over the last two decades. In the case of human genomics, for instance, national politics has been quick to appropriate the power of data to represent heritage, identity, and ethnicity, as exemplified by the controversies over the Human Genome Diversity Project and attempts to nationalize genomic resources as in the case of the "Mexican genome" law for the preservation of genetic sovereignty.[46] Attempts have also been made to nationalize plant data, and particularly the management of data about plant compounds of potential use for drug discovery, as exemplified by the Indian government's development of the Traditional Knowledge Digital Library, a national database collecting and administering access to plant data of relevance to Ayurvedic medicine in order to protect traditional knowledge from exploitation by pharmaceutical corporations abroad.[47] These and many other similarly fraught cases raise questions around how national rights, interests, and claims to ownership may be reasserted in the future in relation to plant phenotypic data such as those circulated by the Crop Ontology, especially at a moment where governmental policies around digital data are growing more sophisticated. There is already extensive debate around how national boundaries could and should be drawn around digital assets—such as, for example, Germany's efforts to nationalize cybersecurity and associated digital infrastructures, which generated controversy as an attempt to nationalize the internet itself.[48]

Furthermore, there are questions around how the Crop Ontology system fares with respect to crops that are already commodified in neoliberal markets and enshrined in national political economies around land management and the distribution of varieties. The case of cassava (and some of the other crops under the remit of the Crop Ontology) is notable for the absence—so far—of a strong commercial interest from the West and of highly centralized governmental control over what is being farmed. In Nigeria, for instance, cassava farming has been largely in the hands of small farming communities supplying local markets. The lack of commercial "value added" in neoliberal market systems has arguably been a major factor accounting for the freedom by researchers and breeders to work with farmers to produce better varieties through transnational sharing and cooperation, and the landscape is now changing rapidly, not least thanks to the agricultural innovations wrought by large-scale data mining. On the one hand, the Crop Ontology has been set up to pick up and emphasize biodiversity, by adapting its standards to the specific traits found in local collections and framing its epistemic economy around the value of local strains. This has arguably worked best when removed from the corporate monocultures and proprietary regimes of food security characterizing much of US agriculture. On the other hand, the Crop Ontology incorporates a whole range of assumptions about what makes a plant trait most valuable and to whom—including an expectation, particularly by funders such as the Gates Foundation, that such wide-ranging data collection will enable the identification of superior varieties or traits that could be commercialized across the globe, in line with previous histories of the commodification of staple crops in the West. As the Crop Ontology becomes ever more embedded in the emerging Platform for Big Data of CGIAR, it is difficult to predict how the inclusivity and effectiveness of data management systems will be interpreted, and with what consequences for agronomic research and agricultural development.

Conclusion: Prospects for the Datafication of Plant Science

The epistemic economy of plant data infrastructures such as the Crop Ontology intersects with national economies in ways that are only partially aligned with global capitalism, and whose impact on local political economies is yet to take a firm shape. The Crop Ontology, and apps such as Nuru, depend significantly on technological developments controlled by the Global North, including most advances in AI, data science, and data semantics. They also feed data collected from local communities into a transnational system with few checks in place against exploitation and appropriation by powerful players.

At the same time, the manner in which plant linkage infrastructures and related applications absorb such knowledge and standards can be reflexive and even critical. For instance, the Crop Ontology's initial skepticism about adopting an OWL-formatted plant ontology—on the grounds that its logical structure would have acted as a straitjacket on the absorption of useful information from users and data producers—signals the willingness to recognize and mark alternative epistemic spaces, thus resisting the overarching theory of value (both economic and scientific) attached to plant data by much AI discourse and, relatedly, the neoliberal market. Nuru, like many other applications of data linkage tools such as the Crop Ontology, has enormous potential to make a difference to national-scale agriculture, by importing transnational knowledge and its peculiar epistemic economy into local fields and smallhold farming. It also signals an appropriation of technology and a relocation or spread of sites of innovation that transcends national borders. While the creators of the app are US researchers, contributors include data providers and users from around the globe. In this sense, this is not—or at least, not only—a postcolonial story about African and Latin American countries functioning as recipients for technical assistance programs framed in Europe and the United States. As Mateos and Suárez-Díaz discuss in their excellent study of the material movements of radiochemistry, cases such as these make it possible "to address the translation of transnational into local interests. Communication practices certainly are an important aspect of scientific exchanges, but they are embedded in the larger question of the materialities of travel."[49] Looking at transnational data travels and the infrastructures that facilitate them puts in sharp relief the importance of *procedural* innovation and understandings of *expertise* that underpin those "materialities of travel." In Krige's words, "knowledge is a national resource, and the art of statecraft lies in defining policies and instruments that help draw the line between what kind of knowledge will be shared with (or denied to) whom."[50] As I hope to have shown, the multiplicity and complexity of ways in which plant data can travel, of potential outcomes of such travels and of the infrastructural, conceptual, procedural, and material components of their journeys, make it difficult for national governments to draw this line effectively and decisively.

The translational epistemic economy of plant data movements promises to forge data systems, and related conceptual approaches, that bridge between biological research on conservation, biodiversity, and food security, and take account of local knowledge and uses of the plants in question. This may be construed as an essential avenue toward the transformation of long-standing knowledge systems to reimagine agriculture away from high-yield monocultures. The Crop Ontology certainly invites consideration of how local biodiversity can boost the sustainability of food production systems, thus framing research in plant science so that conservation efforts and the resilience of local communities are placed at the center of food security. This includes the repositioning of taxonomic efforts around plant traits, as well as data analytics and visualization tools, toward supporting a better understanding of the multiple roles of plants as components of complex ecosystems.

And yet, this same approach can be coupled with highly restrictive and predatory forms of intellectual property, such as the patenting of plant genomic modifications, control over the choice and distribution of varieties, and the tracking of food production cycles (which makes how plants are stored, selected, bred, stocked, commercialized, distributed, and monitored subject to market analysis and further privatization).[51] These can in turn be used to support the establishment of novel monopolies grounded on control over information rather than over genetic materials, thus proving a natural extension to the Green Revolution and the intensification of agriculture, which precision agriculture is meant to perfect and further augment. Relatedly, the dematerialization of biological materials achieved through the triangulation of genomic, phenotypic, and breeding data may further facilitate the exclusion of breeders and farmers from decision making around crop choice and management, creating more inequality and a further breakdown of communication and trust. In closing this chapter, I wish to highlight the extent to which these, as no doubt other, different narratives coexist in relation to plant data movements. Future intersections between national interests and the translational epistemic economy of plant data linkage will no doubt play a crucial role in determining which narratives prevail.

Acknowledgments

This chapter has benefited from helpful discussions with: John Krige, to whom I am particularly grateful for challenging me to write on this theme and providing insightful feedback and encouragement; all the participants in the excellent workshop on "Transnational Transactions" hosted at CalTech in February 2020; my colleagues Hugh Williamson, Rachel Ankeny, Mike Dietrich, and John Dupre; and my collaborators in the field, particularly Elizabeth Arnaud, Afolabi Agbona, and Peter Kulakow. Funding for this research was generously provided by the European Research Council (award numbers 335925 and 101001145), the Australian Research Council (award number DP160102989), and the Alan Turing Institute (EPSRC grant EP/N510129/1).

Notes

1. Marie Bolger, Rainer Schwacke, Heidrun Gundlach, Thomas Schmutzer, Jinbo Chen, Daniel Arend, Markus Oppermann, et al., "From Plant Genomes to Phenotypes," *Journal of Biotechnology* 261 (2017): 46–52.

2. Ted Striphas, "Algorithmic Culture," *European Journal of Cultural Studies* 18, no. 4–5 (2015): 395–412, https://doi.org/10.1177/1367549415577392; Chris Miles, "The Combine Will Tell the Truth: On Precision Agriculture and Algorithmic Rationality," *Big Data and Society* 6, no. 1 (2019), https://journals.sagepub.com/doi/full/10.1177/2053951719849444.

3. The coupling of digital data and material resources is highly valuable in scientific terms but very difficult to obtain precisely because the transfer of material assets is subject to clear international regulations, while digital data—and particularly data deemed not to be sensitive because they do not directly concern individuals—remain in a legal limbo.

4. As documented in this volume by Jessica Wang (chapter 1), Gabriela Soto Laveaga (chapter 6), Tiero Saraiva (chapter 8), and Courtney Fullilove (chapter 9).

5. Mariana Mazzucato, "Preventing Digital Feudalism," *Project Syndicate*, October 2, 2019, https://www.project-syndicate.org/commentary/platform-economy-digital-feudalism-by-mariana-mazzucato-2019-10.

6. United Nations, Department of Economic and Social Affairs, "Sustainable Development Goals," https://sdgs.un.org/goals, accessed August 31, 2021.

7. Jan Ralph Kloppenburg Jr., *First the Seed: The Political Economy of Plant Biotechnology*, 2nd ed. (Madison: University of Wisconsin Press, 2004); Courtney Fullilove, *The Profit of the Earth* (Chicago: University of Chicago Press, 2017).

8. Kloppenburg, *First the Seed*; Sabina Leonelli, "Data: From Objects to Assets," *Nature* 574 (2019): 317–321.

9. Jay Sanderson, *Plants, People and Practices: The Nature and History of the UPOV Convention* (Cambridge: Cambridge University Press, 2014).

10. For example, Juliana Santilli, *Agrobiodiversity and the Law: Regulating Genetic Resources, Food Security and Cultural Diversity* (Abingdon, UK: Earthscan, 2012).

11. Christine Frison, *Redesigning the Global Seed Commons: Law and Policy for Agrobiodiversity and Food Security* (Abingdon, UK: Routledge, 2018).

12. Sabina Leonelli, "Centralising Labels to Distribute Data: The Regulatory Role of Genomic Consortia," in *The Handbook for Genetics and Society: Mapping the New Genomic Era*, ed. Paul Atkinson, Peter Glasner, and Margaret Lock (London: Routledge, 2009), 469–485.

13. For an overview of the significant role played by the Research Data Alliance in international coordination of data resources, see Francine Berman and Merce Crosas, "The Research Data Alliance: Benefits and Challenges of Building a Community Organization," *Harvard Data Science Review* 2, no. 1 (2020), https://hdsr.mitpress.mit.edu/pub/i4eo09f0/release/2; on the bioversity work on crop descriptors, see Biodiversity International, "Developing Crop Descriptor Lists: Guidelines for Developers," *Biodiversity Technical Bulletin* 13 (2007), https://cgspace.cgiar.org/bitstream/handle/10568/74489/Developing_crop_descriptor_lists.pdf?sequence=1&isAllowed=y.

14. Research Data Alliance, https://www.rd-alliance.org/, accessed August 31, 2020.

15. Hans J. P. Marvin, Esmée M. Janssen, Yamine Bouzembrak, Peter J. M. Hendriksen, and Martijn Staats, "Big Data in Food Safety: An Overview," *Critical Reviews in Food Science and Nutrition* 57, no. 11 (2017): 2286–2295; Elizabeth Arnaud, Marie-Angélique Laporte, Soonho Kim, Céline Aubert, Sabina Leonelli, Berta Miro, Laurel Cooper, et al., "The Ontologies Community of Practice of the CGIAR Platform on Big Data in Agriculture: Improving Practices in Developing and Applying Semantics for Data Annotation." *Patterns* 1, no. 7 (2020): 100105.

16. The degree to which the philanthropic work of the Gates Foundation includes a commitment to specific forms of "technological fix" and commercialization of

science (themselves steeped in American perceptions of success and Western approaches to the eradication of disease, following on the footsteps of long-standing charity bodies such as the Rockefeller Foundation) is a matter of ongoing debate. See, for instance, Lindsey McGoey, *No Such Thing as a Free Gift: The Gates Foundation and the Price of Philanthropy* (London: Verso, 2015); Adam Moe Fejerskov, "The New Technopolitics of Development and the Global South as a Laboratory of Technological Experimentation," *Science Technology and Human Values* 42, no. 5 (2017): 947–968; and David Reubi, "Epidemiological Accountability: Philanthropists, Global Health and the Audit of Saving Lives." *Economy and Society* 47, no. 1 (2018): 83–110.

17. When not otherwise indicated, quotations are extracted from transcripts of interviews I held with Crop Ontology and CassavaBase curators in Exeter (UK), Ibadan (Nigeria), and Montpellier (France) between 2016 and 2018. Those interviews that are not bound by confidentiality concerns are available as an Open Data collection on the Exeter Data Studies section of Zenodo, with permission from the interviewees: https://zenodo.org/communities/datastudies/?page=1&size=20, accessed April 1, 2021.

18. Rosemary Shrestha, Elizabeth Arnaud, Ramil Mauleon, Martin Senger, Guy F. Davenport, David Hancock, Norman Morrison, Richard Bruskiewich, and Graham McLaren, "Multifunctional Crop Trait Ontology for Breeders ' Data : Field Book, Annotation, Data Discovery and Semantic Enrichment of the Literature." *AoB Plants* 2010 (2010): plq008, https://doi.org/10.1093/aobpla/plq008; Sabina Leonelli, *Data-Centric Biology: A Philosophical Study* (Chicago: University of Chicago Press, 2016).

19. Rosemary Shrestha, Luca Matteis, Milko Skofic, Arlett Portugal, Graham McLaren, Glenn Hyman, and Elizabeth Arnaud, "Bridging the Phenotypic and Genetic Data Useful for Integrated Breeding through a Data Annotation Using the Crop Ontology Developed by the Crop Communities of Practice," *Frontiers in Physiology* 3 (2012): 1–10, https://doi.org/10.3389/fphys.2012.00326; Arnaud et al., "The Ontologies Community of Practice."

20. Arnaud et al., "The Ontologies Community of Practice."

21. Alessandro Minelli, "The Galaxy of the Non-Linnaean Nomenclature," *History and Philosophy of the Life Sciences* 41, no. 3 (2019), https://doi.org/10.1007/s40656-019-0271-0.

22. Sabina Leonelli, Robert P. Davey, Elizabeth Arnaud, Geraint Parry, and Ruth Bastow, "Data Management and Best Practice for Plant Science," *Nature Plants* 3 (2017): 17086, https://doi.org/10.1038/nplants.2017.86.

23. Generation Challenge Programme, https://www.generationcp.org/aboutus.html, accessed August 31, 2020.

24. Helen Curry and Tim Lorek, *Historical Perspectives on the Consultative Group on International Agricultural Research (CGIAR)* (Cambridge: Cambridge University Press, forthcoming).

25. For other examples of international consortia enacting this vision for data sharing, see Sabina Leonelli, "Scientific Agency and Social Scaffolding in Contemporary Data-Intensive Biology," in *Beyond the Meme: Development and Structure in Cultural Evolution*, ed. Alan C. Love and William C. Wimsatt (Minneapolis: University of Minnesota Press, 2019), 42–63.

26. I was one of the keynote speakers at the most recent such meeting in Montpellier in 2018 and can attest to the variety of stakeholders and initiatives involved from around the globe. While corporations like Syngenta and Bayern, and international agencies like FAO and Bioversity were in attendance, there was no formal representation from governmental bodies.

27. Key components of this network are documented in Hugh Williamson and Sabina Leonelli, eds., *Towards Responsible Plant Data Linkage* (Cham, Switzerland: Springer, 2022).

28. Hanna Ćwiek-Kupczyńska, Thomas Altmann, Daniel Arend, Elizabeth Arnaud, Dijun Chen, Guillaume Cornut, Fabio Fiorani, et al., "Measures for Interoperability of Phenotypic Data: Minimum Information Requirements and Formatting," *Plant Methods* 12 (2016): 44.

29. Plant data are a particularly attractive type of data on which to enact this vision, since it is often claimed that they do not have ethical implications for privacy (at least not in the same way as personal data used in health research) and food security is a widely shared concern. The problems with this view of plant data are discussed in detail by Williamson and Leonelli, *Towards Responsible Data Linkage*.

30. Noe Fernandez-Pozo, Naama Menda, Jeremy D. Edwards, Surya Saha, Isaak Y. Tecle, Susan R. Strickler, Aureliano Bombarely, et al., "The Sol Genomics Network (SGN): From Genotype to Phenotype to Breeding," *Nucleic Acids Research* 43, no. D1 (2015): D1036–D1041.

31. CassavaBase, https://www.cassavabase.org/, accessed August 31, 2020; NextGen Cassava, https://www.nextgencassava.org/what-we-do/our-mission/, accessed August 31, 2021.

32. CassavaBase is not the only international database developed in the last fifteen years that is devoted to cassava data. Among other relevant databases are the Cassava Genome Hub (https://www.cassavagenome.org/) and the Cassava Online Archive (http://cassava.psc.riken.jp/), whose history and relations to the Crop Ontology and CassavaBase I don't have the scope to discuss here.

33. Anna Burns, Roslyn Gleadow, Julie Cliff, Anabela Zacarias, and Timothy Cavagnaro, "Cassava: The Drought, War and Famine Crop in a Changing World," *Sustainability* 2, no. 11 (2010): 3572–3607, https://doi.org/10.3390/su2113572.

34. Béla Teeken, Olamide Olaosebikan, Joyce Haleegoah, Elizabeth Oladejo, Tessy Madu, Abolore Bello, Elizabeth Parkes, et al., "Cassava Trait Preferences of Men and Women Farmers in Nigeria: Implications for Breeding," *Economic Botany* 72 (2018): 263–277, https://doi.org/10.1007/s12231-018-9421-7.

35. Evaluations need to happen at time of flowering, at harvest stage, and at postharvest stage: for example, organoleptic evaluation, where samples of all clones and varieties are boiled and presented on plates, identified only by numbers, and subjected to "blind" tasting, distinguishing the components of appearance, taste, and texture.

36. Gary N. Atlin, Jill E. Cairns, and Biswanath Das, "Rapid Breeding and Varietal Replacement Are Critical to Adaptation of Cropping Systems in the Developing World to Climate Change," *Global Food Security* 12 (2017): 31–37.

37. The longest-running project on genetic gains funded by the Gates Foundation, the Borlaug Global Rust Initiative, is focused on wheat: https://globalrust.org/page/delivering-genetic-gain-wheat (accessed April 1, 2021).

38. It is important to note that such hard-won neutrality constitutes a significant political stance in and of itself, particularly given the commitment to corporate-owner technological fixes permeating the agricultural landscape worldwide.

39. See also Neema Mbilinyi and Babwale Ahmed, "Tensor Flow App for Cassava Disease Diagnosis," Milan Milenovic, YouTube video, September 4, 2017, 4:09, https://youtu.be/479p-PEubZk; "Revolutionary Mobile App for Monitoring Crop Pests and Diseases," CGIAR News, September 19, 2017, https://www.rtb.cgiar.org/news/revolutionary-mobile-app-monitoring-crop-pests-diseases/.

40. Kaushik Sunder Rajan and Sabina Leonelli, "Introduction: Biomedical Trans-Actions, Post-Genomics and Knowledge/Value," *Public Culture* 25, no. 3 (2013): 463–475.

41. David Beer, *Metric Power* (London: Palgrave Macmillan, 2016); Nick Srnicek, *Platform Capitalism* (Cambridge: Polity Press, 2017).

42. Sandro Mezzadra and Brett Neilson, *Border as Method, or, the Multiplication of Labor* (Durham, NC: Duke University Press, 2013), 3.

43. Mezzadra and Neilson, *Border as Method*, 4.

44. Again, according to Mezzadra and Neilson's insightful analysis, "The border is an epistemological device, which is at work whenever a distinction between subject and object is established" (*Border as Method*, 16). Borders are essential to cognitive processes, to "establishing taxonomies and conceptual hierarchies that structure the movement of thought" (16). In my interpretation, it is the absence of discourse around borders that sends a specific signal about the conceptual framework and epistemic approach favored by researchers involved in plant data transnational transactions.

45. John Krige, *Sharing Knowledge, Shaping Europe: US Technological Collaboration and Nonproliferation* (Cambridge, MA: MIT Press, 2016), 165.

46. Jenny Reardon, *Race to the Finish: Identity and Governance in the Age of Genomics* (Princeton, NJ: Princeton University Press, 2005); Robert Benjamin, "A Lab of Their Own: Genomic Sovereignty as Postcolonial Science Policy," *Policy and Society* 28, no. 4 (2019): 341–355.

47. Jean-Paul Gaudillière, "An Indian Path to Biocapital? The Traditional Knowledge Digital Library, Drug Patents, and the Reformulation Regime of Contemporary Ayurveda," *East Asian Science, Technology and Society* 8, no. 4 (2014): 391–415.

48. Norma Möllers, "Making Digital Territory: Cybersecurity, Techno-nationalism, and the Moral Boundaries of the State," *Science, Technology, and Human Values* 46, no. 1 (2021): 112–138, https://doi.org/10.1177/0162243920904436.

49. Gisela Mateos and Edna Suárez-Díaz, "Technical Assistance in Movement: Nuclear Knowledge Crosses Latin American Borders," in *How Knowledge Moves: Writing the Transnational History of Science and Technology*, ed. John Krige (Chicago: University of Chicago Press, 2019), 345–367.

50. John Krige, "Introduction: Writing the Transnational History of Science and Technology," in Krige, *How Knowledge Moves*, 13.

51. Miles, "The Combine Will Tell the Truth," for instance, critiques the rationalization of biology associated with big data analysis, the alienation of farmers from decision making about their harvest, and the economic surveillance and quantification of farming "value" enabled by automated systems of sensing and monitoring from field directly to market and back. He thus points out that the apparently democratic idea of making food chains and processing "transparent" and "traceable" could backfire.

Conclusion

Decentering the Global North

JOHN KRIGE

This collection (along with several others that I have been engaged with recently) takes knowledge and know-how embodied in people, ideas, and things as the object of a transnational approach for historians of science and technology.[1] This is not the only way to characterize transnational history, of course.[2] However, it has a certain primacy for this subdiscipline. For, as historian of science James Secord points out, the question of how and why knowledge circulates is "the central question of our field."[3] It also opens up interesting and important lines of inquiry with major intellectual, policy, and political implications.

Knowledge is an object of desire. What people know empowers them, enabling them to act upon the world to change it. As it moves from one local site to another on its journey, knowledge is appropriated, enriched, and repurposed in successive local contexts, where it acquires new meanings and serves new goals. When knowledge is acquired by others it can also empower *them* to think and act in new ways, ways that can be unexpected and uncontrollable by those who passed it on. From a trans*national* perspective, this entanglement of knowledge with power raises the question of its relationship with national interests, broadly conceived, and asks if and how such interests are engaged, impacted, or jeopardized as knowledge crosses borders into the heads and hands of "others." The nexus of knowledge, power, and "national interest" is a specificity of the genre of transnational history adopted in this volume that is alert to the factors that facilitate, or impede, the movement of knowledge across the borders of the national frame and to the changing power relationships involved.

The empowerment at the "receiving end" of transnational knowledge flows is particularly visible when it has unintended consequences, as we see in Michael Falcone's contribution (chapter 3). Early in World War II, the British realized that they did not have the means to mass-produce penicillin, and shared what they knew with the United States. They were persuaded to bracket discussions over intellectual property so as to facilitate a rapid scaling-up to mass production that was impossible at home under the pressure of war. When questions of ownership eventually arose, the American authorities had vested the rights to the exploitation of penicillin in their pharmaceutical companies like Merck and Pfizer, to the frustration and anger of British firms and the government, who could do little to redress their loss of control over the production of the drug and its markets. Tiago Saraiva (chapter 8) illustrates the same point. When the Portuguese government sent a local colonial administrator, Amílcar Cabral, to participate in a training course organized by the United Nations Food and Agriculture Organization (FAO) in Ibadan, Nigeria, they wanted to secure international respectability by participating in a worldwide project to produce agricultural surveys of Third World countries. Cabral used what he learned to develop proxy measures for colonial exploitation in Guinea-Bissau, and used them in leading a violent revolutionary movement to overthrow Portuguese rule there. Transnational knowledge flows can recalibrate power relationships between the partners, often in unexpected ways.

The chapters in this volume have been organized along two main axes. The first group focuses on the resistance by governments in the Global North to sharing knowledge of military and commercial significance with rival powers, be they allies or enemies. Knowledge flows are regulated to ensure that the competitive advantage that they bestow are not squandered by sharing key or "sensitive" components with another country that has the indigenous capacity to exploit their rich possibilities. The second group focuses on the transnational movement of agricultural knowledge and commodities between the Global North and the Middle East and Global South in colonial and postcolonial contexts. Knowledge flows are fostered by major powers to provide access to world markets, to increase yields of grains and vegetables, and to enhance food security and food sovereignty. Humanistic ambitions to combat poverty are interwoven with political ambitions, even if this means propping up the economy of a colonial dictatorship.

Geopolitically speaking, the chapters in Part I focus on leading global powers in the twentieth century, notably the United States, but include the United Kingdom and China, and are not restricted to Cold War superpower rivalry. The chapters in Part II do not simply consider North-South relations. They introduce the "Global South" as an actor that mobilizes knowledge to achieve political goals in a geopolitical order marked by asymmetrical structures of power and increasing global inequality. Contrast Vannevar Bush, Norman Borlaug, the Royal Society, the Rockefeller Foundation, the British Admiralty, the US national-security state, Merck, the Control Data Corporation, Hughes Aerospace, and the China Great Wall Industry Corporation with Amílcar Cabral, Octavio Paz, a workers cooperative in Hebron, a guerilla movement in Guinea-Bissau, coffee farmers in the cloud forests of Angola, Portuguese agronomists standardizing coffee beans, Mexican and Indian agronomists on farms in India, and cassava growers in Kenya. Negotiations over the transnational flows of knowledge in the Global North are intended to secure political and commercial advantage in a struggle for global dominance. The mobilization of agricultural knowledge to commercial and political ends by actors in the "Global South" can be an act of resistance or affirmation from the "Other," an attempt to engage with the world order being constructed by the major industrialized powers, and to secure national sovereignty, food security, human dignity, and basic rights of access to land. These chapters both include and decenter the Global North, and by introducing an array of actors from the "Global South" gesture toward a global history of science and technology. They also insist that that history be sensitive to the kinds of knowledge that crosses borders (the transnational flow of knowledge about penicillin and high-speed computers is subject to very different logics than that moving hybrid seeds and coffee beans across borders).

The transnational approach explored here poses major intellectual (and not just archival and linguistic) challenges. Saraiva (chapter 8) remarks that the transnational history of statistical methods that he describes calls on the historian to simultaneously engage with (at least) four major historiographies: the history of international institutions (the UN, the FAO, the WHO, etc.), of science and the American state, the history of colonial science, and histories of science from the Global South. My study of US-Chinese space collaboration in the 1990s (chapter 5) also invokes four rather different historical threads. Each of these histories has its own temporalities and its own intellectual agendas; interlaced, they provide thick descriptions of how knowledge moves transnationally.

Incorporating histories written from the point of view of the "Global South" does more than simply enrich existing narratives, it urges us to question the very narratives we have relied on to explain our present.[4] It treats the Global North and the Global South as coproduced in an asymmetrical field of power relations that is structured by the capacity of hegemonic actors to exploit resources in the "Other" to consolidate the projection of their power abroad and the construction of world order. It throws light on the mutual shaping of their historical trajectories and the inequalities it engendered. And it opens the way for alternative histories that highlight "South-South" entanglements. For example, as Gabriela Soto Laveaga (chapter 6) shows, here and elsewhere, there is a distinct Mexican tale to be told of the import of high-yielding wheat into India that is not just a national retelling of storied pasts, but one that foregrounds an interconnected Global South. It begins with the arrival of Indian agronomist Pandurang Khankhoje in Mexico in the 1920s, an individual on the margins of society, an itinerant, and a political refugee who arrived with "a socialist project steeped in anti-(British) imperialist ideology" that was "co-mingled with Mexican revolutionary ideals."[5] The narrative continues through a program of national agrarian reform in the 1930s, and the associated establishment of agronomical research stations by rich farmers in the Yaqui valley dedicated to developing new varieties of hybrid wheat. It describes the links established between Mexico and India soon after independence, and the help given to the Indian Agricultural Research Institute in the early 1960s by providing Mexican seeds for domestic research and experimentation. And it ends with Mexico's ambassador to India and his agronomists out in the fields demonstrating to Indian farmers the advantages of hybrid seeds bred in Mexico. A simplified US-centric variant of the "Green Revolution" locates the export of hybrid strains of Mexican wheat to India in the context of US development aid, Walt Rostow's noncommunist manifesto, and agronomical research by Norman Borlaug in the Yaqui valley supported financially by the Rockefeller Foundation, with some local assistance. It reduces Mexico's contribution to the production of hybrid seeds and India's to one of grateful recognition. The history of the Green Revolution from the point of view of the "Global South" is a history that goes beyond an outstanding "hunger-fighting" American scientist and instead incorporates experienced domestic agronomists in the Yaqui valley, Delhi, and elsewhere as equal protagonists. The narrative written "from below" is a story of interdependence and entanglement that is often erased in the asymmetric relationship of power between Mexico and India, on the one hand, and the United States, on the other.[6]

International alliances and foreign policy agendas provide the necessary backdrop to the transnational flows of knowledge discussed here (as Jessica Wang emphasizes in chapter 1). Many of these stories enrich our understanding of the evolving diplomatic relationships between national governments as expressed through negotiations over knowledge flows. The relationship between the British and the American governments is a major thread in Falcone's chapter, and hovers in the background of Katherine

Epstein's (chapter 2). The export of high-performance computers and the installation of Western monitors at Serpukhov (see Mario Daniels, chapter 4) exploited the window of opportunity opened by a brief thaw in US-Soviet relations in the early 1970s (détente) that deescalated military tension, encouraged commercial ties, and imagined exporting US political values behind the Iron Curtain to transform Soviet society. The agreement to use Chinese launchers for American-made satellites (Krige, chapter 5) only makes sense if we factor in Deng Xiaoping's opening of China to the West, and US-China foreign relations, which changed from Reagan's policy of treating it as a friendly, nonaligned country to Clinton's pursuit of constructive engagement and trade liberalization in an accelerating process of globalization. The travel of hybrid seeds from Mexico to India facilitated by Norman Borlaug was embedded in the Rostowian model of development that propelled Third World countries out of poverty through several stages toward a Western-style modernity, as well as in competition with the Soviet Union for the allegiance of a nonaligned state.[7] Portugal's quest for political legitimacy on the postwar global stage, and especially vis-à-vis the United States, prompted it to seek the implantation of statistical knowledge in the administration of its colonies in Africa (Saraiva) and the export of standardized Angolan coffee beans to American markets (Maria Gago, chapter 7). Cabral's struggle to liberate Guineau-Bissau brings the control of knowledge into the process of decolonization and reminds us of the violence that accompanied it. The quest for food security and food sovereignty by a nongovernmental organization (NGO) in Hebron (Courtney Fullilove, chapter 9) immerses us directly into the Arab-Israeli conflict in the Occupied West Bank and a local political agenda of nation building under the threat of the territorial expansion of Jewish settlements. Transnational histories of knowledge flows necessarily intersect with histories of international diplomatic and political relations.[8]

"Whether" and "how" the nation-state is implicated in the transnational circulation of knowledge is not simply an empirical question to be answered by the historian.[9] It can also be a political question posed and negotiated by national actors themselves who position transnational transactions in global contexts specifically so as to secure their (relative) autonomy.[10] The initiative taken by a Palestinian NGO that focused on peasants growing garden vegetables on drylands in Palestine, and that sought to preserve their seeds in a high-tech seedbank in Hebron, was driven by the quest for both food sovereignty and political sovereignty. The flow of these knowledges across "a jagged circuit of geographic, infrastructural, and political barriers" in the occupied West Bank enrolled local peasants in the process of nation-state building (Fullilove). The place of the state in knowledge circulation was also a major consideration shaping the construction of a platform for sharing mobile, interoperable data on the properties of the cassava plant produced collectively by farmers, plant breeders, and researchers at multiple national sites (Sabina Leonelli, chapter 10). Taking a bottom-up approach that respected the different meanings that cassava had for these social actors, this alliance structured a "border-free" space in which data traveled transnationally and in which national, corporate, and global agendas supported by the Global North were deliberately marginalized in the name of food sustainability in the Global South.

The transformative power of knowledge implicates the state and its interests more or less explicitly in the transnational activities described in this volume. This is not to deny that states can also create "borderless" regions between them where knowledge circulates "globally." It is only to emphasize that the transnational movement of knowledge takes place in a fluid political context of ongoing border-suppressing and border-creating actions. These selectively target specific kinds of knowledge, or bits of knowledge, that are identified as being of "national interest" broadly conceived, and whose transfer into new spaces can be unrestricted and even encouraged, or contentious and bureaucratically regulated. When knowledge moves transnationally, national borders—far from disappearing from view—become the focal point of the historical narrative itself. It is just as important to ask why (it seems that) knowledge moves smoothly out of the national container, as to inquire into the messy complexity of border crossings; to identify the factors that facilitate transnational flows as to tease out the obstacles to cross-border movement.

A great deal of knowledge circulates globally without any apparent resistance at national borders. One indicator of this: a report on the state of US science and engineering in 2020 noted that "the proportion of worldwide articles produced with international collaboration—that is, by authors from at least two countries—has grown from 14% in 2000 to 23% in 2018."[11] All of the major countries of the European Union had international collaboration rates of over 50 percent. In the United States the rate more than doubled, from 19 percent to 39 percent, during the same period, of which just over one in four were coauthored with an entity in China. This vast enterprise, which engages specific modes of knowledge production and circulation, is possible because national governments encourage, facilitate, or at least do not impede it. On the other hand, it would be wrong to see it as inhabiting a parallel universe to that explored in this collection, as I point out in the introduction. The impression given by this data is that knowledge moves "by itself" in a frictionless global space. In fact, it attests to the underlying presence of a specific alliance between national powers and to the associated global circulation of trained human capital made possible by the border-suppressing policies of nation-states in the name of scientific internationalism. The scope of knowledge that circulates freely and the countries that are considered to be acceptable international partners are circumscribed by political decisions of national governments that readily put obstacles in the path of transnational scientific and technological collaboration if they so choose. The recent tensions and constraints surrounding Sino-American cooperation in science and engineering dramatically illustrate this point.[12]

Describing the assembly of actors and institutions that together facilitate or impede knowledge flows across borders is one thing. It is another to explain the sources of their cohesion, to understand how this collective mobilization is possible. National institutions representing commercial interests, foreign policies, and geopolitical ambitions, along with ideological commitments (including scientific internationalism) fuse to construct the transnational network. For Fullilove and Leonelli, by contrast, major social actors are united in their determination to build collaborative structures that *resist* the imposition of state (and corporate) power, relying on bottom-up cooperation, NGOs, and the exploitation of sophisticated technological tools (a seed bank, the Crop Ontology data linkage tool) to build a community of knowledge producers that maintain control over their agricultural ecosystems. The positioning of nonstate transnational actors vis-à-vis the state is a particularly interesting empirical question that is not addressed systematically here but that can further enrich answers to the questions of whether and how to engage with the nation-state in transnational history.

Transnationalism as an approach to writing history has political effects. Scientific and technological achievements are tied up with national economic and military power, and with national pride. They can also feed nationalist sentiments of superiority in a world marked by intense competition between different states and social systems. Treating the nation-state as situated in an interconnected, interdependent web of relationships, rather than as a self-contained, autonomous, innovative social actor, necessarily dilutes the grandeur, the "sublimity" (to quote David Nye), of spectacular national feats of science and engineering.[13] Comedian Tom Lehrer's song, written in the early 1960s to deflate American pride in the Apollo program, is a case in point. In that song Lehrer slyly reminded his listeners that the ex-Nazi rocket engineer Wernher von Braun had played a key role in the successful "American" moon landing. In an age when major and spectacular scientific and technological achievements have become heavily charged as symbols of modernity, and instrumentalized as tools of nation building, a transnational approach to knowledge flows is necessarily subversive of nationalist ideology.[14]

David Edgerton has given multiple examples of transnational knowledge flows from the Global North that were constitutive of major domestic technological achievements but that were "written out" of national histories and of official narratives.[15] He does not address the political ramifications of his emphasis on foreign borrowings. We may laugh with Tom Lehrer when he highlights the influx of technology, knowledge, and know-how, notably from Nazi Germany, to prick the national pride of the leaders and citizens of a global power like the United States. It is far more controversial when such an approach undermines the technologically based sense of achievement of a newly emerging nation. The immense sensitivity shown by the Chinese authorities to claims that their successful nuclear and space programs benefited from foreign inputs (see chapter 5) is a somewhat extreme example of this, exacerbated by accusations of theft, but it also attests to deep-rooted resentments going back a century and more.

Michael Adas points out that, since the nineteenth century, the mastery of science and technology has been a marker of modernity and of "progress" for countries struggling to throw off the yoke of colonialism. It has freed them from the humiliating discourse of the "civilizing mission" and empowered them to take control of their own destinies.[16] At the same time it is difficult for national actors aspiring to modernity to avoid assessing their "progress" without using the norms espoused by global scientific and technological leaders. Francesca Bray writes that "the history of science or technology of anywhere or any group in the 'modern' or contemporary world is inevitably comparative," and "has to be contextualized within a broader flow of what has become a power-laden transnational field of knowledge production within which accuracy and truth, efficiency and originality are expected to be assessed."[17] It is not only historians, however, who make such comparisons: many national leaders feel obliged to do so as well. For them, too, scientific and technological originality and innovation are hallmarks of modernity. To highlight the borrowings and dependencies on knowledge that flowed into the country from abroad, as one does in a transnational approach, can easily be interpreted as parading the power of technologically advanced nations, and as belittling the successes of emerging countries. It is tantamount to reproducing neocolonial attitudes and discourses that have humiliated those countries for centuries and denying the national sovereignty that they have won in their anti-imperial and anticolonial struggles.

In a recent analysis of the state of global history today, Jeremy Adelman emphasizes that global narratives that highlighted interconnection and integration often meant losing contact with "'deep stories' of resentment about loss of and threat to local attachments." "The older patriotic narrative," Adelman goes on, "had tethered people to a sense of bounded unity," and had secured "heartlanders' ties to a sense of place in the world."[18] How are transnational historians to avoid demeaning this sense of national belonging, especially when writing from a privileged social position in the "Global North"? One way forward, emerging from this volume, is to suggest that a nation's sense of identity and of "modernity" be built not on its scientific and technological self-sufficiency but on its ability to successfully merge cross-border flows of knowledge into national technoscientific projects. It lies in the ability to build national competencies that are mobilized by local actors to use technology and knowledge from abroad in innovative ways to meet domestic political, industrial, and cultural exigencies.[19] A history that throws light on the strategies, policies, and practices invoked by national actors who position themselves transnationally, and who draw on the global pool of knowledge production and circulation to advance their prestigious national projects, respects the pride felt by leaders and citizens in their country's achievements.

That being said, it is not by framing the terms of the debate in advance, as I am doing here, but by participating in a transnational scholarly community that addresses these intellectually and politically complex issues that we can hope to promote a liberating transnational history. As it embraces regions beyond the "advanced" industrialized world, that history will surely engage with what Boaventura de Sousa Santos calls epistemologies of the Global South, neglected ways of knowing born in repression and opposition that are embedded in a cognitive framework that aspires to "the mutual enrichment of different knowledges and cultures."[20] The disciplinary and national heterogeneity of the contributors to this volume takes one small step in this direction, though all are located institutionally in the "Global North." It is distressing that, with the intensification of national rivalry and protectionism today, even this limited mode of transnational cooperation cannot be taken for granted any more.

Notes

1. John Krige, ed., *How Knowledge Moves: Writing the Transnational History of Science and Technology* (Chicago: University of Chicago Press, 2019); Axel Jansen, John Krige, and Jessica Wang, eds., "Empires of Knowledge," special issue, *History and Technology* 35, no. 3 (2019).

2. Zuoyue Wang, for example, defines transnational science as dealing with "the movements of scientists, scientific institutions, practices, instruments and ideologies across national boundaries," showing how such movements "interacted with indigenous traditions and contexts within any particular nation-state," and shaped national and international scientific developments; Wang, "The Cold War and the

Reshaping of Transnational Science in China," in *Science and Technology in the Global Cold War*, ed. Naomi Oreskes and John Krige (Cambridge, MA: MIT Press, 2014), 343–370, at 344. Simone Turchetti and his coeditors summarize various transnational approaches to the history of science as involving the "integrated study of different forms of global circulation of scientific knowledge and products, including the construction and functioning of international institutional and professional spaces devoted to science": Simone Turchetti, Néstor Herran, and Soraya Boudia, "Introduction: Have We Ever Been 'Transnational'? Towards a History of Science across and beyond Borders," in "Transnational History of Science," ed. Turchetti, Herran, and Boudia, special issue, *British Journal for the History of Science* 45, no. 3 (2012): 319–336, at 330.

3. James Secord, "Knowledge in Transit," *Isis* 95, no. 4 (2004): 654–672, at 655.

4. Thanks to Gabriela Soto Laveaga for this formulation.

5. Gabriela Soto Laveaga, "*Largo dislocare*: Connecting Microhistories to Remap and Recenter Histories of Science," in "Thinking with the World: Histories of Science and Technology from the 'Out There,'" ed. Gabriela Soto Laveaga and Pablo F. Gómez, special issue, *History and Technology* 34, no. 1 (2018): 21–30. See also her introduction with Pablo F. Gómez in the same special issue of *History and Technology*, 5–10.

6. Hiromi Mizuno, "Mutant Rice and Agricultural Modernization in Asia," *History and Technology* 36, no. 3–4 (2020): 360–381, describes a parallel and quite different dynamic. See also Tao-Ho Kim, "Making Miracle Rice: Tongil and Mobilizing a Domestic 'Green Revolution' in South Korea," in *Engineering Asia: Technology, Colonial Development and the Cold War Order*, ed. Hiromi Mizuno, Aaron S. Moore, and John DiMoia (London: Bloomsbury Academic, 2018), 189–208.

7. David C. Engerman, *The Price of Aid: The Economic Cold War in India* (Cambridge, MA: Harvard University Press, 2018).

8. Jeffrey James Byrne "Reflecting on the Global Turn in International History or: How I Learned to Stop Worrying and Love Being a Historian of Nowhere," *Rivista Italiana di Storia Internazionale* 1, no. 1 (2018): 11–42, emphasizes the interdependence of these different genres of historiography. For a brief "theoretical" discussion of this theme, see John Krige, "Technodiplomacy: A Concept and Its Application to U.S.-France Nuclear Cooperation in the Nixon-Kissinger Era," *Federal History* 12 (2020): 99–116.

9. See Patricia Clavin, "Defining Transnationalism," *Contemporary European History* 14, no. 4 (2005): 421–439.

10. It inspired the analysis by John Krige and Sabina Leonelli, "Mobilizing the Transnational History of Knowledge Flows: COVID-19 and the Politics of Research at the Borders," *History and Technology* 37, no. 1 (2021), https://www.tandfonline.com/doi/full/10.1080/07341512.2021.1890524.

11. National Science Board, National Science Foundation, *Science and Engineering Indicators 2020: The State of U.S. Science and Engineering* (Alexandria, VA: National Science Board, 2020), 12–13 and fig. 23, https://ncses.nsf.gov/pubs/nsb20201/.

12. Described in Mario Daniels and John Krige, *Knowledge Regulation and National Security in Postwar America* (Chicago: University of Chicago Press, 2022), epilogue.

13. David E. Nye, *American Technological Sublime* (Cambridge, MA: MIT Press, 1994).

14. Michael J. Neufeld has surveyed the extent of the phenomenon in one major field: Neufeld, "The Nazi Aerospace Exodus: Towards a Global Transnational History," *History and Technology* 28, no. 1 (2012): 49–67.

15. David E. H. Edgerton, "The Contradictions of Techno-Nationalism and Techno-Globalism: A Historical Perspective," *New Global Studies* 1, no. 1 (2007): 1–32.

16. Michael Adas, "Modernization Theory and the American Revival of the Scientific and Technological Standards of Social Achievement and Human Worth," in *Staging Growth: Modernization, Development, and the Global Cold War*, ed. David C. Engerman, Nils Gilman, Mark H. Haefele, and Michael E. Latham (Amherst: University of Massachusetts Press, 2003), 25–43.

17. Francesca Bray, "Only Connect: Comparative, National, and Global History as Frameworks for the History of Science and Technology in Asia," *East Asian Science, Technology and Society* 6, no. 2 (2012), 233–241, at 236.

18. Jeremy Adelman, "What Is Global History Now?," *AEON*, March 2, 2017, https://aeon.co/essays/is-global-history-still-possible-or-has-it-had-its-moment.

19. Lissa Roberts, inspired by Edgerton's argument, claims that, globally speaking, "Most scientific and technological knowledge has, in fact, historically involved the local adaptation and use of extant knowledge, procedures, apparatus and artefacts"; Roberts, "Situating Science in Global History: Local Exchanges and Networks of Circulation," *Itinerario* 32, no. 1 (2009): 9–29, at 18. This writes innovation entirely out of the picture for "most" knowledge, and it ignores the global inequalities between countries that innovate to secure scientific and technological advantage and those that can often do little more than "adapt and use" that knowledge as best they can.

20. Boaventura de Sousa Santos, *The End of the Cognitive Empire: The Coming of Age of Epistemologies of the South* (Durham, NC: Duke University Press, 2018), 275. See also Gabriela Soto Laveaga, "Moving From, and Beyond Invented Categories: Afterwords," *History and Theory* 59, no. 3 (2020): 439–447.

Acknowledgments

The chapters in this volume were first presented in preliminary form at a workshop sponsored by the Francis Bacon Foundation as part of the biennial Francis Bacon Prize in the History and Philosophy of Science and Technology, awarded to John Krige in 2019. It was held at the Department of Humanities and Social Sciences at the California Institute of Technology in Pasadena under the watchful eye of Professor Jed Buchwald, Doris and Henry Dreyfuss Professor of History. The event was attended by about twenty-five people. Professors Ted Porter (UCLA) and Steve Usselman (Georgia Tech) graciously agreed to share their comments throughout the meeting and at a dedicated session on the last afternoon. Fran Tise took care of organizational matters with diligence and patience. Karen Merikangas Darling at the University of Chicago Press has accompanied this project from the outset: my thanks to her and her associates for their outstanding support.

John Krige
Paris

Contributors

MARIO DANIELS is the Deutscher Akademischer Austauschdienst (DAAD)-Fachlektor at Duitsland Instituut Amsterdam. He holds a PhD from the University of Tübingen, taught at the Universities of Tübingen and Hannover, and was twice a research fellow at the German Historical Institute in Washington, DC. From 2015 to 2020 he was the DAAD Visiting Professor at the BMW Center for German and European Studies at Georgetown University. His latest book, *Knowledge Regulation and National Security in Postwar America*, coauthored with John Krige, will be published by the University of Chicago Press in 2022.

KATHERINE C. EPSTEIN is associate professor of history at Rutgers University–Camden and the author of *Torpedo: Inventing the Military-Industrial Complex in the United States and Great Britain*. She studies technology transfer, the intersection of national security and intellectual property regimes, and the political economy of power projection.

MICHAEL A. FALCONE is Chauncey Postdoctoral Fellow in the Brady-Johnson Program in Grand Strategy and International Security Studies at Yale University. His forthcoming book project, tentatively titled *The Rocket's Red Glare: Transatlantic Technology and the Rise of American Global Power*, analyzes the role of science, technology, and diplomacy in the construction of US hegemony and the contemporary world order. In 2019–2020 he was postdoctoral fellow in US foreign policy at the Dickey Center for International Understanding at Dartmouth College. He received his PhD from Northwestern University in 2019.

COURTNEY FULLILOVE is an associate professor of history, environmental studies, and science in society at Wesleyan University. She researches the history of practices now gathered under the rubric of world development: sustainability, biodiversity, intellectual property law, traditional knowledge, and cultural heritage. She is the author of *The Profit of the Earth: The Global Seeds of American Agriculture*, which received an Honorable Mention for the 2018 Frederick Jackson Turner Award by the Organization of American Historians. She is currently working on a book about biodiversity preservation.

MARIA GAGO is currently employed at the Institute of Contemporary History, NOVA School of Social Sciences and Humanities, Portugal. She is a historian of science and technology interested in the global history of crops, notably coffee. Her doctoral dissertation in history at the Institute of Social Science, University of Lisbon, brought together the history of science and technology, environmental history, and imperial history to produce a nuanced narrative of Robusta coffee and Portuguese colonialism in Angola. Gago is currently a Max Weber Fellow for 2020–2022 at the European University Institute in Florence, Italy.

JOHN KRIGE is the Kranzberg Professor Emeritus in the School of History and Sociology at the Georgia Institute of Technology. He researches at the intersection of the history of science, technology, and (American) foreign policy. His recent work has focused on export controls as instruments to regulate transnational knowledge flows and on the transnational approach as method. He is the editor of *How Knowledge Moves: Writing the Transnational History of Science and Technology* (University of Chicago Press, 2016) and coauthor, with Mario Daniels, of *Knowledge Regulation and National Security in Postwar America* (University of Chicago Press, 2022).

SABINA LEONELLI is professor of philosophy and history of science at the University of Exeter, where she codirects the Centre for the Study of the Life Sciences and leads the governance strand of the Institute for Data Science and Artificial Intelligence. She is a fellow of the Alan Turing Institute, the Academia Europaea, and the Académie Internationale de Philosophie de la Science; editor-in-chief of *History and Philosophy of the Life Sciences*; associate editor of *Harvard Data Science Review*; and twice a European Research Council grantee. Her books include *Data-Centric Biology: A Philosophical Study*, *Data Journeys in the Sciences*, and *Data and Society: A Critical Introduction*.

TIAGO SARAIVA is associate professor of history at Drexel University; coeditor of the journal *History and Technology*; and author of *Fascist Pigs: Technoscientific Organisms and the History of Fascism*, winner of the 2017 Pfizer Prize awarded by the History of Science Society. He is coeditor of *Nature Remade* and coauthor of *Moving Crops and the Scales of History*. His work intersects the history of science and technology, political history, and global history.

GABRIELA SOTO LAVEAGA is professor of the history of science and Antonio Madero Professor for the Study of Mexico at Harvard University. Her first book, *Jungle Laboratories: Mexican Peasants, National Projects and the Making of the Pill*, won the Robert K. Merton Best Book prize in Science, Knowledge, and Technology Studies from the American Sociological Association. She is completing two book manuscripts: one on doctors as agents of social unrest, and her third book, which examines agricultural science exchange between India and Mexico. Two recent publications include a coedited issue with Warwick Anderson on Decolonizing Histories in Theory and Practice," in *History and Theory*, and "The Socialist Origins of the Green Revolution," in *History and Technology*.

JESSICA WANG is professor of US history at the University of British Columbia. Her publications include *American Science in an Age of Anxiety: Scientists, Anticommunism, and the Cold War* and *Mad Dogs and Other New Yorkers: Rabies, Medicine, and Society in an American Metropolis, 1840–1920*. Her current research focuses on tropical agriculture, interimperial collaboration, and American empire in the early twentieth century.

Index